Boston Studies in the Philosophy and History of Science

Volume 326

More information about this series at http://www.springer.com/series/5710

Christophe Bouton • Philippe Huneman
Editors

Time of Nature and the Nature of Time

Philosophical Perspectives of Time in Natural Sciences

Springer

Editors
Christophe Bouton
Département de Philosophie/SPH,
UFR Humanités
Université Bordeaux Montaigne
Pessac Cedex, France

Philippe Huneman
Institut d'Histoire et de Philosophie des
Sciences et des Techniques (IHPST)
CNRS/Université Paris 1 Panthéon-Sorbonne
Paris, France

ISSN 0068-0346 ISSN 2214-7942 (electronic)
Boston Studies in the Philosophy and History of Science
ISBN 978-3-319-53723-8 ISBN 978-3-319-53725-2 (eBook)
DOI 10.1007/978-3-319-53725-2

Library of Congress Control Number: 2017939580

Printed on acid-free paper

This Springer imprint is published by Springer Nature
The registered company is Springer International Publishing AG
The registered company address is: Gewerbestrasse 11, 6330 Cham, Switzerland

Acknowledgments

This publication is supported by the Agence Nationale de la Recherche (ANR) Project 13 BSH3 0007 “Explabio”; the Laboratoire International Associé (LIA) CNRS Paris IHPST-Montréal “Epistemic and Conceptual Issues in Evolutionary Biology” (ECIEB); the research center “Sciences, Philosophie, Humanités” (SPH, EA 4574) of the Université Bordeaux Montaigne; and the Institut Universitaire de France (IUF).

The editors of the book thank Pascal Duris (SPH, University of Bordeaux), Jean Gayon (IHPST, CNRS/Université Paris 1 Panthéon-Sorbonne), Valéry Laurand (SPH, Université Bordeaux Montaigne), Pascal Weil (LaBRI, University of Bordeaux), the “Institut d’Histoire et de Philosophie des Sciences et des Techniques” (IHPST, UMR CNRS 8590), the “Laboratoire Bordelais de Recherche en Informatique” (LaBRI) of the University of Bordeaux, the “Maison des Sciences de l’Homme d’Aquitaine” (MSHA), and the région Aquitaine for their long-lasting support. They also thank anonymous referees for their constructive comments on the manuscript.

Contents

About the Editors and Authors

About the Editors

Christophe Bouton is professor of philosophy at the Université Bordeaux Montaigne and member of the research center "Sciences, Philosophie, Humanités" (SPH EA 4574). His research area is focused on the issue of time in the contemporary philosophy. He was fellow of the Institut Universitaire de France (Research Project 2008–2013: "Time: Philosophical and Scientific Approaches") and visiting scholar at the Centre for Time of the University of Sydney (April–July 2012) and at the NYU Department of Philosophy (April–July 2015). His principal publications on the issue of time are *Time and Freedom* (Northwestern University Press, SPEP Series, 2014) and "The Emergence of Time: Kant, Bergson, and Modern Physics," *KronoScope* 13, 1, 96–111 (2013).

Philippe Huneman is a researcher (directeur de recherche) at the Institut d'Histoire et de Philosophie des Sciences et des Techniques, CNRS/Université Paris 1 Panthéon-Sorbonne. After having worked and published on the constitution of the concept of organism and Kantian metaphysics, he currently investigates philosophical issues in evolutionary theory and ecology, such as the emergence of individuality, the relations between variation and natural selection in evolutionary theory, the role of the concept of organism, or the varieties of explanations in ecology. He has been visiting professor at the University of Chicago (2005). His research has been published in *Erkenntnis*, *Synthese*, *Philosophy of Science*, *Biology and Philosophy*, etc., and as editor, he published *Functions: Selection and Mechanisms* (Synthese Library, Springer, 2012), *From Groups to Individuals: Evolution and Emerging Individuality* (MIT Press, 2013, with F. Bouchard), *Handbook of Evolutionary Thinking in the Sciences* (ed. with T. Heams, G. Lecointre, M. Silberstein, Springer, 2015), and *Challenging the Modern Synthesis: Development, Adaptation and Inheritance* (with D. Walsh, Oxford University Press, 2017). He is series coeditor of "History, Philosophy and Theory in the Life Sciences" (Springer).

About the Authors

Vincent Ardourel has a PhD in philosophy of science from Paris 1 Panthéon-Sorbonne University, on discrete mathematical models in physics, under the supervision of Jacques Dubucs and Anouk Barberousse. He is currently postdoctoral researcher at the Université catholique de Louvain. His research interests are the philosophy of physics and the philosophy of applied mathematics. His recent publication on time and discrete mechanics is "A discrete solution of the paradox of Achilles and the tortoise," *Synthese* 192, 2843 (2015).

Anouk Barberousse is professor of philosophy of science at the Université Paris-Sorbonne. Her principal fields of research are the philosophy of physics (and in particular problems arising from models, idealization, and computer simulations), the philosophy of probability, and the epistemology of scientific pictures. She is the author of *La physique face à la probabilité* (Vrin, 2000) and editor of (with M. Morange and T. Pradeu) *Mapping the Future of Biology: Evolving Concepts and Theories* (Springer, Boston Studies in the Philosophy and History of Science, 2010) and of a special issue of *Synthese* on models and simulations (2011).

Joseph Berkovitz is a professor at the Institute for Philosophy and History of Science and Technology at the University of Toronto. He was previously associate professor at the University of Maryland and a visiting professor in several places, including the Centre for Time at the University of Sydney and at the London School of Economics. His research is in philosophy of science, especially the philosophy of physics, philosophy of probability, and philosophy of economics. In his work, he has used philosophical analyses of causation and probability in order to clarify a variety of issues in science, epistemology, and metaphysics. In the philosophy of physics, he has studied the curious nature of causality and probability in the quantum realm and randomness and chaos in the classical realm. He has contributed to the Stanford Encyclopedia of Philosophy, and his work has been published in philosophical journals such as *Erkenntnis, Mind, Philosophy of Science* and *Studies in History and Philosophy of Modern Physics*.

David Braddon-Mitchell is professor of philosophy at the University of Sydney (Centre for Time). His research interests are philosophy of mind and metaphysics; his area of competence includes philosophy of science, philosophy of biology, and epistemology. He is author, with Frank Jackson, of the *Philosophy of Mind and Cognition* (Blackwell, 1996).

Armand J. de Ricqlès is a professor of historical biology and evolutionism at the Collège de France in Paris. A world expert in vertebrate paleohistology, he initially worked on the functional significance of extant histodiversity and applied this newly gained knowledge in paleobiological inferences. He was visiting professor of the Department of Anatomy at the University of Chicago and visiting curator of the Department of Geology at the Field Museum of Natural History, Chicago (1979–1980). He had published over a hundred scientific papers in biology and paleontology journals.

Jean Gayon is a professor of philosophy at the Université Paris 1 Panthéon-Sorbonne and chair of the Institut d'Histoire et de Philosophie des Sciences et des Techniques (CNRS/Université Paris 1 Panthéon-Sorbonne), as well as a senior fellow of the Institut Universitaire de France. His research addresses history and philosophy of evolutionary biology and genetics and deals with issues such as the species concept, chance in evolution units of selection, or questions raised by human enhancement. He is the author of *Darwinism's Struggle for Survival* (transl. of *Darwin et l'après-Darwin*) (Cambridge University Press, 1998) and editor of many books such as (with A. de Ricqles and M. Mossio) *Les fonctions: des organismes aux artefacts* (PUF, 2010), (with C. Malaterre and M. Morange) *Defining Life* (Springer, 2010), and (with A. Brenner) *French Studies in the Philosophy of Science* (Springer, 2009). He published papers on genetics, evolutionary theory, eugenics, and history of philosophy of science in major history and philosophy of science journals.

John Huss is an associate professor at the University of Akron, Ohio. He is a philosopher of biology, interested in paleontology and evolution, as well as in philosophical issues raised by metagenomics and by personal medicine. He did his PhD at the University of Chicago on philosophy of evolutionary biology with a focus on paleobiology (2004). His last publications include "The MBL Model and Clade Shape" in *The Paleobiological Revolution* (Hull and Sepkoski eds., University of Chicago Press, 2009) and, as an editor, *Planet of the Apes and Philosophy* (Open Court, 2013).

Jenann Ismael is full professor at the University of Arizona (Department of Philosophy). Her research interests are the philosophy of physics and topics related to mind, cognition, and phenomenology. An important part of her publications is focused on the issue of time (she was visiting fellow at the Centre for Time of the University of Sydney from 2005 until 2010). Her selected publications include *Essays on Symmetry* (Garland, 2000), *The Situated Self* (Oxford University Press, 2006), and *How Physics Makes Us Free* (Oxford University Press, 2016).

Annick Lesne is a directeur de recherches of the CNRS (full professor) at the Laboratoire de Physique de la Matière Condensée, Université Pierre et Marie Curie, Paris. Trained as a physicist, her research addresses complex systems modeling in general – in physics and biology. Her papers addressed the issues of robustness in complex systems and of randomness and determinism in nonlinear dynamics, and she has been especially interested in understanding and modeling multiscale dynamics. Her books include *Chaos and Coarse Graining in Statistical Mechanics*, along with P. Castiglione, M. Falcioni, and A. Vulpiani (Cambridge University Press, 2008), and *Scale Invariance*, along with M. Laguës (Springer, 2011). She published papers about metabolic networks, multiscale ecological modeling, and genetic networks in medicine, ecology, biology, computational biology, and neuroscience journals, as well as general papers on robustness in biological and physical dynamics and multiscale structures (e.g., "Multiscale analysis of biological functions," *Acta Biotheoretica* 61, 2013; "Robustness: confronting lessons from physics and biology," *Biological Reviews* 83, 2008).

Kristie Miller is associate professor at the University of Sydney (Department of Philosophy) and currently joint director of the Centre for Time of the University of Sydney. Her research interests lie primarily in metaphysics, in particular in philosophy of time, mereology, composition, and persistence. Her selected publications include *Issues in Theoretical Diversity: Persistence, Composition, and Time* (*Philosophical Studies* series, Volume 106, Springer, 2006), and, with R. Ciuni and G. Torrego, New Papers on the Present: Focus on Presentism (Philosophia Verlag, 2013).

Maël Montévil is a theoretical biologist. He is a contractual researcher in the Matière et Systèmes Complexes laboratory, Université Paris Diderot (Paris 7), and associated to Institut d'Histoire et de Philosophie des Sciences et des Techniques. His work focuses on the relationship between biology and mathematical methods and objects, in particular in the book coauthored with Giuseppe Longo, *Perspectives on Organisms: Biological Time, Symmetries and Singularities* (Springer, 2013).

Antonine Nicoglou defended her thesis in philosophy at the University Paris 1 Panthéon-Sorbonne on November 19, 2013, under the supervision of Professor Jean Gayon and Philippe Huneman. She also graduated from the University Paris Diderot (Paris 7) in cell biology and physiology. Her dissertation concerned the notion of plasticity in life sciences from early embryology until the recent problematics of the extended synthesis. Her interests include questions on plasticity but also on theoretical problems in developmental biology, embryology, evolution, and the synthesis of these disciplines. She is currently a postdoctorate fellow in the laboratory of excellence "Who Am I?" at the University Paris Diderot where she is working on the nature-nurture debate and its implication in different fields including experimental and theoretical biology, medicine, and philosophy. She will publish in 2017 a book in history and philosophy of science about plasticity in the life sciences.

Huw Price is the Bertrand Russell professor of philosophy and a fellow of Trinity College at the University of Cambridge. He was previously ARC Federation fellow and Challis professor of philosophy at the University of Sydney, where from 2002 to 2012 he was founding director of the Centre for Time. He is a fellow of the British Academy, a fellow and former member of the Council of the Australian Academy of the Humanities, and a past president of the Australasian Association of Philosophy. He made widely discussed contributions to the philosophy of physics, to pragmatist epistemology, and to the question of causation. His books include *Facts and the Function of Truth* (Blackwell, 1988; 2nd. edn. Oxford University

Press, forthcoming), *Time's Arrow and Archimedes' Point* (Oxford University Press, 1996), and *Naturalism Without Mirrors* (Oxford University Press, 2011). He is also coeditor (with Richard Corry) of *Causation, Physics, and the Constitution of Reality: Russell's Republic Revisited* (Oxford University Press, 2007).

Carlo Proietti is a postdoctoral researcher at Lund University. He did his PhD in philosophy on the problem of the contingent futures and Fitch paradox (Paris 2008), and his current research areas are philosophical logic and epistemology, with a focus on modal logics and its applications. He is currently working on a project to develop "Logical Modeling of Collective Attitudes and Their Dynamics" financed by the Swedish Research Council (2013–2015). His most recent publication includes "The Abundance of the Future. A Paraconsistent Approach to Future Contingents" *Logic and Logical Philosophy*, 2013, with R. Curl.

Pascal Tassy is a professor at the Museum d'Histoire Naturelle de Paris. He investigates the phylogeny of mammals as well as paleobiogeography and biostratigraphy, and he also addressed issues in the history of paleontology and the methodology of systematics. He has published, among papers in biology journals (*PNAS*, *Nature*), *La Reconstruction phylogénétique* (with P. Darlu, Masson, 1993), *L'arbre à remonter le temps* (Christian Bourgois, 1991), and *L'invention du mastodonte: aux origines de la paléontologie* (Belin, 2009).

Ken Wharton is professor of physics and astronomy at San José State University. His research is in the field of quantum foundations. He is particularly interested in time-symmetric and causally neutral approaches, working toward the ultimate goal of reframing quantum theory in a manner fully compatible with general relativity. His specific research projects concern simple Lagrangian-based classical models that exhibit quantum-like behavior and exploring basic deviations from strict classical theory that might lead to realistic explanations of quantum phenomena. His most recent publications include "A Novel Interpretation of the Klein-Gordon Equation," *Foundations of Physics* 40, 2010, and, with D. J. Miller and Huw Price, "Action Duality: A Constructive Principle for Quantum Foundations," *Symmetry* 3, 2011.

Chapter 1
Introduction: Time Between Metaphysics and Natural Sciences: From Physics to Biology

Christophe Bouton and Philippe Huneman

Abstract Although each scientific discipline has a specific approach to time and distinctive methodologies for implementing time in their models, pervasive issues about time still arise in all sciences, like the reality of time, the measurement of time, the definition of irreversibility and reversibility (time's arrow), the status of the past, the notion of timescale, etc. In this introductive chapter, we claim that a comparative and interdisciplinary approach to time as it is used and represented in the natural sciences would be the most appropriate way to deal with these issues in order to provide a philosophical understanding of the time of nature. In the natural sciences, time is a crucial dimension of physics. However, we stress that "time in physics" is something of an abstraction, since "the physics" itself is nowadays a set of different disciplines working with heterogeneous models, assumptions, and experimental settings. Because of this disparity, several perspectives on time and time paradoxes emerge from various fields in physics. In addition to the controversies about the time in physics, we argue that geology, paleontology and biology should be included in the philosophical assessment of the nature of time: within these sciences, time indeed displays different properties, is investigated using very different methods and tools, and raise specific problems, many of them due to the fact that evolutionary theory is the current framework for much of biological investigation. Finally, this interdisciplinary approach leads to the question of the univocal or pluralist nature of the concept of time, which is raised in the conclusion of the chapter.

C. Bouton (✉)
Département de Philosophie/SPH, UFR Humanités, Université Bordeaux Montaigne, Domaine Universitaire, 19 esplanade des Antilles, 33607 Pessac Cedex, France
e-mail: christophe.bouton33@orange.fr

P. Huneman
Institut d'Histoire et de Philosophie des Sciences et des Techniques (IHPST), CNRS/Université Paris 1 Panthéon-Sorbonne, 13 rue du Four, 75006 Paris, France
e-mail: philippe.huneman@gmail.com

C. Bouton, P. Huneman (eds.), *Time of Nature and the Nature of Time*, Boston Studies in the Philosophy and History of Science 326,
DOI 10.1007/978-3-319-53725-2_1

This interdisciplinary volume intends to shed light on the nature of time as it appears in natural science. It considers several natural sciences with the intention of constructing a comparative framework that would explore physics, the heart of current philosophical accounts of time, and other sciences. It confronts the notion of time in several branches of physics – the theory of relativity, quantum physics, and discrete mechanics – since there is no a priori warrant that all these notions are identical, and then extends this comparison to geology and to evolutionary and developmental biology. Several disciplines of evolutionary biology, i.e. paleontology and theoretical and experimental population genetics, are examined. The epistemological issues raised by modeling time in special sciences are confronted. Several topics related to natural time are explored: in particular, the reality of time, the measurement of time, the univocal or pluralist nature of the concept of time, the notions of irreversibility and reversibility. Therefore, this volume aims to cultivate a discussion between philosophy and the natural sciences, and presents an up-to-date account of the question of time from the viewpoints of a philosophy fueled by the natural sciences.

1.1 The General Issue Addressed by the Book

Time is a longstanding vexed question in philosophy, and metaphysicians since Aristotle have made important contributions to the study of time. As it has become apparent, time can most generally be approached from two perspectives: the time of nature, for example when Aristotle relates time with the concept of motion, and time as an essential structure of consciousness or thought, as illustrated by the Kantian thesis of time as an a priori form of intuition, or later by Husserl and his phenomenology of inner time. One of the latest systematic and general philosophical studies of time, Paul Ricoeur's *Time and Narrative*, (Ricoeur 1984, 1985, 1988) concluded that the two main approaches, the time of nature and the subjective time of lived experience, are irreducible to one another. The present volume addresses the question of time from the perspective of the time of nature, intending to provide some insights about the nature of time on the basis of the scientific uses of the concept of time. While it may not ultimately assess Ricoeur's claim of irreducibility, at least it clarifies the natural side of this dichotomy, possibly challenging his assumption of a unity of the "time of nature."

In the natural sciences, time *prima facie* is a crucial dimension of physics. The first philosophical sophisticated theory of time, understood as the "number of motion according to the anterior and the posterior", occurs in Aristotle's book on *Physics* (Part IV, chap. 10–14) as a systematic examination of the major concepts of the science of nature. And Aristotle seems to have captured a major feature of this concept: the fact that time is measured by regular natural processes, such as the motion of planets and, more recently, the radioactive decay of a cesium atom, indicates that physics – traditionally understood as the science of matter and its motions – plays an essential role in our understanding of time. Models in physics generally investigate trajectories of systems along time, time being a crucial variable for all models. But more generally, major advances in physical theories arguably have been reflecting and rethinking the meaning and dimension of time: Newton's views on absolute space and absolute time, as well as Einstein's formulation of space and time solidarity and their co-constitution of Minkowski's space-time, or the recent questioning of the discrete vs. continuous fundamental micro-structure of space-time, easily illustrate this assertion.

Physicists since Galileo seem indeed to have constantly faced major questions about the nature of time, and their theoretical and modeling advances often included tentative answers to such questions. For instance, Newton and Leibniz were concerned with the issue of the relativity or the absoluteness of time – as the correspondence between Leibniz and Clarke attests (Leibniz 1998). And the arguably major mathematical advance in theoretical physics in the early modern age, namely the calculus, was intended to mathematically capture motion in time – an achievement that needed the best mathematicians of their time, culminating with Leibniz and Newton (Boyer 1949). Later on, Einstein's special relativity put into question, as many philosophers will notice, one of the most intuitive notions on which any theoretical and mathematical concept of time relies, namely *simultaneity* (Sklar 1974, see in this volume Bouton Sect. 6.2.3). As one knows, special relativity makes time into a magnitude relative to the reference-frame of the observer, a view that required major modifications to the mathematical apparatus used to formulate physical theories. Within the theory of general relativity, Riemannian spaces were indeed required to replace Euclidean space, and Newton's theory in Euclidean space appeared then as a mere approximation of the true, relativistic theories, plausible only when the systems move much slower than the speed of light.

Yet in another context, a second intuitive notion constituting our general understanding of time as human beings, namely its *irreversibility*, or in more technical terms the dissymmetry between past and future, or what some call the "arrow of time", becomes a problem in theoretical physics. Actually, unlike what happened to simultaneity, the problem itself did not emerge with modern physics, but is already present in classical, Newtonian physics: its equations are time–reversible, and it seems that the descriptions of nature given by this science of nature are not capable of making sense of a quasi-natural feature of time, that is, irreversibility. On the other hand, thermodynamics, as it was developed in the second half of the nineteenth century especially after the works of Clausius and Carnot, is a time-oriented theory (see for a survey North 2011; Chang 2004), because it includes a principle – namely, the "second principle", about the necessary increase of entropy – that confers a direction to the time variable. Theoreticians of far-from-equilibrium thermodynamics, such as Prigogine (1961, 1977), Kauffman (1993) or philosophers, like Feigl, insisted on this gap between classical mechanics and the thermodynamics regarding the direction of time (see also Chibbaro et al. 2014). But the justification of the second principle is a hotly debated issue among physicists and philosophers of physics, and, *pace* Boltzmann's sophisticated elaborations aimed at grounding this principle (Boltzmann 1964, 1966), for more than a century no consensus has been found regarding an apodictic justification of this principle

(Uffink 2007). Therefore, physics does not unambiguously provides us with the bases of a theory of time that would make sense of its most common intuitive features, reduce them to theoretically indubitable primitives, or explain them away in an uncontroversial fashion.

Granted, physics is still crucial for our understanding of time, and has been taken as such by philosophers in the tradition. However, "time in physics" is an abstraction, since "physics" itself is nowadays a set of different disciplines working with heterogeneous models, assumptions, and experimental settings. The Large Hadron Collider in particle physics, the percolation models of statistical mechanics, or the Navier-Stokes equations of hydrodynamics are indeed very different, and could arguably be seen as equally distant from each other than are statistical physics and evolutionary biology. Because of this disparity proper to the field – which proves problematic for any philosophical conceptions that just opposes "the physics" to anything else. . . – several perspectives on time and issues about time emerge from various fields in physics.

In general, from Hans Reichenbach (1920, 1928, 1956), to Lawrence Sklar (1974, 1985, 1993) or Huw Price (1996), most of the philosophy of science books on time in the twentieth century focused on physics. More recently, several edited volumes have presented the issue of time in the framework of metaphysics (e.g. assessments of McTaggart on the "unreality" of time, the question of presentism, issues of time's passage, etc.), physics (thermodynamics, special and general relativity, quantum mechanics, quantum gravity) and to a lesser extent of cognitive neuroscience.[1] However the present book is not a handbook dealing exhaustively with the philosophy of time. It is a series of studies on various approaches of time of nature related to the natural sciences. These approaches focus much more on the dialogue between philosophy of science and the sciences, and much less on metaphysics. They include geology, paleontology and biology in the assessment of the philosophy of time: within these sciences, time indeed displays different properties, is investigated and represented using very different methods and tools, and raise specific problems, many of them (but not all) due to the fact that evolutionary theory is the current framework for much of biological investigation.

In the nineteenth century, Charles Lyell (although he had his predecessors), in geology elaborated on the notion of "deep time", the fact that earth has a nearly unimaginably long history, far longer than that recounted in the Bible (see Rudwick 2005 and Ricqlès' chapter 11 in this volume). This is important because it constitutes the framework of evolutionary theory. The timeframe of "deep time" in

[1]See e.g. Thomas Gold and Herman Bondi (eds.), *The Nature of Time*, 1967, Raymond Flood and Michael Lockwood (eds.), *The Nature of Time,* 1986, Dennis Dieks (ed.), *The Ontology of Spacetime*, 2 Vol. 2006, 2008, and most recently Craig Callender (ed.), *The Oxford Handbook of Philosophy of Time*, 2011, Adrian Bardon, *The Future of the Philosophy of Time*, 2012, Adrian Bardon and Heather Dyke (eds.), *A Companion to the Philosophy of Time,* 2013, Abhay Ashtekar, and Vesselin Petkov (eds.), *Springer Handbook of Spacetime*, 2014 and L. Nathan Oaklander (ed.), *Debates in the Metaphysics of Time*, 2014. See also Dean Rickles and Maria Kon (eds.) *Annals of the New York Academy of Sciences,* Vol. 1326, Special Issue on "Flow of Time", October 2014.

geology, paleontology and evolutionary theory rests on evidence that are vestiges from the past, which raises specific epistemological issues about our knowledge of the past, and our epistemic access to past events and processes through present facts. Genes or fossils inform us about the past, but such information is not given immediately, and information retrieving raises epistemic questions such as estimating information loss along time. Then, measuring time is a crucial aspect of the epistemological issues raised by time in specific sciences like geology or evolutionary biology, as exemplified by the fact that time measurement methods sometimes conflict. For instance, the time of the human-chimp split has been sometimes estimated to be 25 million years old, sometimes 8 or, more recently, 5–7 million years old or 13 million years old. Each of these estimations relies on specific methodological and theoretical commitments about genetics and paleontology. Measuring and representing time is therefore an important aspect of the problem of time in biology and geology, though it is left out in most metaphysical approaches of time in the last decades.

In biology, the concept of time is required at each level of its structure: most notably, in evolutionary theory, which makes it clear that all traits in the living world are historical. A major consequence of Darwin's theory has indeed been a *historical or temporal reinterpretation of major biological concepts or categories*. "*Adaptation*", the fit between organisms an their environments that have bewildered natural philosophers and theologians alike for centuries, is now understood as a result of cumulative natural selection for very long periods. The *diversity* of species on Earth is the result of a 4 billion year-long non-directed history, whose main turns may have been contingent (Gould 1989), or not (Dawkins 1982). The *classification* of living creatures, paradigmatically achieved by naturalists such as Linnaeus in the eighteenth century, should in turn be interpreted as the trace of history: distance measured in terms of the amount of intermediaries within the classification, indicates a historical distance; and similarities indicate common ancestry (in general).

Granted, as Lovejoy (1936) forcefully argued a long time ago, transformist theories that were frequent after 1750, such as De Maillet's, Erasmus Darwin's, Lamarck's or Herder's, were "temporalizing the great chain of being". But this was just about interpreting a chain (which is linear) into a temporal line. Darwinism, on the other hand, introduced time in a classification that was already tree-like – a hierarchy, mathematically speaking, promoting a branching, non-oriented, history. Temporalizing a linear chain is one thing, historicizing taxonomic distances, as Darwin did, is another thing, and it is part of what is often meant when biologists claim that Darwinism is the unanimous general framework for biology. This pervasive role of the dimension of time in the living world and in our understanding of it diffracts into various aspects and philosophical issues.

As Aristotle definitively formulated it in his *Physics*, time is the dimension of change, and as a matter of fact biological entities do change. *Evolution* is the change of species. It concerns only populations, as Darwin claimed, and the specific, non-random, population-level process called "natural selection" is the only process that accounts for adaptive evolution (much evolution is indeed non adaptive, led by stochastic processes such as the one called "genetic random drift" since Wright 1932). Individuals in contrast, never evolve: the individual organisms' change in time, from the zygote stage to the adult stage (when they are metazoan organisms) is called *development*.

Interestingly, the word "development" in the seventeenth century named precisely the development of organisms in the framework of the theory called "preformationist", which opposed "epigeneticism" (Roger 1963; Hoffheimer 1982; Roe 1980). Preformationists were scientists who claimed that the individual somehow already exists within the zygote, as a miniature or a set of dispositions; pure mechanical laws support the development of this preexisting form, which was in those times called "evolution", and opposed to a process led by interactions with the environment, sometimes called "eduction". Later on, through an interesting semantic switch, in the end of the nineteenth century and with the rise of transformist theories (especially Darwin's), evolution came to name the process of transforming species that Darwin studied (Bowler 1971; Richards 1992).

Developmental theory therefore investigates the change of an individual organism through time. Historians are still debating the exact inception of modern developmental theory: its conditions included overcoming the debates between preformationist and epigeneticists, ovists (i.e., among preformationists, the ones who thought the preformed individual was in the egg) and animalculists (opponents of ovists); acquiring experimental techniques to manipulate the egg; and then accessing the first and foremost cellular divisions, namely the early embryogenesis, by contrast with organogenesis, which was studied by eighteenth century embryologists, and especially in Wolff's groundbreaking book *Theoria generationis* (1758, see Dupont 2007; Huneman 2007; Goy 2014). Von Baer's *Entwicklungsgeschichte der Thiere* (1828) is a landmark in this story, as Darwin himself acknowledged by considering Von Baer the most important nineteenth century biologist. It provided the general framework for embryology until its merging with the late nineteenth century cellular theory (Duchesneau 1987; Lenoir 1982; Oppenheimer 1967) and then the rise of the mechanicist materialist *Entwicklungsmechanik* developed by Wilhelm His (Hopwood 2000) and Wilhem Roux (Kearl 2009), which moved embryology from description to experimentation.

All these theories assumed various concepts of time, because development is obviously a process in time, whose regularities or laws were investigated by developmental theorists during this period. Moreover, given that developmental theory includes a crucial dimension of comparative anatomy or embryology, many questions concern the differences and commonalities between developmental episodes across species or lineages. As it appears, some stages of development are shorter or longer in a species than in another, or they may be inverted, or some developmental stages can be deleted in a species evolved from a former one. Those switches and shifts have been captured through various concepts such as heterochrony, paedomorphism or neoteny, which biologists started to forge in the last nineteenth century with Haeckel, and that S.J. Gould systematically assessed and revisited in its groundbreaking book *Ontogeny and phylogeny* (1977). This book was the first attempt to conceptually and historically interpret the emerging field of evolutionary developmental biology (or Evo-devo), which nowadays continues to shake up the accepted framework of evolutionary biology (e.g. Müller and Pigliucci 2011; Huneman and Walsh 2017). But overall, such conceptual elaborations indicate that time itself is a crucial notion for develomental biology: what is a temporal stage? What is a stage switch, how should one define the relation between stages orders and duration, etc.? Could one use the distinction set by Kant in the *Critique of Pure Reason* between "order of time" and "duration of time" to capture those subtle concepts? What does "stage" or "phase" mean, and how could they be identified across distinct species or families that share little evolutionary history? And moreover, *how does developmental time relate with evolutionary time*?

The latter question is indeed central to any general philosophical interrogation about time in biology. Evolutionary time is underpinned by a "process" called inheritance (though the extent to which it's a *bona fide* process is debatable). While development is the process that leads from zygote to adult stages of an individual, inheritance names the patterns of similarity that relate individuals of various generations, and eventually characters (or genes) of those individuals to characters of ancestor species (on "character" see Barriel 2015). Inheritance is character-based and intergenerational, whereas development is organism-focused and intragenerational.

This is taken for granted by current biologists, yet it has not been always the case: pre-darwinian biologists were often likely to conflate development and inheritance as two intergenerational individual-based processes (Amundson 2005). Supporting this conflation, a very famous way of understanding the relation between evolutionary time and developmental time has indeed been provided for decades by the so called "recapitulation law", according to which ontogeny "recapitulates" phylogeny, meaning that the stages of the developmental process of an individual of a given species mimics the types of the species that preceded it in phylogeny. Interestingly, this "law", that the great popularizer of evolutionary biology Ernst Haeckel first formulated in those evolutionary terms under the name "biogenetic law" in 1889, emerged before Darwinian biology. It was first formulated in a fixist framework by the morphologists Etienne Serres and August Meckel in the early nineteenth century (they were some of the first followers of embryologist Caspar Wolff) (Canguilhem et al. 1962). The recapitulation achieved by an embryological process, according to them, would not represent evolutionary stages but species that precede the species under focus in a logical order, that is, within the classification of species. In order to find his biogenetic law, Haeckel therefore only had to reinterpret classification in evolutionary terms – as is entailed by Darwin's theory itself.

However, what is most striking is that Von Baer himself had already refuted the law of recapitulation (or "law of Meckel-Serres" at those times). He has shown that the "stages" recapitulated in embryology were not adult *stages* of other species, but *embryonic* stages of those species. And in evolutionary biology, it is easy to make sense of that once one considers that a species shares the first stages of its embryogenesis with its ancestor species. But the appeal of the recapitulation law, namely its simplicity, and most probably, the easy way it provides us with a clear link between evolutionary time and developmental time, led Haeckel to forget or reject Von Baer's sound refutation. In the end, though rejected as such by most scientists, the recapitulation law is still an idea that may have an illuminative potential regarding some interesting cases in evolutionary developmental theory, as many Evo-Devo researchers acknowledge.

On the other hand, two decades after Haeckel, the field of *inheritance* studies has been founded and systematized by Mendelian genetics, through the first statistical concepts of heritability and measures of heritability due to Pearson, Galton and the biometricians in early twentieth century (Gayon 1998) and finally by population geneticists like Fisher in the 1930s. Thus, for the most part of the twentieth century heredity meant something clear at least for evolutionary biologists. Moreover, Franklin's, Crick's and Watson's discovery of the structure of DNA allowed grounding the notion of heredity on a materially accessible and empirically manipulable substrate, made of nucleotides. Yet this whole field is currently shaken by empirically documented claims about the existence of non-genetic inheritance, such as epigenetic inheritance and cultural inheritance (Danchin et al. 2011; Jablonka and Lamb 2005). Biologists are therefore debating about the actual extant of those inheritances and the changes their acknowledgment should bring into evolutionary biology (e.g. Scott-Philips et al. 2011, vs. Mesoudi et al. 2013).

In this context, some authors tried to distinguish between various senses of inheritance, for instance by differentiating inheritance as intergenerational *patterns* of similarity and heredity as *processes* that underlie those patterns (Danchin and Pocheville 2014), attesting to the fact that clear conceptual demarcations are what is badly needed in the field, now that empirical evidence accumulates. But, notwithstanding the legitimacy of those claims and conceptual switches, what strikes us about those controversies is that the question of the relationship between inheritance, development, their timescales and more generally the related articulation between evolutionary and developmental times remains an open question that the current controversies about major concepts and systems in evolutionary biology is metaphysically reviving.

Thus, one sees that time in the natural sciences raises more philosophical questions than the questions about nature, relativity, essentiality, existence or the subjective vs. objective character of time: it raises questions about modeling, representing, evaluating, about criteria of explanations and timescales delimitations, etc., i.e. questions proper to philosophy of science. Finally, a more general problem emerges from those sketchy considerations regarding time in the natural sciences: are all these notions of time in various sciences identical? And in this case, is it because they are all reducible to the same physical time? Or because each of them instantiate the same general concept of time? In the latter case, then, how are they related, given the assumption that in all scientific models are involving something common that we want to call time? A philosophical investigation of time in distinct disciplines of the natural sciences would provide preliminary considerations to formulate and suggest hypothetical solutions to such questions. This is the endeavor of the present book.

1.2 Structure of the Book

Although each scientific discipline has a specific approach to time, specific tools to measure time and ways to define timescales, and distinctive methodologies for implementing time in their models, pervasive issues about time still arise in all sciences: this volume relies on the idea, shared by the authors, that the general metaphysics of time may have overlooked specificities of time as it is dealt with in the special sciences, and finally, that a comparative approach to these pervasive issues regarding time in the sciences would be the best way to provide a general philosophy of time with a philosophical understanding of what time in nature is for those who strive to understand nature. In addition to the controversies about the passage of time in physics and metaphysics, it will study issues of time that have come recently under focus in distinct areas of science and philosophy of science: the non-standard logics that address time differences (past, present and future), the various timescales used for modeling natural systems, the retro-causality issue in quantum physics, the controversies about dating last common ancestors in specific families such as primates in biology, the problems of articulating macroevolution and microevolution in evolutionary in the face of recent advances by developmental theory, paleontology and molecular phylogenetics, etc. The structure of the book, as well as its interdisciplinary scope, which mixes philosophers' and scientists' approaches, reflects these shared commitments, and tries to define a common language for difficult philosophical questions ordinarily approached from distinct perspectives.

The first section, entitled "*The experience of time and the scientific framing of time*", explores the experience of time in traditional philosophy and its relation to time in nature, with a set of chapters that bring together what human experience, metaphysical questioning and physics (especially relativity theory) enable philosophers to say about time. Ismael's chapter sketches the difference between our lived experience of time as flowing and time measured in a scientific context, as it appears in our best current physics. She deflates the apparent tension between the phenomenology of time and the 'block universe' view of physics (studied in Bouton's chapter), by arguing that they are two different temporal perspectives that depend on the considered reference frame. Aiming to construe a very general philosophical framework for expressing time, the second chapter, by Proietti, moves to another level of abstraction and presents attempts to formulate time in the formal framework of tense logic, as it arose after Arthur Prior's early achievements. Here time perception is confronted with the most formal science, that is to say, mathematical logic, which solves two paradoxes of time from the theory of relativity: not only the paradox of the twins, but also the one of time travel in the past. In her chapter, Lesne begins with a statement attributed to Poincaré: "Time is what a clock measures", which could be the basic definition of time in natural sciences. She shows that beyond this definition, the different sciences employ different representations of time (quantitative, qualitative, continuous, discrete, etc.) relative to their various purposes, so that time in this context seems to be an operational concept rather than a metaphysical one. Lesne also addresses the problems of the multiplicity of timescales at which a system should be considered, the selection of the most adequate timescale (and time variable) and the switching between timescales. This notion of timescale is crucial not only in physics, but also in paleontology and evolutionary biology, as shown in Ricqlès' and Huneman's chapters. The chapter by Braddon-Mitchell and Miller examines the issue, raised by Lesne's chapter, of multiplicity and equivocation of time as it is handled by various sciences. They suggest that the special sciences – biology, evolutionary theory, geology, cognitive science, ecology and so forth – perform a "subject setting role" for lower-level disciplines such as physics, which establishes whether there is anything like time in the world and what its hidden nature is. The interdisciplinary research on the nature of time can be fruitful despite the fact that the disciplines in question have different approaches, since the diversity of explanatory projects contributes to the overall project of locating and describing the phenomenon called time.

The second section, entitled "*Time paradoxes in physics*", studies some puzzling paradoxes about time raised by the physical sciences.

The first one comes from the theory of relativity. Proietti's chapter deals with the twins paradox and the grandfather paradox from a logical point of view. These issues lead to a more general paradox: physics, which could be defined as the study of matter and motion in space and time, tells us that there is no time. This claim is often related to the special theory of relativity. It has been indeed stated by the physics of special relativity that the non existence of an absolute present may entail the fact that all events, be they past, present or future, are on a par, which means that the difference between non-present and present that seems to be constitutive of time is just a subjective aspect of reality. This is a prominent metaphysical view of time inspired by physics, called the "block universe", i.e. a general four-dimensional space-time in which all events are positioned and where there is no 'flow' of time. The "block universe" favored by many physicists and philosophers of physics (such as Eddington 1990; Weyl 1949; Gödel 1949; Grünbaum 1963; Putnam 1967; Price 1996) is rather counterintuitive, and a discussion of the difficulties it raises and of its relation to the special theory of relativity is provided by Bouton's chapter. Bouton unpacks the notion of a "block universe" into different claims and interpretations. He then considers three main arguments in favour of the implication of the block universe view by relativity and argues that some of the arguments are incomplete or fallacious. Even the strongest argument (the "Rietdijk-Putnam argument") does not absolutely imply the block universe. The second paradox concerns quantum mechanics. More recently, the issue of the arrow of time has been raised from a new perspective in quantum physics: here the controversies focus on "retrocausation" or "backward causation", i.e. the possibility of changing past quantum states of the world. The reader is then introduced to fascinating paradoxes about retro-causality in quantum mechanics, where it is thought future events can cause past ones. Retro-causality is introduced here in order to solve one of the puzzling consequences of quantum mechanics, which had been pointed to in the famous Einstein-Podolsky-Rosen thought experiment (published in 1935). Two particles that have interacted with each other are always correlated thereafter, so that a measurement on one of the two particles influences the state of the other particle. This phenomenon called "quantum entanglement" or "non-separability" seems to imply an action at a distance ("non-locality"), in the sense that measuring a particle immediately determines the state of the other, regardless of the distance between them in the universe. The entangled particles "behave synchronously with no intermediary; it is nonlocal." (Albert and Galchen 2009, p. 32) This non-local action contradicts the special theory of relativity, which states that two events cannot have a causal relationship between them if they are "space-like" separated in space-time, that is, if they cannot be connected by at least one light beam. Beyond this limit, no interaction is possible. This "spooky action at a distance" is the reason why Einstein was dubious, until the end of his life, about quantum mechanics, considering that it was an incomplete description of reality. He thought that there should be "hidden variables" that explain in a deterministic and local way the strange behavior of the quantum particles. Yet Bell's theorem established in the 1960s, and the experiments to which it gave rise confirmed the "non-locality". Bell showed how even a realistic interpretation of quantum mechanics involving hidden variables – what Einstein had in mind – implies non-locality.

In their chapter, Price and Wharton point out that this reasoning nevertheless requires a specific assumption, namely that retrocausality is impossible. They show that such notion of retro-causality can be exemplified by a phenomenon Einstein discovered in 1905 along with the theory of relativity and before quantum mechanics: the quantization of light, namely the discovery that light is absorbed in discrete "quanta." Their new argument, based on the polarization of light, states that the hidden variables theory could escape the objection of "non-locality" through the concept of "retrocausality". This argument would have been available to Einstein after 1905, and would have allowed him an escape route from the 'spooky' action at a distance in quantum mechanics.

Berkovitz addresses this same issue of retrocausality but in the framework of the Bohmian model of quantum mechanics, which, in contrast with the Copenhagen interpretation introduced by Niels Bohr and assumed in many discussions, offers a deterministic, local 'hidden variables' model of the Einstein-Podolsky-Rosen experiment. He shows that by postulating backward causation, retro-causal interpretations of quantum mechanics could explain away the non-local influences and accordingly reconcile quantum mechanics with relativity. In this framework, the measurement on a particle entangled with another particle does not influence "at a distance" the state of the other particle but it influences, by a process of "retrocausality", the past initial state of the two particles system. This interpretation of quantum mechanics raises a new challenge for the issue of causation, and in particular, for Reichenbach's principle of the common cause. More generally, the concept of "retrocausation" challenges the idea of a causal arrow of time, according to which the direction of time would be given by the anteriority of the cause to its effects. To that extent, this concept can provide grist to the mill of those who support the block universe view, since in this model of the universe, there is no objective direction of time.

Finally, from the epistemic side, Barberousse and Ardourel investigate the general issue of studying and measuring time, which is assumed to be continuous, through models with discrete intrinsic time such as computer models, addressing in particular the recent paradigm of "discrete mechanics" in physics. Discrete mechanics uses consistent models where time is represented as discontinuous, as if this property of time was not intrinsic but relative to the theory used to describe physical phenomena. To that extent, the question of whether the structure of time is continuous, and more generally the question concerning the nature of time itself, is beyond the scope of physical theories. This chapter can illustrate Lesne's idea that what matters to natural science is not the metaphysical nature of time, but the more practical and operational issues of how to represent time and to choose an appropriate time variable.

These four chapters suggest that the questions of the existence, the nature and the properties of time as it is represented in physics depend on the field of physics that is considered. And within each area, there may be variations between special and general relativity, among the various interpretations of quantum mechanics (with or without hidden variables, deterministic or indeterministic), and even between different versions of the classical mechanics (e.g. discrete or continuous). Any search for a unitary concept of time in physics must take account of this plurality of approaches and contexts, which does not exclude the possibility of a common core to all these physical conceptions of time (as succession for example). This consequence challenges the intuitive notion of "physical time" as one single phenomenon, which is often assumed in philosophical discussions about time. Such a result therefore illustrates the way an investigation of time as handled by distinct sciences may directly impinge upon the metaphysical examination of the nature of time.

The third and fourth sections concern biology and realize the ambition of the book to disentangle philosophy of time and issues in physics.

As biologist Theodosius Dobzhansky famously said: "nothing in biology makes sense except in the light of evolution" (1973). Therefore the distinctive problems and conceptions of time in the life sciences have to be first sought in an investigation of the epistemological and metaphysical aspects of evolutionary biology that concern time. Evolutionary biology brought to the fore several peculiar ways of handling time scientifically. In this framework, the living beings are understood historically, which means that they are carrying information about their past, and that their nature is to some extent dependent upon the contingencies of their past history. This raises questions about the specific evidences regarding the knowledge about the past. Moreover, within evolutionary biology, explanatory goals are multiple and varied, which means that timescales are different – the timescale of molecular phylogenetics, for example, is not the timescale of population genetics or behavioural ecology –, and this entails issues about their connection and compatibility. The last problem is raised by the fact that biology is, as Ernst Mayr famously remarked, divided between the biology of "proximate causes", considering functioning of individual organisms and often asking how-questions, and the biology of "ultimate causes", that is mostly evolutionary, i.e. answers questions by considering past populations of a species, and considers mostly why-questions (Mayr 1961).

This division may plausibly entail major differences in the way time is, in each case, modelled, represented, assumed, or just handled.

The third section of the book, entitled "*Dealing with deep time (Paleontology)*", directly addresses what has been called "deep time" in evolutionary theory, namely the fact that it deals with a time incommensurably longer than the ordinary human life. This section is therefore devoted to deep time in the sciences that are part of evolutionary biology or related to it: phylogeny, paleontology, and geology. Deep time may confront us with interesting methodological peculiarities and challenges. Philosopher John Huss's chapter provides a perspective on the epistemology of knowledge of the past by addressing the specific issue of epistemic limits to our reconstruction of past species from presently existing traces. Huss argues that the knowledge of the remote past is not as imperfect as it seems. Absolute dating, based on radioactive decay, can complete relative dating. We find here again a reflection about the notion of timescale – considered in the first part of the volume – as it is used in paleontology, addressed through a comparison between relative timescale and absolute timescale. Huss's chapter also includes a case study of the color of dinosaurs, and shows that, in a counterintuitive manner, this quest is not as desperate as it would initially seem. It was once thought that this would be forever epistemically inaccessible to us, but recent advances in the understanding of fossilized melanosomes allow paleontologists to reconstruct the colors of some dinosaurs. Discussing Darwin's book metaphor, whereby the missing pieces in the fossil records are compared with missing chapters, pages, lines, and words in the book of the history of Earth, he suggests in his conclusion that the absence of evidence of life may in fact be evidence of absence of life.

Then two biologists, De Ricqlès in functional paleomorphology and Tassy in systematics, address the peculiarities of such kind of time, which lies at the intersection of geology and biology, and the issues it raises for both classification and paleontological enquiry. In his chapter De Ricqlès points out the differences between the timescales appropriate for the paleontologist and for the biologist and draws the consequences that derive from the difference between considering extinct species and considering extant species – especially in terms of epistemic constraints. Tassy compares the linear time of geology with the branching time of living beings, especially as it has been understood and modeled through the powerful and now dominant paradigm of classification, namely cladism, stemming from Willi Hennig's view of taxonomy and definition of genuine phylogenetic categories.

The fourth section, entitled "*Time of the Evolution (Evolutionary and Developmental Biology)*" deals with the biology of the processes of change, and the kind of knowledge it provides. Population genetics – the science of the variation of gene frequencies in populations under the action of natural selection – has been the core of the so-called Modern Synthesis of the 1930s–1950s that synthesized Mendelian genetics and Darwinian evolutionary theory. In this section, two chapters are concerned with specificities of time in population genetics, as a straightforward way of modeling the process of natural selection. Importantly, some issues appear to resonate with issues seen in the philosophy of physics: reversibility and irreversibility are a central problem in evolutionary theory, as made clear by Gayon's and Montévil's chapter on reproduction and reversibility in theoretical population genetics, experimental population genetics and phylogeny – here too, like in physics, reversibility, as one assumes it in theoretical population genetics modeling, coexists with irreversibility, as some so-called laws in paleontology (e.g. Dollo's law) tended to emphasize. Yet in its conclusions the chapter undermines the usual parallels between population genetics and classical mechanics or thermodynamics, on the grounds that the reversibility issues are precisely not structured in the same way.

Then the internal complexity of evolutionary biology, and its pluralistic understanding of time, is brought to the fore in Huneman's chapter, which illustrates once again the importance of taking timescale into account when addressing issues in evolutionary biology: within this science, there exists discontinuities in time as one goes from small timescales (e.g. population genetics) to large timescales (e.g. history of the succession of clades, mass extinctions, etc.). Following a distinction made by mathematician Mandelbrot, Huneman distinguishes between "mild randomness" and "wild randomness", and argues that at very large timescales randomness is more likely to be wild than mild. For this reason it is not possible to extrapolate from microevolution to macroevolution.

Finally, the last chapter on biology by Nicoglou concerns "functional biology", or biology of the proximate mechanisms. Development, that is, the trajectory of an individual from the zygote stage to the adult stage, is a process in time, whose timescale is decoupled from the evolutionary timescales, and which may include various processes at distinct space-scales and time-scales (molecular, cellular, etc.). Developmental theory therefore has to integrate these multiple processes, each of them likely to be modeled in its own right. The time of development results from this integration and therefore displays much specificity. Nicoglou's chapter addresses these peculiarities of developmental timing, and its relation to other timescales. The whole section therefore sets the stage for a systematic apprehension of the articulation of the specific times proper to each level of biological change.

One of the conclusions of this book is that the relationship of natural sciences to the nature of time can take very different forms. In particular, one may distinguish ontological versus epistemic approaches, or scientific realism versus scientific instrumentalism. On the one hand, one may think that the science of nature, reduced to physics, is able to discover the hidden nature of time, to reveal what it is in itself (even if this truth is that time, at least the time as we perceive it, does not exist). But on the other hand, we can consider that the natural sciences do not claim to answer the question of the nature of time itself, they simply give us some information on the nature of time as they use it and represent it: time as a variable 't', continuity versus discontinuity, quantitative versus qualitative succession, irreversibility versus reversibility, absolute present versus relative present, A series-relations (past, present and future) versus B series-relations (succession and simultaneity), etc. This information varies depending on the methods applied, the role assigned to the time, time and space scales implemented, theoretical and experimental contexts used. From this epistemic perspective, "The physicist means only by 'fundamental' 'our theory doesn't say anything more about the nature of these things beyond what they do in our theory'", as Braddon-Mitchell and Miller put it (Chap. 5, Sect. 5.3.3.1). In this sense, the natural sciences show us many aspects of the concept of time as an instrument of knowledge, and raise several questions about that to which they do not give a univocal answer: how are these multiple aspects of time mutually compatible? Do they refer to a common structure of time? Or is such a unity of 'time', which is not guaranteed *a priori*, an illusion? The general issue of reductionism (to fundamental physics' time) vs. pluralism lurks under such problems. We don't have to decide on this issue here, but we feel that it could not be properly addressed without patiently investigating the peculiarities, specificities, features and difficulties of the nature of time in the special natural sciences.

Given the diversity of approaches to the problem of time presented in this volume, it may seem like a collection of papers with nothing in common except the word 'time' in the title: Braddon-Mitchell and Miller in their chapter raise this important methodological issue that any interdisciplinary inquiry has to face. But they suggest that the range and variety of approaches both to the study of time and the use of the concept of time, offer genuine novel insights into the unifying theme and contribute in diverse ways to a better understanding of the nature of time. In conclusion, we would like to highlight this point by a few examples of transversal topics that can illustrate the connections between the chapters of this volume. We select three key notions (but the reader will easily find other examples): the status of the past, the notion of timescale and time's arrow.

It is often said that in physics, the distinction between past, present and future is not fundamental, as it does not appear in the equations that express the laws of nature. However, the notion of past, which is investigated in various forms in this volume, seems to be significant for the other domains. It concerns not only psychology (see the experience of the past described in Ismael's chapter), but also paleontology and evolutionary biology, which cast a crucial light onto the semantics of the past in the natural sciences. And even in physics, in relativity theory, the notion of the past seems to be used in a specific way, through the concept of 'backward light cone'. These different approaches show that such a notion of "past" cannot be easily reduced or eliminated from the conceptual framework of the natural sciences – even though arguments from fundamental physics would dismiss the objective reality of "past". Chapters on evolutionary biology, on biology and paleontology show here that the notion of a difference between extinct and extant, past and present, plays a structuring role in the theoretical frameworks elaborated in those fields. Whether or not this leads to a unified notion of "past", one is nevertheless reminded here that it takes more than a thorough analysis of theoretical physics to motivate an eliminativist view regarding the past. The additional step one has therefore to take is to decide whether the philosopher's task from this point on consists in the endeavor to elaborate a unified theory of time across fields, or whether, inversely, she rather has to design a coherent pluralistic science-inspired theory of time. To this extent, the present collection of chapters contributes to the recently defined field of "metaphysics of science" (Bird 2007; Ladyman and Ross 2009; Maudlin 2007) – namely, the exploration of fundamental metaphysical concepts based on what our best science tells us – by sketching an alternative between two avenues of development.

Another transversal notion of this volume is that of timescale. The notion of scale, which originally comes from cartography and architecture, is widespread in the natural sciences. At a certain speed scale, close to the speed of light, the laws of Newtonian mechanics must be replaced by those of relativity. At a certain size scale, at the atomic level, the laws of quantum mechanics apply. Spatial scale and time scale are related, for example "in physics, fast processes generally correspond to spatially microscopic processes, and slow ones to spatially macroscopic ones" (Lesne's chapter, Sect. 4.4). Similarly, at a long time scale, the temporal characteristics of evolution change. On the scale of the individual organism, the timing of development is different again. With respect to the considered time scale, nature seems not to obey the same laws – at least within the models used to account for it. A possible lesson of the compared analyses included in this volume would therefore be the following: the notion of timescale – involving possibilities of timescale coupling and decoupling – is constitutive of the complete notion of time, as it is found in the theoretical activities of the natural sciences. Thus, a philosophical inquiry into the nature of time according to the metaphysics of science project should include a reflection of the notion of timescales, its role in various ontological fields, its significance relative to the measures of time, and the meaning involved, for the actual world, of the actual decoupling of timescales across and within scientific fields. It is not the place here to develop a study on the concept of timescale: suffice it to say that its study cannot be carried out without an interdisciplinary approach that confronts the uses of this notion not only in the various fields of physics, but also in developmental biology, evolutionary biology and paleontology.

A similar lesson holds for the notion of time's arrow. Most often, debates about the 'arrow of time' focus on physics. This question is traditionally related to the increase of entropy according to the Second Law of Thermodynamics. In this framework, the apparent inconsistency between the irreversibility of phenomena at the macro-level and the reversibility of molecular dynamics at the micro-level is also a matter of scales (see Lesne's chapter, Sect. 4.9). Another approach attempts to complete the thermodynamic arrow of time with a causal arrow of time (Reichenbach 1956): in this view, the direction of time is based on the anteriority of the cause to its effects. The cause-effect relationship always seems to be aligned with the 'past to future' order, and such an approach interprets this alignment in a foundational way. In this volume, we can see how the validity of the causal theory of time varies with respect to the field of physics that is considered. The special theory of relativity doesn't conflict with this theory: relationships of succession are absolute between two events whose space-time interval is time-like or light-like, meaning that they can be linked by a causal relationship. However, Gödel (1949) pointed out that the general theory of relativity allows possible worlds in which it is possible to travel in any region of the past, suggesting that the 'earlier-later' relationship therefore appears non objective (see Bouton's chapter Sect. 6.4). As seen above, the causal theory of time is also challenged by more recent interpretations of quantum mechanics that appeal to the notion of retrocausality (see Price's and Wharton's chapter and Berkovitz's chapter). To shed new light on this debate on time's arrow, it is useful to consider other natural sciences, as it's done in the fourth section of this volume – for instance evolutionary biology, since one commonly assumes that "evolution is unique and irreversible". In their chapter on theoretical population genetics, Gayon and Montevil distinguish three meanings of reversibility, which are often entangled. Retrodictibility is the property of a physical system according to which it is possible to infer the past states from the present state; 'mathematical' reversibility is understood as time-symmetry, the property of equations of being invariant if the sign of time is reversed, like the fundamental laws of mechanics; and 'physical' reversibility means that a physical system can spontaneously return to a prior physical state. Contrary to the fundamental laws of mechanics, the laws of population genetics are not always reversible, and exhibit an evolutionary time's arrow in some cases. Gayon and Montevil explain that retrodictability seems obvious for deterministic models of population genetics, while it does not apply to stochastic models. On the contrary, time reversibility in the mathematical sense is instantiated in some important stochastic models of population genetics, whereas basic deterministic models do not exhibit this property. Concerning 'physical' irreversibility, they study Fisher's attempt to draw certain formal analogies between entropy in the mechanistic models introduced by Boltzmann to analyse physical systems, and the evolutionary models proposed by Darwin to explain adaptation in biological systems. Even if they are skeptical regarding this comparison between thermodynamics and deterministic models of population genetics, their reflection shows how the concept of reversibility/irreversibility can circulate from physics to biology, and generate representations of time that vary within the same area. This excursion into evolutionary biology indicates that the general notion of "time's arrow" is not exhausted by a thorough examination of the subdisciplines of physics – even though this shows how problematic such a common-sensical notion is. It is therefore a safe bet to think that a dialogue between the various theoretical stances regarding time's arrow across various empirical sciences will eventually provide a clearer idea of what it is, how it should appear to us in distinct ontological regions, and why this is the case.

As one might guess, the contributors of this volume would not all concur on the same answers to these questions. Yet they share the idea that no answer can be reached without starting with a thorough examination of time as it is for the many natural sciences. Even if the task of comparing and articulating the results of those investigations is left open, the authors wanted to contribute materials that would bring together lessons from the philosophy of science and the metaphysics of time, in a project that could be appropriately named the "metaphysics of the science of time".

Acknowledgements The authors are grateful to Andrew McFarland (The City University of New York) for a thorough language check of the text.

References

Albert, D., and R. Galchen. 2009. A quantum threat to special relativity. *Scientific American* 300: 32–39.

Amundson, R. 2005. *The changing role of the embryo in evolutionary thought: Roots of Evo-Devo*. Cambridge: Cambridge University Press.

Ashtekar, A., and V. Petkov, (eds.). 2014. *Springer handbook of spacetime*. Berlin : Springer.

Baer, K.E. 1828. *Über die Entwicklungsgeschichte der Thiere*. Königsberg: Bornträger.

Bardon, A. 2012. *The future of the philosophy of time*. New York: Routledge.

Bardon, A., and H. Dyke, (eds.). (2013). *A companion to the philosophy of time*. Chichester: Wiley-Blackwell.

Barriel, V. 2015. Character. In *Handbook of evolutionary thinking in the sciences*, ed. T. Heams, P. Huneman, G. Lecointre, and M. Silberstein, 115–140. Dordrecht: Springer.

Bird, A. 2007. *Nature's metaphysics: Laws and properties*. Oxford: Oxford University Press.

Boltzmann, L. 1964 [1896]. *Lectures on gas theory*. Trans. Stephen G. Brush. Berkeley: University of California Press.

———. 1966 [1877]. On the relation of a general mechanical theorem to the second law of thermodynamics. In *Kinetic theory. Volume 2: Irreversible processes*, ed. S.G. Brush, 188–193. Oxford: Pergamon Press.

Bowler, P. 1971. Preformation and pre-existence in the seventeenth century: A brief analysis. *Journal of the History of Biology* 4 (2): 221–244.

Boyer, C. 1949. *The history of Calculus*. New York: Dover Publications.

Callender, C. (ed.). (2011). *The Oxford handbook of philosophy of time*. Oxford: Oxford University Press.

Canguilhem, G., G. Lapassade, G. Piquemal, and J. Ullman. 1962. *Du développement à l'évolution au 19ème siècle*. Paris: PUF.

Chang, H. 2004. *Inventing temperature: Measurement and scientific progress*. New York: Oxford University Press.

Chibbaro, S., L. Rondoni, and A. Vulpiani. 2014. *Reductionism, emergence and levels of reality*. Dordrecht: Springer.

Danchin, E., A. Charmontier, F.A. Champagne, A. Mesoudi, B. Pujol, and S. Blanchet. 2011. Beyond DNA: Integrating inclusive inheritance into an extended theory of evolution. *Nature Reviews Genetics* 12: 475–486.

Danchin, E., and A. Pocheville. 2014. Inheritance is where physiology meets evolution. *Journal of Physiology* 592: 2307–2317.

Dawkins, R. 1982. *The extended phenotype*. Oxford: Oxford University Press.

Dieks, D. (ed.). (2006). *The ontology of spacetime*. Amsterdam: Elsevier.

———. (ed.). (2008). *The ontology of spacetime II*. Amsterdam: Elsevier.

Dobzhansky, Th. 1973. Nothing in biology makes sense except in the light of evolution. *American Biology Teacher* 35 (3): 125–129.

Duchesneau, F. 1987. *Genèse de la théorie cellulaire*. Paris: Vrin.

Dupont, J.-C. 2007. Pre-kantian revival of epigenesis: Caspar Friedrich Wolff's *De formatione intestinorum*. In *Understanding purpose: Kant and the philosophy of biology*, ed. P. Huneman, 37–49. Rochester: University Rochester Press.

Eddington, A. 1990. *Space, time and gravitation. An outline of the general relativity theory*. Cambridge: Cambridge University Press.

Flood, R., and M. Lockwood, (eds.). 1986. *The nature of time*. Oxford: Blackwell.

Gayon, J. 1998. *Darwinism's struggle for survival: Heredity and the hypothesis of natural selection*. Trans. M. Cobb. Cambridge: Cambridge University press.

Gödel, K. 1949. A remark about the relationship between relativity theory and idealistic philosophy. In *Albert Einstein: Philosopher-scientist*, ed. P.A. Schilpp, 557–562. La Salle: Open Court.

Gold, T., and H. Bondi, (eds.) (1967). *The nature of time*. Ithaca: Cornell University Press.
Gould, S.J. 1977. *Ontogeny and phylogeny*. Cambridge, MA: Harvard University Press.
———. 1989. *Wonderful life: The burgess shale and the nature of history*. New York: W. W. Norton.
Goy, I. 2014. Epigenetic theories. Caspar Friedrich Wolff and Immanuel Kant. In *Kant's theory of biology*, ed. I. Goy and E. Watkins, 45–60. Berlin/New York: De Gruyter.
Grünbaum, A. 1963. *Philosophical problems of space and time*. New York: Alfred A. Knopf.
Hoffheimer, M.H. 1982. Maupertuis and the eighteenth century critique of pre-existence. *Journal of the History of Biology* 15 (1): 119–144.
Hopwood, N. 2000. Producing development: The anatomy of human embryos and the norms of Wilhelm His. *Bulletin of the History of Medicine* 74(1): 29–79.
Huneman, P. 2007. Reflexive judgement and wolffian embryology: Kant's shift between the first and the third Critique. In *Understanding purpose ? Kant and the philosophy of biology*, Publication Series of the North American Kant Society, ed. P. Huneman, 75–100. Rochester: University of Rochester Press.
Huneman P., and D. Walsh, (eds.) (2017). *Challenging the modern synthesis: Development, inheritance, adaptation*. Oxford: Oxford University Press.
Jablonka, E. (with Lamb, M.). 2005. *Evolution in four dimensions*. Cambridge: MIT Press.
Kauffman, S. 1993. *The origins of order: Self organization and selection in evolution*. New York: Oxford University Press.
Kearl, M. 2009. Wilhelm Roux (1850–1924). *Embryo project encyclopedia* (2009–07-22). ISSN: 1940–5030. http://embryo.asu.edu/handle/10776/1753
Ladyman, J. (with Ross, Don). 2009. *Every thing must go: Metaphysics naturalized*. Oxford: Oxford University Press.
Leibniz, G.W. 1998. *The Leibniz-Clarke correspondence*, ed. H.G. Alexander. Manchester: Manchester University Press.
Lenoir, T. 1982. *The strategy of life. Teleology and mechanism in nineteenth century German biology*. Dordrecht: Reidel.
Lovejoy, A. 1936. *The great chain of being: A study of the history of an idea*. Harvard: Harvard University Press.
Maudlin, T. 2007. *The Metaphysics within physics*. Oxford: Oxford University Press.
Mayr, E. 1961. Cause and effect in biology. *Science* 134 (3489): 1501–1506.
Mesoudi, A., S. Blanchet, A. Charmentier, et al. 2013. Is non-genetic inheritance just a proximate mechanism? A corroboration of the extended evolutionary synthesis. *Biological Theory* 7 (3): 189–195.
North, J. 2011. Time in Thermodynamics. In *The Oxford handbook of philosophy of time*, ed. G. Callender, 312–350. Oxford: Oxford University Press.
Oaklander, L. N. (ed.). (2014). *Debates in the metaphysics of time*. London: Bloomsbury.
Oppenheimer, J. 1967. *Essays in history of embryology*. Cambridge: MIT Press.
Pigliucci, M., and G. Müller, (eds.) (2011). *Evolution: The extended synthesis*. Cambridge: MIT Press.
Price, H. 1996. *Time's arrow and Archimedes' point*. New York/Oxford: Oxford University Press.
Prigogine, I. 1961. *Introduction to the thermodynamics of irreversible processes*. New York: Interscience.
Prigogine, I. (with Nicolis, G.). 1977. *Self-organization in non-equilibrium systems*. New York: Wiley.
Putnam, H. 1967. Time and physical geometry. *The Journal of Philosophy* 64 (8): 240–247.
Reichenbach, H. 1920. *Relativitatstheorie und Erkenntnis a priori*. Berlin: J. Springer.
———. 1928. *Philosophie der Raum-Zeit-Lehre*. Berlin and Leipzig: Walter de Gruyter.
———. 1956. *The direction of time*. Berkeley and Los Angeles: University of California Press.
Richards, R.J. 1992. *The meaning of evolution. The morphological construction and ideological reconstruction of Darwin's theory*. Chicago: Chicago University Press.

Rickles, D., and M. Kon, (eds.) (2014). Flow of Time. *Special issue of the annals of the New York academy of sciences*, 1326, 1–96.

Ricoeur, P. 1984. *Time and narrative*. vol. 1. Trans. K. McLaughlin and P. Pellauer. Chicago: University of Chicago Press.

———. 1985. *Time and narrative*. vol. 2. Trans. K. McLaughlin and P. Pellauer. Chicago: University of Chicago Press.

———. 1988. *Time and narrative*. vol. 3: *Narrated time*. Trans. K. McLaughlin and P. Pellauer. Chicago: University of Chicago Press.

Roe, S. 1980. *Matter, life and generation. Eighteenth century embryology and the Haller-Wolff debate*. Cambridge: Cambridge University Press.

Roger, J. 1963. *Les sciences de la vie dans la pensée française au 18ème siècle*. Paris: Albin Michel.

Rudwick, M.J.S. 2005. *Bursting the limits of time: The reconstruction of geohistory in the age of revolution*. Chicago: University of Chicago Press.

Scott-Phillips, T.C., et al. 2011. Evolutionary theory and the ultimate-proximate distinction in the human behavioral sciences. *Perspectives on Psychological Science* 6 (1): 38–47.

Sklar, L. 1974. *Time, space, and space-time*. Berkeley: University of California Press.

———. 1985. *Philosophy and space-time Physics*. Berkeley: University of California Press.

———. 1993. *Physics and chance. Philosophical issues in the foundations of statistical mechanics*. Cambridge: University of Cambridge University Press.

Uffink, J. 2007. Compendium to the foundations of classical statistical mechanics. In *Handbook for the philosophy of physics*, ed. J. Butterfield and J. Earman, 924–1074. Amsterdam: Elsevier.

Weyl, H. 1949. *Philosophy of mathematics and natural science*. Princeton: Princeton University Press.

Wright, S. 1932. The roles of mutation, inbreeding, crossbreeding and selection in evolution. *Proceedings of the VI International Congress of Genetics* 1: 356–366.

Part I
The Experience of Time and the Scientific Framing of Time

Chapter 2
Passage, Flow, and the Logic of Temporal Perspectives

Jenann Ismael

> *Let us hug to us as closely as we like that there is real succession, that rivers flow and winds blow, that things burn and burst, that men strive and guess and die. All this is the concrete stuff of the manifold, the reality of serial happening, one event after another, in exactly the time spread which we have been at pains to diagram. What does the theory allege except what we find, and what do we find that is not accepted and asserted by the theory? Suppose a pure intelligence, bred outside of time, instructed in the nature of the manifold and the* design of the human space-time worm, *with its mnemic organization and the strands of world history which flank it, and suppose him incarnated among us: what could he have expected the temporal experience to be like except just about what he actually discovers it to be? How, in brief, could processes which endure and succeed each other along the time line appear as anything other than enduring and successive processes?*
>
> D.C. Williams, "The Myth of Passage"

Abstract In this paper, an attempt is made to inject a little formal precision into the discussion of passage. Instead of focusing on the quality of temporal experience, we talk about the content, and we argue that a good many of the issues can be resolved with an examination of the logic of temporal perspectives.

J. Ismael (✉)
University of Arizona, Tucson, AZ, USA
e-mail: jtismael@u.arizona.edu

C. Bouton, P. Huneman (eds.), *Time of Nature and the Nature of Time*, Boston Studies in the Philosophy and History of Science 326,
DOI 10.1007/978-3-319-53725-2_2

2.1 The Topic

One of the most persistent debates both inside and outside of philosophy about the nature of time concerns the relationship between the manifest, flowing time of everyday experience and time as it appears in our best current physics, as one dimension in a static manifold of events.[1] The debate about whether these conflict and, if so, which has priority is deeply hindered by the fact that descriptions of temporal experience tend towards the metaphorical. In describing the experience of time, people often speak of the quality of passage and the felt whoosh of experience.

It's hard to get any traction without an exact, literal account of the content of temporal experience. As John Norton has said: "The real and troubling mystery lies in asking what more can be said. ... It is not at all clear how properly even to describe the passage of time in precise terms."; "We usually end up describing passage with metaphors that prove circular and then, in desperation, gestures." (Norton 2010, p. 23)

I'm going to inject a little formal precision into the discussion of passage. I'm going to suggest that we can resolve some of the issues with an examination of the logic of temporal perspectives. Much has been written about the human experience of time. From Proust to Eliot and Joyce to Khayyam, there is no shortage of literary attempts to capture the complexity of our temporal experience. The goal here to something quite different, viz., to strip away the complexity and get a schematic formal understanding of the logical contours of that experience. We want to obtain a kind of logical skeleton, or schema, of the structure of the experience of beings that process temporal information the way we do.[2]

The first two sections will be descriptive phenomenology. In the third section I'll offer the interpretation of flow and passage, and in the final sections I'll draw some general lessons for metaphysics.

2.2 The Perspective *at* a Time (TEMP)

At different points in our histories, we experience time from different perspectives in the way that we experience a spatially located object, or space itself from different perspectives. What I mean by 'the logic of temporal perspectives' is the formal relations between the contents of representations of time from different perspectives.

---

[1]It is not unlikely that significant departures are needed to attain a theory of quantum gravity. For the controversy about whether physical time is Parmenidean, see Bouton (Chap. 6, this volume), also Price (1996) and Markosian (2010). And for discussion of time in quantum gravity, see Butterfield (1999).

[2]By 'experience', I don't just mean the raw sensory phenomenology. I mean the full evolving contents of consciousness, including the thoughts, emotions, and feelings that make up the everyday flux of mental life in both quality and content.

We start with the Temporally Embedded Momentary Perspective (TEMP), which is a representation of time relativized to a particular moment in a psychological history: a snapshot of time, taken from the here and now. A simple and natural view would have it that the sensory surfaces register information about the environment and relay it to the mind where it produces experience in the way that a video camera registers and relays information to a screen, so that we have real-time covariation of states of the world and states of the screen. Over time, one representational state replaces another, each reflecting the more or less occurrent state of the environment. Although a person watching the screen will remember the passing images and piece them together in his mind to arrive at an idea of how the screen changed over time, there is no representation *of time* on the screen itself, no *accumulation of information* on the screen, the content displayed on the screen doesn't have a dimension representing time.

That's not how it actually works. The brain is integrating information over an interval so that the representational content has temporal breadth (just as the representational content of every visual experience has spatial breadth). There is dispute about how exactly to characterize that content. Almost no-one working scientifically in the field denies that we directly perceive motion, and the received view is that the immediate, uninferred contents of perception span a finite temporal interval. Instead of thinking of perception as generating representations of the momentary state of the environment, cognitive scientists have begun to "reconceive the job of the perceptual system as producing representations that attempt to capture temporally extended *processes* (or, synonymously, *trajectories*) in the environment." (Grush 2007, p. 38)[3] Trajectory estimates don't only represent parts of processes that are already completed. There is a forward-looking, anticipatory component to their content that allows us to prepare reactions to meet expected circumstances. The perceptual system of the baseball player running for a fly ball makes a computation that anticipates where the ball will be. He *sees* (non-inferentially perceives) what his brain anticipates, and so he sees literally where it is going.

Now embed perceptual consciousness in a psychological stream whose full description includes the contents of memory. The temporal content of perceptual states spans a short interval, but they are embedded in a psychological context with very long scope, both in the forward and backward direction.

And that, in its turn, is embedded in an impersonal history that is unbounded in both directions. What we do at the personal level in forming a conception of history to some extent mirrors what perceptual processing does on a very small scale in forming a conception of change over the interval of a specious present.

Like perception, memory is a tremendously complex, and very incompletely understood. In its most general meaning, memory is the process in which information is encoded, stored and retrieved. There are many different processes in the brain that satisfy this description, and scientific questions about what these

[3]The *locus classicus* of the view that the contents of perceptual states have temporal extension, see James (1890).

processes are and how they work are in early stages of understanding. At the conscious level, things are almost as complex, but there are two recognizably different types of memory – known, respectively as episodic and autobiographical memory – that are especially important for our purposes. Episodic memory allows information from past experience to collect in the mind by making records of past events as you experienced them, often in the form of recollected images, sights, sounds and smells. Autobiographical memory gives that information autobiographical form, summarizing, constructing, interpreting, and condensing life experiences, to produce a more or less coherent narrative sense of a personal past.[4]

Memory is notoriously reconstructive and famously selective. Each moment is only very partially and selectively reified in the next, and reification doesn't merely copy, but transforms its objects. We forget, fill in, and embellish. But we do represent our pasts, and re-represent them with every passing moment, reexamining, reevaluating, and reorganizing them in an ongoing process of self-definition. Some of us do this more than others, but all of us do it to some degree. The notion of veridicality that is relevant to assessing autobiographical memory is not straightforward. In some cases, we can assess memories for truth or falsity. If I remember going to the store on a day in which I didn't, I remember falsely. But generally, we need a notion more general than forensic accuracy. An honest or fair representation employing thick ethical concepts is a more subtle matter than a bare transcription of fact. The sorts of mistakes that we are apt to make in memory tend to be self-serving: e.g., remembering a fight or disagreement in a self-flattering way. Search the Times book review for critical discussions of autobiographies to see the complex judgments rendered in critical discussions of autobiographies: Remembering the early stages of a failed marriage as an unending series of attacks on you, for example, or remembering someone as having been a closer friend than they were. The way that we paint our pasts can be misleading or revealing in all of the ways that a portrait can be false or faithful to its subject.

The contents of memory grow by 'accretion of fact' moving up the temporal dimension of a psychological history with the addition of new memories. We do forget things, but if memory is working properly, we remember more than we forget. The result is an asymmetric arrangement, with information accumulating in memory along the temporal dimension, representation and re-representation of the same events in every momentary cross section of experience. It is very easy to fail to see the temporal depth of even momentary states of consciousness, i.e. to forget that it isn't just distributed lengthwise along our world-lines, but contained in every temporal cross section, like a series of Russian Dolls, each sandwiched between others, and each containing internally a representation of those that precede it in the sequence to which every stage adds. Compare this structure with the representational states of a system without a memory, e.g., a guided

[4]There is also procedural memory and semantic memory, which are exercised respectively when you remember a learned skill like riding a bike and hen you consciously recall impersonal factual information as in studying for a test, again. These are heuristic divisions only and theyare not exhaustive. For something much more nuanced, see Sutton (1998) and also http://www.maccs.mq.edu.au/members/profile.html?memberID = 237/

system navigating by an internal map of space, but not representing its past or future. The epistemic states of such a system follow one another in an ordered sequence, but the representational content does not have a temporal dimension. There is no explicit retention of information in the content of the states. Each replaces the next and there is no representation of time at any point in its history, no representational state whose content takes them all into view and integrates them into a history in which they are simultaneously represented in a temporally ordered form. There is not a well-defined internal *point of view* at any point in the system's history whose content spans past, present and future.

Here, schematically, is how to build a system whose representational states have this structure: start with a system that is receiving perceptual input from the environment, add a temporal dimension to its representational states, allow memory to work recursively on those states, first storing them as snapshots of passing experiences (not just images of what was seen, but the full lived content of the experience with all of the thoughts and feelings that were present) and then organizing them into an explicit autobiographical history whose contents are always being reconsidered and retrospectively modified. Now take a temporal cross section of the resulting stream and you have the contents of the TEMP. We have the perceived present, which is itself (recall) a state with some temporal breadth, with the nested memories lining the context in which the perceived present is encountered. The opening lines of Eliot's *Four Quartets,*[5] and the whole of the poem itself are a haunting meditation on the depth of that momentary content, but there is nothing better than Proust's *A la Recherche du Temps Perdu* for conveying the full richness of temporal content (see Proust 2002). A normal person's thought at any given moment contains not just their current experiences, but memories of earlier ones, with an order imposed by autobiographical memory, embedded in the larger narrative of history. And all of this structure – the occurrent thoughts and experiences, the episodic memories, the personal history, and the impersonal history in which it is embedded – is present (in a more or less definite, more or less explicit form) in every momentary part of the psychological life of a normal human adult.[6] This structure is not typically part of the foreground of thought, but it

[5] Time present and time past
Are both perhaps present in time future,
And time future contained in time past. (Eliot, T. S. "Four Quartets 1: Burnt Norton." *Poetry X*. Ed. Jough Dempsey. 13 Jul 2003).

[6]To say that this structure is present in a more or less definite, more or less explicit form in every momentary part of the psychological life of a the life of a consciousness with autobiographical memory is (emphatically) *not* to say that it is fully present in an entirely explicit form in every waking moment of our conscious lives. The psychological presence of this sort of autobiographical content in day-to-day activity is rare and the degree to which it is present at all varies from one person to the next. Some people do carry a self-concept of Proustian complexity, but others have a relatively thin sense of the story of their lives. For strong claims, see Bruner (1987), Dennett (1988), Macintyre (1981). For critique of the strong claims, see Strawson (2004). The claim here is simply the weak and uncontroversial claim this structure is implicit in the ability to think about ourselves as autobiographical subjects and the representational forms we employ in practical deliberation.

is present in a form that allows it to be accessed more or less on demand. The contents of memory are like psychological time capsules, providing each momentary cross section of an evolving consciousness with a compact, backward-looking representation of its own past.

But we don't just represent the past. We also represent the future. We have expectations and plans and fears and hopes. Those are tacked on to our representations of the past, and they too evolve over time, as we learn more about the world and about ourselves. The right way to think of one's view of time is as an evolving representation of history as a whole, with past and future portrayed in different epistemic castes. The events of one's life are encountered from multiple perspectives, first in anticipation, later *in praesentia*, and finally in retrospect. And it is not just the events, but also the perspectives themselves that we represent. Later perspectives have earlier perspectives as constituents. Earlier ones have later ones as constituents. You remember anticipating what it would be like to look back on your early years. You know what it is like to feel sadness at the memory of years of regret attached to expectations for a relationship in light of what actually came to pass. States like this have an exceedingly complex temporal structure of iterated nesting. That kind of structure isn't easy to come by. It has to be built up by passing through those stages; anticipating them, experiencing them, viewing them in retrospect, through all of one's changing fortunes. And you can probably recognize that a lot of the more interesting emotional phenomenology attaches to states with this kind of complex content. If we took a temporal cross section of the contents of a person's consciousness at a particular point in time it would be this rich in temporal content. It would be a view of history, centered on a moment separating remembered history from anticipated future.

2.3 The Perspective *over* Time (TEVPoV)

So much for the TEMP. We obtain the Temporally Evolving Point of View (TEvPoV) – that is, not just the view from a particular moment in a life, but the progression of views over a life as a whole – by stringing together temporally embedded points of view in an order defined by their frame-defining temporal parameter. This is just like taking snapshots of time as it appears at different moments and then stringing them together in temporal order and running the string through a film projector. The contents of TEvPoV capture how time appears over the course of a day, a year, or a life. Formally, the contents of the TEMPs relate to one another as representations *of* time *from* different static frames. The content of the TEvPOV represents the view of time from an *evolving* frame. Compare TEvPoV with the view of space from the perspective of a moving frame of spatial reference. As you move through space, you see an object – a table, say – from different perspectives. Changes in the content of experience as one moves around the object are explained by changes in the frame. The only difference between the

TEvPoV and a moving frame of reference is that the spatial dimension is replaced with time.

Some important features to note. First, we noted that perception isn't like a film camera that projects an image of instantaneous state of the environment on the screen of consciousness. It integrates information over a temporal interval so that the *content* of any perceptual state spans a finite region of both space and time. What you see (using that locution to refer specifically to the contents of the perceptual representations generated by your brain) is movement and change, which belong not to points, but *intervals* of time and *regions* of space. There is no more basic experience from which these are pieced together as the experience of a month is pieced together from experiences of days. There is some dispute among cognitive scientists about whether that content includes a future component and whether acts of awareness are themselves extended.[7] I have sided with Grush on both points because I find his account phenomenologically convincing and the data from temporal illusions persuasive. But we can slot in an extended account model or deny that there is any future temporal content without changing much. These are issues about how the brain processes temporal information, and however they get sorted out, there is nothing metaphysical at stake. The phenomenology of flow – whether it is properly captured by the continuous, direct, experience of flux as I have portrayed it, or one of these other models – is perfectly compatible with a Parmenidean metaphysics, as we shall see below.

Second, over time, as one viewpoint is exchanged for another, asymmetries in long-range temporal vision appear. There is an epistemic asymmetry: the past is remembered whereas the future is inferred. And there is a closely related practical asymmetry. Inject the capacity for decision-governed behavior into a system that is representing the world and you have a system that is not just a consumer of information, but a *producer*. By transforming information into decision and decision into action, the mind becomes not just a passive observer in history, but also an active participant. Decision becomes a source of information that is neither perceptual nor memory-based. And because our beliefs about the future hinge in part on the outcome of our own deliberations, our beliefs about the future can't be stabilized until our own decisions are. We experience the world as an open system with a fixed past and a future that (in the literal sense) awaits decision.[8]

Once we get the process up and running, we can add supervising processes that generate further structures: A continuous cycle of reflective representation and re-representation, preconceiving our histories, planning, acting… and so on. We add differences in emotional attitudes to past and future, and all of the narratively structured emotions like surprise and regret and hope and fear. And we begin to get something close to the full-felt quality and content of our inner lives.

---

[7]See for example Grush (2005) and Dainton (2008). See also Clark (2013), Chuard (2011) and Phillips (2014).

[8]See Joyce (2002), Ismael (2011a), Ramsey (1978), and Price (1992).

2.4 Formal Relations

So now we have a schematic description that captures the temporal contours of our own experience, with the important asymmetries in place and their connections to perception and action. The next task is to find an interpretation for flow in content-level structure. Here we want to look at the content both at and over time. At a time: The fact that the content of experience at any instant spans a finite interval means that the world is perceived at any moment *as* moving, changing, constantly in flux.

Over time: The fact that the TEvPoV is evolving, in the precise sense that the frame is centered *on* different point of time *at* different times gives us a formal interpretation of McTaggart's Moving Now. McTaggart's argument was, simply put, that there is no way to consistently integrate the B-series and the A-series, no way to consistently identify points across the two-series at all times. But there is. The trick is simply to recognize that there is an A-series associated with every moment *t* in time and that 'now' is the name for the fixed point in the function that maps the A-series representation associated with *t* into the B-series. Since the fixed point has different values for the A-series associated with different moments, the now will appear to be evolving, or 'moving through' the B-series.[9]

I prefer to put it in terms of reference frames. There is a frame-dependent representation of time associated with every point along a psychological history of a situated being, and the frame evolves in the sense that it is centered on different points at different moments in that history. This can be described non-question-beggingly in either atemporal or dynamic terms.

The personal experience of time over history relates to the physical conception of time as a frame-dependent to a frame independent one. If we take what is invariant under transformations among all of the temporal perspectives within a life, we get the everyday public image of time; i.e., calendar time, the temporal analogue of the view from nowhere. The only way of integrating the TEMP's located along a psychological history is to see the viewpoint itself as evolving. At any given moment, what we call the present is the moment on which a calendar is centered. It 'changes' in the sense that it has different values (corresponds to different dates) at different times.[10]

[9]The remarks here are a generalization of Chap. 10 in Ismael (2007), integrating the A-series with the B-series by a transformation from a frame-dependent to a frame-independent one where the frame is defined as above; an evolving frame, or if you like a frame centered on the Moving Now, which is interpreted in four-dimensional terms as a fixed point in the mapping between the invariant representation and the various temporally embedded perspectives on time embodied in the different stages of the psychological history of a human being.

[10]From the point of view of the special theory of relativity, the common present of calendar time is local and inherently approximate notion, contingent on our being sufficiently close to one other and moving with relative velocities much less than that of light? The personal present of an individual human at any point along its world-line is the moment on which its representation of time is centered. It 'changes' in the sense that it has different values at different points along its world line.

2.5 The Generator of a Point of View

Neither Newtonian physics nor relativity recognizes a fundamental temporal asymmetry in nature.[11] So the physical basis of the epistemic and practical symmetries have come under close scrutiny. The epistemic asymmetry is relatively well understood.[12] More work needs to be done to understand the nature of the practical asymmetry, but within the framework described here both asymmetries get explained together in a story that is part physical science and part cognitive science. The physical sciences tell us how a certain kind of system – a system that gathers information, stores it, and uses it to guide its behavior – arises in nature. The cognitive sciences tell us how the experience of such a being is structured by the epistemic and practical asymmetries introduced by its point of view.[13] Let's call the processes that structure the experience of a particular kind of being, the generator of a point of view. A point of view is a generalization of the notion of a reference frame. If a frame-dependent representation represents a space in terms that are implicitly relativized to a location, orientation, or state of motion, the generator of a point of view represents a space in terms that are organized around distinctions that have epistemic or practical significance to the systems that use those representations, but do not represent invariant structures of the landscape represented.[14]

We can find in the literature, a useful, schematic rendering of the generator of a point of view for a being whose epistemic and practical situation mirrors our own in a device that James Hartle introduced into the discussion of time called an information gathering and utilizing system (an IGUS, for short).[15] An IGUS is a physical system that has sensors of its environment and uses the information coming through its sensors to navigate. We can describe its workings in mechanical terms, but we are interested here in the formal relations between the contents of perception and the absolute structures on which those contents operate, so we describe its workings in computational terms.

---

[11]Before we can say whether quantum mechanics incorporates such an asymmetry, we need a solution to the measurement problem and an understanding of the ontology. Even if quantum mechanics does impart a temporal asymmetry to nature, it is likely to be washed out by decoherence at the level at which cognitive processes operate.

[12]There is lively dispute about the details, the bulk of the scientific community thinks that an explanation of this asymmetry based on the thermal gradient and additional contingent will be correct. See Albert (2000) and the review by Nick Huggett in *Notre Dame Philosophical Reviews*, http://ndpr.nd.edu/news/23401-time-and-chance/

[13]For a start at this part of the story, see Clark (2013).

[14]Except, of course, insofar as the systems that use those representations are themselves part of the landscape. They appear in the absolute structure of the world in a form that is explicitly relativized to the situation of their users.

[15]See Hartle (2005, p. 101). Hartle himself took the device from Murray Gell-Man, though it has become more closely associated with his own name.

Hartle's example of an IGUS works as follows.[16] It has $n + 1$ memory locations $P0$, $P1$,... Pn which we call 'registers' comprised of a time series of (coarse-grained) images of its external environment. At times separated by intervals t* the image in register Pn is erased and replaced by the image in $Pn - 1$. This is then erased and replaced by the image in $Pn - 2$, and so on. For the last step, the robot captures a new image of its external environment from its sensors and stores it in register $P0$. So, at any given moment, the robot possesses a coarse-grained representation of history extending over a time $(n + 1)t^*$, which it uses to make predictions about its environment at times to the future of the data in $P0$ and direct its behavior based upon these predictions. It does this in two steps employing two different, ongoing processes of computation: U and C.[17] U uses sensory information to update the image in $P0$. C uses the image in $P0$ as the setting for a decision procedure. If we pay attention just to the changing contents of register $P0$ and the processes that operate on those contents, we get something that corresponds to a stream of consciousness. We see an evolving image of the world used to fuel a decision process that guides the gross movements of the robot.

The only addition I made in my discussion was to fill out the internal representational environment a little and emphasize that in us, it is not just the image displayed on the canvas of consciousness (the content of the register $P0$, perhaps with some auxiliary records of stored in back registers about earlier contents), on which decision operates. In us, those images are used to fuel an evolving autobiography, which adds a further layer of quite complex processing that involves much more than passive recording. There is a very complex form of self-definition involved in autobiographizing. This evolving autobiography – which takes the lion's share of conscious mental energy, most ordinary people spend their conscious waking lives, aside from the immediate demands of attention rehearsing and reinterpreting their pasts, planning and reconsidering plans, reflecting on their stories and planning their future – is what really guides decision. This degree of explicit temporal content makes every appearance of being a uniquely human cognitive adaptation. Other animals have internal information-bearing states that mediate stimulus and response. What our minds do is add a temporal dimension to those states with enough room to explicitly represent our whole histories, and let deliberative processes work on those states to select action. And it is in the space opened up by that added dimension in the contents of our representational states that the concept of self as a thing with a past and future, and plans and projects, takes shape. That explicit temporal content is cumulative, evolving, and continuously subjected to the ongoing process of reflective self-definition.

The device is – as Hartle notes – simple enough to be easily analyzed, but complex enough to generate representational states with some of the realistic

[16]Notions of representational content invoked in the description of an IGUS is one that applies to robots and computers as surely as humans.

[17]The letters C and U are chosen to signal the corresponding unconscious and conscious processes in us.

complexity of our own experience. And it can act here as a formal schema that stands in for the much more complex cognitive processing that goes on in us. That formal schema is what relates the Heraclitian character of experience to the Parmenidean character of Time itself. It is what plays the role of the transformation between a frame-dependent and frame –independent representations of time.

2.6 What Time Is *Really* Like

Suppose that you think that from formal point of view, we have a resolution of the apparent conflict between two ways of representing time; time represented in a frame-independent way and time represented relative to a certain kind of reference frame. It is very tempting here to raise the question of which of these two views represents how time *really* is. Defenders of the priority of the phenomenological conception insist that the atemporalist vision is just an abstract representation of a 'really' or 'objectively' or 'mind-independently' flowing time. According to George Ellis, for example, the history of Being is "ongoing irreversible development of time itself." (2008) And according to John Norton: "The passage of time is a real, objective fact that obtains in the world independently of us. . . . It has all the marks of an objective process whose existence is independent of the existence of we humans." (2010, p. 31) A defender of the Parmenidean conception holds that time really is static; any impression of flow is an illusion, an artifact of perspective. So here is Gold: "We ought to eliminate this flow idea from the real picture, but before we can eliminate it we ought to understand how it arises. We should understand that there can be a self-consistent set of rules that would give a beast this kind of phony picture of time." (1967, p. 182) The embedded agent sees a world in progress, unfolding, full of possibility and unrealized potential. But really, according to Gold, this is all illusion, i.e. an artifact of ignorance.

I fail to see how there could be any substantive question about which of these is correct. From the frame-dependent perspective of TEvPoV, the atemporal representation is a static representation of an evolving reality. From an atemporal perspective, TEvPoV is an evolving representation of a static reality. There is a perfect symmetry. Each is implicit in the other, obtainable from it by a transformation akin to the transformation between a frame-dependent and frame-independent representation of space. The idea that one of these is more 'real' than the other is just a mistake. There *is* a well-defined sense in which the TEvPoV is less objective; viz., the TEvPoV represents time in a frame-dependent manner and the specifically Heraclitian features of time (passage and flow) turn out to be frame-dependent.

The TEVPoV contains more information of direct practical relevance to the situated agent, because it represents the events in history at large to those in the evolving here-now of the agent's internal time, and the 'here-now' of the agents own internal time has direct (representationally unmediated) connections to perception and action.[18] Time passes from the evolving point of view of an IGUS in the same way that the rows of houses pass by the passengers on a train. It is an illusion only if we think that it is time itself, rather than our perspective on time, that changes over a life, which is to say only if we think that we have a fixed perspective relative to the Absolute structure of time. I see no reason to convict our ordinary ways of thinking and speaking of that mistake. Indeed, whatever we say in philosophical moods, we all have a good working knowledge of how to integrate temporal information across perspectives that shows that we are *not* making such a mistake.

We have become accustomed to the fact that physics represents the world at an increasingly fine level of resolution, with the result that a lot of what is simple at the level of experience turns out to have a hidden substructure, and gets reconstructed from more fundamental structures. But physics doesn't just go deeper. There is also a progression towards an increasingly Absolute conception of the world, one that aims to capture the intrinsic structure of Being. This progression works by taking what is invariant under transformations between perspectives, where 'perspective' is now widely construed to include relations to any feature of ourselves, our sensory states, our situation, or our local environment. And it is part of the logic of that progression that increasing amounts of the structure that is at the forefront of human thinking about the world gets reconstructed as implicitly relativized to our situation in the world.[19] It should not be surprising that structure that has the most immediate epistemic and practical import for us should be at the forefront of human thinking about the world, or that structure that has the most immediate epistemic and practical import for us should end up being, in this sense, perspectival. We care directly about what is happening nearby and soon. Notions of absolute location in space-time come later, and only as hooks on which to hang information about places we are not at.

There is a tradition in physics that dismisses frame-dependent or coordinate-dependent structure as *unreal*, but that tradition has to be very delicately interpreted. It is dismissed as unreal in cases in which the frame is a virtual object with no physical significance, and the distinction between frame-dependent and frame-independent content is a separation of structure in the mathematical object used to represent from what it represents. But when reference frames represent real features of the situation of viewing agent, the frame-dependent content has physical significance as implicitly relating the objects of representation to his situation. There is nothing illusory about the fact that Paris is far away, or quinoa is healthy.

---

[18]From a four-dimensional point of view, that extra information is reflexive information, and it has a quite complex logical structure. See Perry (2001) and Ismael (2011b).

[19]It is sometimes alleged that the scientific vision of the world loses contact with what matters from a human perspective. There is some justification for this, but it is not quite accurate. What is true is that if we equate the scientific view of the world with fundamental physics, the fundamental structures are very far removed from anything that has immediate epistemic or practical significance for human beings. These appear at somewhat higher levels in the edifice of the scientific view of the world.

But from a more absolute perspective these get reconstructed as relations to a location and particular kind of physical constitution, respectively. What happens in the progression towards an increasingly absolute conception of the world is that implicitly relational structures are explicitly relativized. What we did here was to reconstruct the dynamic features of time as frame-dependent. We found that when we move to a level of description that captures only what is invariant under transformations between temporal perspectives; the dynamic features of time are reconstructed as relations to the evolving point of view of the agent.

I used to think that there is something that is missing in this way of resolving the conflict, i.e. that it defines away a substantive issue. I now think (to use a lovely phrase from P. F. Strawson) that there is a sense in which the problem arises because the solution exists. And the more interesting question is why the problem proves so psychologically resilient. The apparent conflict between the familiar, flowing time of everyday experience and the static time of the Block Universe has a stubborn way of reasserting itself as a substantive and all—important metaphysical disagreement, even in my own mind. It is a reminder of the constant tension in the human between the transcendent and embedded viewpoints, which is in its turn the product of the peculiarly human form of mindedness.

2.7 Generalizing the Lessons

This paper has been about sorting out some of the formal relations between time as it appears from the evolving point of view of experience and time as it appears from the Absolute (or temporally transcendent) perspective adopted by physics.

Once we have a handle on the formalities, we apply the distinction between the Absolute structure of time and those features of our experience of time that turn out to be frame-dependent, and it emerges that flux and flow and all of the so called 'dynamic properties of time' turn out to be frame-dependent. There is nothing illusory about the frame-dependent content. Changes in the point of view from which time is viewed over the course of a life are as real as changes in the point of view from which space is seen as one moves through it. It is just that the only invariant structures are the static relations (such as the order of succession) captured in the Block Universe.

What this reconstruction does is prompt a generalization of the everyday notion of perspective to something with the logical complexity of the generator of a point of view; no longer a static transformation, but a process. And in so doing, it quells the dual tendencies to either reify or reject features of temporal experience as illusory that are actually implicit in the physical conception, as the way time would appear from the perspective of (a certain kind of) embedded creature.

There are some lessons here about how to address arguments that physics is inadequate to some aspect of experience. Philosophers sometimes speak as though the problem is that if one lays the atemporal view of the universe alongside a representation of temporal experience, one doesn't find anything in the atemporal representation that *corresponds* to flow. But this is the wrong account of how physics relates to phenomenology. What physics owes to phenomenology is a non-reductive reconstruction of the contents of the point of view of the agent that tells us how the representational states of an evolved system with a particular combination of epistemic and practical needs would be organized.

To transform an account of the Absolute structure of time into a view of time from the perspective of an embedded agent, we have to understand the perspective from which time is viewed in experience both at a time and over the course of a personal history. The transformation is not a static mapping, but a generative procedure that produces a stream of content in which time is represented from a point of view that evolves relative to the fixed background of absolute time. Formally, the case of time precisely mirrors that of space, with a single dimension in place of space's three dimensions and with the added complication that experience is *extended* in the very dimension *represented* in its content.[20] Once we add the epistemic and practical asymmetries that a creature storing information about its past and making decisions about its future bears to time at different points in its life, we have something that begins to look recognizably like our own temporal experience. The asymmetries are introduced by the agent's perspective. They are asymmetries in her practical and epistemic relations to events, not asymmetries in the intrinsic nature of the events being represented. They vary from one person to the next, and from one moment in time to the next.

There is an assumption built into many of the current practices in analytic metaphysics that for every element of belief or experience, there should be a corresponding element of Being, i.e., that we should be able to take our account of what there is in a perspective-free or Absolute conception of the world and find something that corresponds to every structure that is present at the level of experience. But there are all kinds of reasons that we shouldn't expect extensional correspondences that are that are readily apparent when you look at the transformations wrought by the generative procedure. Relating the world as it appears in physics to phenomenology requires nothing less in general than the generative procedure, iterated over time, to allow structure to (i) train up the schema, (ii) accumulate in the content, and (iii) evolve over time. The training up period builds a lot of structure into our models of the world in the form of invariant patterns stabilized out of experience in the training period (expressed in a mature theory as about dispositions, capacities, chances, and dispositions) used to guide expectation and action. Information about the way things generally hang together gets built into our representations of the way they are in the here and now, so that information gets all mixed up and spread around in a manner that precisely confounds those nice, discrete mappings. The accumulation of content means that even momentary

[20]In physical terms, space-time is more fundamental than either space or time, so that we should really be talking about the relationship between the Absolute structure of space-time and the view from a frame of reference that is moving relative to the fixed structure of Absolute space-time.

representations have content that spans all of history. And the evolution means that a static domain is experienced as in flux.

Separating of ‘the products of the processing’ from the absolute structures on which that processing operates gives us an account of the phenomenology that neither reifies the features that turn out to be frame-dependent, nor dismisses them as illusory. The picture we end up with is very close to one that Frank Ramsey defended. What changes or evolves is not time, or Being, but a probability distribution and an evolving catalogue of practical possibilities defined over Being that represents the changing epistemic and practical relations that an IGUS bears to the events in history as we trace a path up its world line.[21] The mistake that opponents of the atemporal conception make is to reify, in the absolute structure of time, features that belong properly to the point of view of such a being.

References

Albert, D.Z. 2000. *Time and chance*. Harvard: Harvard University Press.
Bruner, J. 1987. Life as narrative. *Social Research* 54: 11–32.
Butterfield, J. 1999. *The arguments of time*. Oxford: Oxford Clarendon Press.
Chuard, P. 2011. Temporal experiences and their parts. *Philosopher's Imprint* 11: 1–28.
Clark, A. 2013. Whatever next? Predictive brains, situated agents, and the future of cognitive science. *Behavioral and Brain Sciences* 36 (3): 181–204.
Dainton, B. 2008. Sensing change. *Philosophical Issues* 18: 362–384.
Dennett, D. 1988. Why everyone is a novelist. *Times Literary Supplement*, September 16–22.
Ellis, G. 2008. On the flow of time. arXiv:0812.0240.
Gold, Th. (Ed.). (1967). *The nature of time*. Ithaca: Cornell University Press.
Grush, R. 2005. Internal models and the construction of time: Generalizing from state estimation to trajectory estimation to addresses temporal features of perception, including temporal illusions. *Journal of Neural Engineering* 2: 209–218.
———. 2007. Time and experience. In *Philosophie der Zeit*, ed. Th. Müller, 27–44. Frankfurt: Klosterman.
Hartle, J. 2005. The physics of now. *American Journal of Physics* 73 (2): 101–109.
Ismael, J. 2007. *The situated self*. Oxford: Oxford University Press.
———. 2011a. Decision and the open future. In *The future of the philosophy of time*, ed. A. Bardon, 149–168. Oxford: Oxford University Press.
———. 2011b. Reflexivity, fixed points, and semantic descent; How I learned to stop worrying and love reflexivity. *Acta Analytica* 26 (4): 295–310.
———. 2016. *How physics makes us free*. Oxford: Oxford University Press.
James, W. 1890. *The principles of psychology*. New York: Dover.
Joyce, J. 2002. Levi on causal decision theory and the possibility of predicting one's own actions. *Philosophical Studies* 110: 69–102.
Macintyre, A. 1981. *After virtue*. London: Duckworth.
Markosian, N. 2010. Time. In *The Stanford encyclopedia of philosophy*, ed. E.N. Zalta. (Winter 2010 Edition). URL = http://plato.stanford.edu/archives/win2010/entries/time/

[21]On the status of these practical possibilities, see Ismael (2016). If anything flows in this picture, it is information, flowing in through perception, over the field of belief, and into the decision procedures and back out into the world in the form of action.

Norton, J. 2010. Time really passes. *Humana Mente: Journal of Philosophical Studies* 13: 23–34.
Perry, J. 2001. *Reference and reflexivity*. Stanford: CSLI Publications.
Phillips, I. 2014. The temporal structure of experience. In *Subjective time: The philosophy, psychology, and neuroscience of temporality*, ed. V. Arstila and D. Lloyd, 139–238. Cambridge, MA: MIT Press.
Price, H. 1992. The direction of causation: Ramsey's ultimate contingency. In *PSA 92 (Proceedings of the Biennial Meeting of the Philosophy of Science Association. Vol. 1992)*, ed. D. Hull, M. Forbes, and K. Okruhlik, vol. 2, 253–267. Chicago: The University of Chicago Press.
———. 1996. *Time's arrow and Archimedes' point: New directions for the physics of time*. Oxford: Oxford University Press.
Proust, M. 2002. *In search of lost time*. London: Allen Lane. (6 vols).
Ramsey, F.P. 1978. General propositions and causality. In *Foundations: Essays in philosophy, logic, mathematics and economics*, ed. D.H. Mellor, 133–151. London: Routledge & Kegan Paul.
Strawson, G. 2004. Against narrativity. *Ratio* (new series) XVII, December 4.
Sutton, J. 1998. *Philosophy and memory traces: Descartes to connectionism*. Cambridge: Cambridge University Press.

Chapter 3
Time of Logics and Time of Physics

Carlo Proietti

Abstract The history of reasoning about time is filled with paradoxes and conundra; contemporary physics is no exception. Logics in general, and more specifically temporal logics, represent a rigorous formal tool in order to solve or clarify problems of this kind. In what follows we will first explain, from the point of view of philosophical logic, what is a paradox and what should count as a solution to it. After that we will illustrate A.N. Prior's formalization of the traditional paradox of *future contingency* and *determinism*. Then we will focus on two modern paradoxes – the *twin paradox* and the *time travel paradox* – and show how an adequate temporal logic can help their framing and understanding.

3.1 Introduction

Temporal logics (or tense logics) have a relatively recent history as a formal discipline, dating back to a pioneering essay of Hans Reichenbach (1947) which predates the fundamental and extensive work of Arthur Norman Prior in the 1950s and the 1960s (see Prior 1957, 1967, 1968). However, logic and time share a long intertwined history in western philosophy and theology. Indeed, logical arguments were the most relevant tool of medieval schoolmen to try and reconcile *divine foreknowledge* with the *contingency of future events*, the latter being a fundamental prerequisite of human *free-will*. This problem is strongly connected with an older one: the Aristotelian problem of future contingency and necessitation (Aristotle 1941) and the *Master Argument* by Diodorus Chronus. The link is an easy one. If God knows, for any event, whether or not it will happen then (a) every proposition concerning future events is already true or false.[1] According to both Aristotle and Diodorus, from (a) it follows that (b) every proposition concerning future events is

[1]Knowing that A implies that A is true and knowing that not-A implies that A is false. This is an intuitive property of knowledge that the medievals expressed with the *motto nihil scitum nisi verum*.

C. Proietti (✉)
University of Lund, Lund, Sweden
e-mail: carlo.proietti@fil.lu.se

C. Bouton, P. Huneman (eds.), *Time of Nature and the Nature of Time*, Boston Studies in the Philosophy and History of Science 326,
DOI 10.1007/978-3-319-53725-2_3

either *necessary* or *impossible* and therefore assuming (a) implies that there is no contingency concerning future events.[2] A. Prior was the first to frame this problem in formal logical terms (chapter VII of Prior 1967) and to "solve the paradox" or at least to clarify the preconditions and the import of the previously purported solutions.

Every paradox, if properly framed, has two essential features: (1) a number of intuitive (and seemingly innocent) propositions are assumed and (2) a counterintuitive proposition or a plain contradiction – such as "*p* and not-*p*" – is derived from the assumptions by simple inferential steps. The derivation of a counterintuitive conclusion may be called a paradox in the *weak sense*. Indeed, counterintuitiveness is not a problem per se and is often explained away by adopting a more adequate background theory. Deriving a contradiction constitutes instead a paradox in a *strong sense* and is the symptom of a deeper problem. Indeed, the derivation of a contradiction indicates that something must have gone wrong in our reasoning: either some assumption is *false* or some inferential step is not *sound*. Therefore, solving a paradox in the strong sense is only done by amending an inferential flaw or an invalid assumption. To solve a paradox in the weak sense, we have instead the additional possibility of explaining away the conclusion while accepting all the inferential steps leading to it.

It is no coincidence that temporal logics originated, in Prior's work, to solve a paradox. Rather, this is a characteristic of the whole history of formal logics: formal theories are created to solve the problems carried by an intuitive or semiformal approach. The most famous example of such a dynamics is probably *Zermelo-Fraenkel set theory*, built up to answer the paradoxes of set theory at its naive stage. In tense logics, as in set theory, paradoxes arise from the consequences of seemingly innocent general assumptions about our notion of tenses. Just as set theory, temporal logics (ideally) play a "hygienic" role: they should help us by discovering flaws in, and possibly amending, our (common sense) understanding of time.

Our reasoning about time, including the formal logical one, builds upon a mathematical/geometrical representation of it.[3] Contemporary physics provides us a new geometrical interpretation of time, framed in the theory of *space-time relativity,* which is very different from the traditional one – we may call it "Newtonian" – where the old paradoxes were discovered. Space-time relativity also comes with a number of new paradoxes. Among the most famous we may count the *twin*

[2]Of course, (a) must hold for the argument to be conclusive and that is where Aristotle and Diodorus diverge in their analysis. Most commentators read the whole chapter IX of Aristotele (1941) as an attempt to consistently reject (a) and therefore discard the Diodorean argument.

[3]Our object of analysis here is the context-independent (or absolute) notion of time, as opposed to the context-dependent (or phenomenological) one (see Ismael, Chap. 2 of this volume). The latter brings in additional elements of complexity, although, as argued by Ismael, it can be framed in rigorous terms.

paradox[4] and the *paradox of time travel* or *grandfather's paradox*.[5] As we will see, Prior's formal analysis of the old paradoxes, and the logics built after it, is formulated in a Newtonian framework, where time is constituted of moments, each moment "containing" all the simultaneous events happening in space.[6] In order to understand and analyze the paradoxes of space-time relativity a different logical apparatus is needed which is still under construction.[7]

The aim of the present chapter is to show how some of the new paradoxes can be analyzed in a logical framework. To make our explanation more vivid we shall define temporal logics by their *model-theoretic semantics*. The latter indeed provide a pictorial representation that better illustrates our analysis. As a preliminary step we need to understand what should count as a solution for a paradox in the weak logical sense. To this end, in Sect. 3.2 we will illustrate Prior's approach to the problem of future contingency. This has the twofold utility of exemplifying a successful logical treatment of a paradox and laying the ground for understanding along which lines rigorous solutions to paradoxes in contemporary physics can succeed. To this end, in Sect. 3.3 we shall present the twin paradox and the time travel paradox. The former is usually deemed as a pseudo-paradox, while the latter as a genuine one. We shall see that both are indeed paradoxes in the weak sense and explain what their difference consists of.

Building a full logic for space-time relativity is an ongoing collective enterprise and lies far beyond the scope of the present introductory work. For our purposes it is

---

[4]The twin paradox dates back to the very early days of relativity theory. Einstein himself stressed as a "peculiar consequence" of the theory the fact that "if at the points A and B of K there are stationary clocks which, viewed in the stationary system, are synchronous; and if the clock at A is moved with the velocity v along the line AB to B, then on its arrival at B the two clocks no longer synchronize, but the clock moved from A to B lags behind the other which has remained at B" (in Einstein 1905 § 4, see also Einstein 1911). This scenario was popularized by Langevin (1911) with the more vivid description of a travel back and forth from Earth in a projectile. Finally, it was Weyl (1918) who upgraded the example with two twins aging differently, one staying on Earth and one travelling in the projectile. However, neither Einstein nor Langevin nor Weyl called it a paradox, since the situation is perfectly consistent with the theory. The alleged paradoxality originated mostly in philosophical discussion, due to the objection that if motion is relative then the situation should be symmetric – from the local perspective of each twin, the other is the one moving – and therefore there seems to be no reason why the twins should age at a different pace. The reader may consult During (2014) for a detailed historical account of how the example developed and crystallized into a paradox.

[5]The original source of the paradox, and of its name, is the science fiction novel by Barjavel (1944) whose main character, an imprudent time traveller, kills his grandfather before the latter meets the time traveller's grandmother.

[6]This was not because Prior was unaware of special and general relativity, the point is that a Newtonian framework is detailed enough for dealing with traditional paradoxes and conundrums. Indeed, when dealing with a paradox, a good logical analysis consists first and foremost in simplifying the picture in a way that is fine-grained enough for the problem to arise and hopefully to articulate a solution. In the case of determinism and free-will the further complexity induced by relativistic space-time is to a large extent superfluous and therefore negligible.

[7]For a more advanced reading see Belnap (1992), McCall (1994) and Strobach (2007).

enough to explain how the syntax and semantics of contemporary tense logics may help us to understand and frame the aforementioned puzzles. As far as we know, nobody has previously analysed the time travel paradox by means of temporal logics, which therefore constitutes the most original contribution of the present chapter.

3.2 The Problem of Future Contingents

As we mentioned in the introduction, one of the most urgent problems of medieval philosophers was to reconcile divine foreknowledge and human free-will. A fundamental prerequisite for free-will is the contingency of the future: if everything happens by necessity there is no place for a real free choice. Divine foreknowledge nevertheless implies that future tensed propositions – such as "Peter will deny three times before the rooster crows"- are true (respectively false) before the event, i.e. before Peter's denial. It seems to follow from the latter that future tensed propositions are true (or false) by *necessity* and that therefore future events are either necessary or impossible and, by consequence, predetermined.

The necessitation of future contingent propositions was a well-known problem since Aristotle and Diodorus. However, according to Prior, the most rigorous construction of a paradoxical argument is due to Peter de Rivo (see Baudry 1950). The argument, as presented by Prior in chapter VII of Prior (1967), has the following structure.

1. Whatever has been the case necessarily (now-unpreventably) has been the case (general assumption)
2. Therefore, if it has been the case that it will be, it necessarily has been the case that it will be
3. Of what will be, it has been the case that it will be (general assumption)
4. Therefore, by 2 and 3, of what will be, it necessarily has been the case that it will be
5. If necessarily, p implies q, and necessarily p, then necessarily q (general assumption)
6. Necessarily, if it has been the case that it will be, it will be (general assumption)
7. Therefore, by applying 5–6, if necessarily it has been the case that it will be, necessarily it will be.
8. Finally, by 4 and 7, if it will be then necessarily it will be

Step number 8 is the fatalistic conclusion at stake. It should be stressed that the concept of necessity (or *settledness*) that appears in this argument is a specific one: it is the necessity belonging to past and present events because they are no more modifiable. Medievals called it *necessitas per accidens*.

As a first step of logical analysis, Prior formalizes de Rivo's argument by means of a language for propositional logic with temporal and modal operators added. The language is recursively defined over a set of *atomic formulas* $p, q, r...$ (intuitively representing immediate present tense propositions like "the cat is on the mat") by using boolean operators $\neg$ ("not"), $\wedge$ ("and"), $\vee$ ("or") and $\rightarrow$ ("if... then") with the further addition of temporal operators F_n ("in n steps it will be the case that"), P_n ("it has been the case that n steps ago") and a necessity operator $\Box$ ("it is necessary that"). Later on we will also use the operator $\Diamond$ ("it is possible that") which is defined as the concatenation $\neg\Box\neg$. Recursive definitional clauses for the language are:

- If φ and ψ are formulas so are $\neg\varphi$, $\varphi\wedge\psi$, $\varphi\vee\psi$ and $\varphi \rightarrow \psi$, $F_n\varphi$ and $\Box\varphi$

De Rivo argument is then formalized as follows.

1. $P_m p \rightarrow \Box P_m p$ (PNP)
2. $P_m F_{m+n} p \rightarrow \Box P_m F_{m+n} p$ (from 1 by substitution of $F_{m+n} p$ for p)
3. $F_n p \rightarrow P_m F_{m+n} p$ (PR)
4. $F_n p \rightarrow \Box P_m F_{m+n} p$ (from 2 and 3 by transitivity of $\rightarrow$)
5. $\Box(p \rightarrow q) \rightarrow (\Box p \rightarrow \Box q)$ (K)
6. $\Box(P_m F_{m+n} p \rightarrow F_n p)$ ($\Box$-CPR)
7. $\Box P_m F_{m+n} p \rightarrow \Box F_n p$ (from 5 and 6 by substitution and modus ponens)
8. $F_n p \rightarrow \Box F_n p$ (from 4 and 7 by transitivity of $\rightarrow$)

Framing the argument in this way helps us clarifying its assumptions and spotting its problematic steps. Step 1 formalizes the so-called *principle of necessity of the past* (PNP), that is a specific feature of necessity *per accidens*. Step 2 follows from Step 1 by simply substituting $F_{m+n} p$ for p in 1 (*uniform substitution* is an admissible inference rule in most logical systems). The *principle of retrogradation* (PR) intuitively says that present events should be regarded as future happenings from a previous time perspective. (PR) is a common assumption about temporal operators (its non-metric version $p \rightarrow PFp$ is actually an axiom of so-called minimal temporal logics). Step 5 is instead the principle of distribution of necessity over implication, which is known in contemporary modal logics as axiom K (for Saul Kripke) and is valid in all systems of *normal* modal logic. Step 6 consists of the necessitation of the converse of 3 and is also a common assumption. 7 is a logical consequence of applying 5–6 as an antecedent by *modus ponens*. Finally, 8 can be derived from 4 and 7 by the transitivity of $\rightarrow$. As, mentioned, 8 is not (nor it entails) a contradiction, and this is why this is a paradox only in the weak sense.

Once the structure of the argument is clear it is possible to find the weak spots in it. Given that most of the logical steps are unimpeachable we are essentially left with two possible lines of response.

(i) Step 2 substitutes a statement which somehow concerns the future ($P_m F_{m+n}$) as an instance of PNP. However, PNP intuitively holds only when applied to propositions that concern the "pure" past or present, viz. their verification does not involve any reference to some point in the future. Therefore the substitution $F_{m+n} p$ for p in PNP may be questioned.

(ii) Assumption 3 in its general form $p \rightarrow P_n F_n p$ implies that we can derive "it was the case that it will be the case that p" from "it is now the case that p".

However, assuming that a statement which is now true might be evaluated as previously true in the future is objectionable.

Ockham was the philosopher who most famously raised an objection of type (i). Objections of the kind of (ii) have instead been raised by Aristotle,[8] Aquinas,[9] Peter de Rivo[10] and, more recently, by C.S. Peirce (at least according to Prior's reading). The originality of Prior consists in the construction of two alternative and complete logical systems, one for each solution: the *ockhamist logic* (Ock) and the *peircean logic* (Peirc).

We shall define both Ock and Peirc by their respective semantics, also known as *tree-like semantics for branching time.*[11]

In mathematical terms, a *tree-like structure* T like the one in Fig. 3.1 is a pair $(T,<)$, where T is a set of moments $m,m'\ldots$ and $<$ is a strict partial ordering relation (i.e. irreflexive, transitive and asymmetrical), where the $<$−predecessors of any point m are *totally ordered* by $<$. A *history* h is a *maximal chain* in T for the relation $<$ (e.g. in Fig. 3.1 h_1 is the chain containing m_{-1}, m_0 and m_1). Given a moment m, H_m designates the set of all histories containing it. $H(T)$ is instead the set of all the histories in our structure. We may notice that if $m < m'$ then $H_{m'} \subseteq H_m$. Given a structure, an *evaluation* is a function V assigning to every atomic formula p a set of pairs (m,h) – a moment and a history -, intuitively saying which basic propositions are true at a given pair (m,h). For example, in Fig. 3.1, the atomic proposition p is true at (m_1,h_1) and false anywhere else. An important requirement is that, given a moment m, V does not vary with the different histories in H_m, i.e. we have

$$(\textbf{Uniqueness})\ (m,h)\in V(p) \text{ if and only if for all } h'\in H_m, (m,h')\in V(p)$$

We may interpret Fig. 3.1 as follows. At some point m_0 (say the present) we have only one possible course of past events and many possible courses of future events. This is represented by the fact that m_0 belongs to two different histories h_1 and h_2 (chronicles in Prior's terminology) that run together up to m_0 and diverge later on. Now, for some histories (h_1) the proposition p is true at the next moment (m_1), while in some others (h_2) it is not (m_2).

A history h should represent a specific and well determined course of events where every formula should be assigned a *truth value* at a given (m,h). To this end, the Ockhamist semantics Ock is completed with a definition of truth for our propositions. The definition is a recursive one, based on V, and is given by the following clauses.

[8]See *De Interpretatione* chap. IX.

[9]See among others *Summa theologiae* I.14.13 (Aquinas 1964) and *De Veritate Q. 2* (Aquinas 1949).

[10]See Baudry (1950).

[11]Our approach diverges from Prior's insofar as his construction was mostly syntactic and employed semantics just as an auxiliary tools. Prior had a specific philosophical motivation for this. However, a semantic approach fits better with our explanatory purpose.

Fig. 3.1 A tree-like model

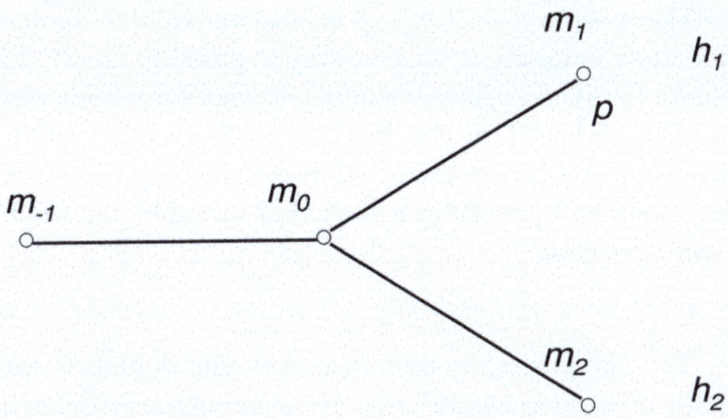

- p is true at (m,h) if and only if V assigns p to (m,h).
- $\neg\varphi$ is true at (m,h) if and only if φ is not true at (m,h)
- $\varphi \wedge \psi$ is true at (m,h) if and only if both φ and ψ are true at (m,h)
- $\varphi \vee \psi$ is true at (m,h) if and only if φ is true or ψ is true at (m,h)
- $\varphi \rightarrow \psi$ is true at (m,h) if and only if φ is not true at (m,h) or ψ is true at (m,h)
- $F_n\varphi$ is true at (m,h) if and only if φ is true at (m',h), where m' is n moments after m in h w.r.t. the order $<$
- $P_n\varphi$ is true at (m,h) if and only if φ is true at (m',h), where m' is n moments before m in h w.r.t. the order $<$.
- $\Box\varphi$ is true at (m,h) if and only if φ is true at (m,h') for every history h' containing m.

From the truth clause of $\Box$ and $\neg$ we can automatically derive the clause for $\Diamond$ (defined as $\neg\Box\neg$), i.e.

- $\Diamond\varphi$ is true at (m,h) if and only if φ is true at (m,h') for some history h' containing m.

A structure with an evaluation is called a *model* and it provides a full interpretation of the formulas of our language. An example of a model is given in Fig. 3.1. A formula is *valid* if and only if it is true for every possible (m,h) in every possible model given every possible structure. Most importantly, in the model-theoretic understanding, a *contradiction* corresponds to a formula φ whose negation $\neg\varphi$ is valid. Otherwise said, we can read a contradiction as a formula which no (m,h) in any model makes true.

Given these clauses we may look at Fig. 3.1 and ascertain that $P_1F_2p \rightarrow \Box P_1F_2p$ is false at (m_0,h_1) and is therefore not valid. In fact, the antecedent P_1F_2p is true – one step behind and two steps forward in h_1 leads to (m_1,h_1) which verifies p – but the consequent is false since h_2 falsifies p at (m_2,h_2). Therefore, (PNP) is invalid in Ock and our argument is blocked at step 2. We may however verify that (PNP) holds for a restricted class of formulas – e.g. for p, P_np or P_nF_m (where $n > m$) – whose truth does not depend on some moment which is future w.r.t. the moment of evaluation. Therefore our Step 1 was correct, but the inference to Step 2 was not.

The schema $p \rightarrow P_n F_{m+n} p$ is however valid in the semantics of Ock, which does not allow to falsify (PR). One way to possibly falsify (PR) is to revise the semantic clause for the F_n operator and this is what Prior does with Peirc. Peircean semantics are always based on tree-like structures and the notion of evaluation remains the same but, contrary to Ock, truth is defined only with respect to a moment m and with no regard to a particular history. For example, the truth clause for atomic propositions now reads

- p is true at m if and only if V assigns p to (m,h) for some h.

The Uniqueness condition makes this definition unambiguous. Also truth for boolean operators and P_n may be unambiguously defined in this way, e.g. the clause for negation will read

- $\neg\varphi$ is true at m if and only if φ is not true at m

What changes in Peirc is the meaning of F_n which now reads as follows

- $F_n\varphi$ is true at m if and only if φ is true n moments after m for every history h' containing m.

We can easily verify that our model of Fig. 3.1 now falsifies $p \rightarrow P_1 F_1 p$: p is true at m_1 but if we go one step "back" (P_1) we see that $F_1 p$ is false at m_0 since p is false at m_2 and therefore the consequent is false at m_1. Indeed, in the semantics of Peirc, F_n means the same as "necessarily it will be the case that p in n moments" in Ock. Such a reading of the operator F_n is based on the idea that it does not make sense to refer to a specific course of events when talking about the future: all possible future histories are on a par. Thus, when we say something like "it will be the case that p", this can only be true because p is already "present in its causes" and therefore cannot be falsified in any possible future history. Therefore, Peirc gives an alternative way of undermining De Rivo's argument, i.e. by invalidating (PR).

In sum, Prior's logical analysis provides two possible solutions for the future contingents problem. They are based on two different intuitions but both of them are, in different ways, compatible with our understanding and use of tenses in temporal reasoning. With this semantic picture in mind we may proceed to analyze the case of space-time relativity and its puzzles.

3.3 Relativistic Space-Time

As mentioned before, Priorean semantics captures a Newtonian and pre-relativistic concept of time. The building blocks of our structures are indeed *moments*, that is a particular kind of entity which encompasses everything that simultaneously happens in space. A history is conceived as a succession of moments: time is like a carpet and space walks over it. This picture does not fit with contemporary physics where space and time are interdependent entities. According to the theory of *space-time relativity* there is no such thing as absolute simultaneity: "being at the same

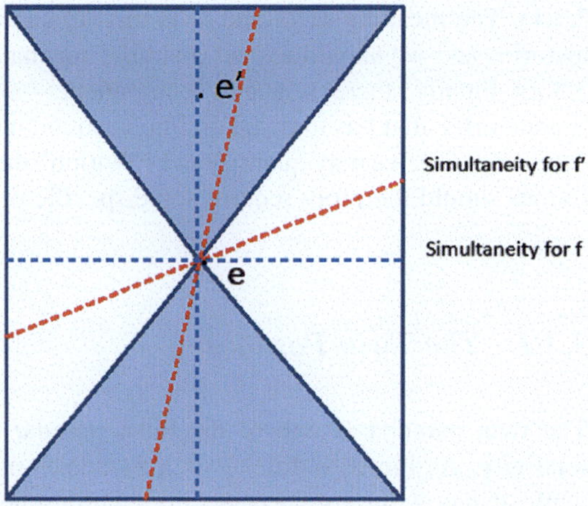

Fig. 3.2 Minkowski space-time. The reference frame *f* is represented by the *blue-dotted* coordinate axes while the reference frame *f'* by the *red dotted* ones (in oblique)

time" – but also "being at the same place" – is relative to a specific *frame of reference*. As a consequence, the notion of a history, intended as a well determined possible course of events, has to be revised. The geometrical representation provided by special relativity is that of a *Minkowski space-time* (see Fig. 3.2) and an adequate semantics for a temporal logic of relativistic space-time should be consistent with it. Accordingly, we should think of a history as a full *block universe* (see Bouton, Chap. 6 of this volume) where all spatio-temporal events are specified. Moreover, in order to have indeterminism in our semantics, we need to allow different possible histories and a notion of branching.[12]

Belnap (1992) provides an overall picture of *branching space-time* as a semantics for temporal logic.[13] The basic items here are *point events* (we will name them *e, e', e"*). According to the Minkowskian picture point events are related to each other not by a temporal but by a *causal order*. A history is now a Minkowski space-time where the causal order among point events is determined by their *past* and *future lightcones* (the blue colored areas in Fig. 3.2). We have that *e* is before *e'* in the causal order (shortly $e << e'$) if and only if *e'* is in the future light cone of *e* or, equivalently, if and only if *e* is in the past light cone of *e'*.

In the Priorean semantics the meaning of the temporal operators P_n and F_n is expressed by quantification over moments, e.g. $F_n p$ is true at *m* if *p* is true for some *m'* which lies *n* steps after *m* w.r.t. <. Analogously, any temporal operator defined in this semantics should be based on << and the *metrics* defined over it (see Sect.

[12]Allowing multiple histories is not merely a logician's trick: in *quantum mechanics* a *many-world interpretation* is actually provided which allows parallel universes as in our scenario.

[13]In what follows we will not fully adhere with Belnap's presentation. For example, Belnap does not provide truth clauses for temporal operators and does not introduce operators for possible reference frames. However, to understand some puzzles of space-time relativity, we need to give at least a partial account of how these operators should work. For this purpose we will (freely) borrow from Strobach (2007).

3.3.1). The meaning of □ and ◇ is instead captured by Prior by quantifying over histories and we should expect a similar mechanism to be in place here (see Sect. 3.3.2). Finally, there is a new component – which is absent in the Newtonian framework – that has to be taken into account, namely the many possible *reference frame sf,f',f''*... we may quantify over "within" the same history. Quantification over frames should therefore require some specific operator (see Sect. 3.3.1).

3.3.1 The Twin Paradox

The twin paradox is one of the most popular introductory examples of special relativity. As hinted before, it is a paradox only from a Newtonian perspective, while it has a rigorous explanation within special relativity. The story runs as follows. We have twins, Alvin and Bert, who both live on planet Earth. At some point in (space)time Alvin decides to stay on Earth while Bert travels away in a spaceship at a speed of 0.8c. After 3 years Bert changes direction and moves back to Earth at the same speed. When they meet again Alvin is 10 years older and Bert is only 6 years older.

To put this puzzle in geometrical terms, we may consider two points *e* (our initial point where Bert leaves Earth) and *e'* (the final point where Alvin and Bert meet again). Our story presupposes that it is possible to reach *e'* from *e* in both 10 (Alvin's life path) and 6 (Bert's life path) steps forward. This amounts to saying, loosely, that both (i) "possibly in 6 steps *e'*" and (ii) "possibly in 10 steps *e'*" should somehow be verifiable at *e*. In the Priorean semantics the distance between any two points is unique and it is thus not possible to reach a future moment via two alternative forward paths. Otherwise said, any possible formulation of "(i) and (ii)" would lead to a contradiction.

However, by assuming Minkowski space-times as a new basis of our semantics the paradox is easily dissolved, though this requires some additional machinery. First of all, our truth clauses should be relativized to triples *(e,h,f)* where the new component *f* stands for a reference frame. Relatedly, as suggested by Strobach (2007), we may extend our semantics by adding a new operator +, intuitively reading "for some alternative reference frame" whose truth clause is the following.

- $+\varphi$ is true at *(e,h,f)* if and only if φ is true at *(e,h,f')* for some *f'* alternative to *f* at *e*.

We should then define a specific causal future operator, say F^*_n, whose meaning is "it is the case *n* steps forward in the causal future". The distance here is given by a metric *d* defined over a Minkowski space-time and which is frame relative. The truth clause for our operator can then be defined as follows.

- $F^*_n\varphi$ is true at *(e,h,f)* if and only if φ is true at *(e',h,f)* for some *e'* such that $e << e'$ and $d(e,e') = n$ w.r.t. the frame *f*.

Fig. 3.3 The twins' travel. Different colors represent different frames in the Minkowski space-time. There are two paths from *e* to *e'* which take different frames although they lie on the same history

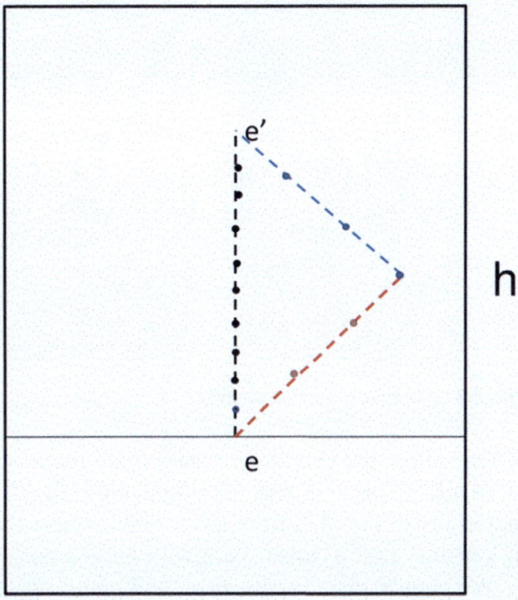

Given these definitions we can find a way to express our "(i) and (ii)" in a consistent way. The formula $F^{*}{}_{10}e' \wedge + F^{*}{}_{3} + F^{*}{}_{3}e'$ can do the job. This formula is indeed true at some *(e,h,f)* in some model: Fig. 3.3 schematically illustrates one. The key point is that we can reach point *e'* from *e* either by ten steps in the initial frame (the black path in Fig. 3.3) or by three plus three steps in two different frames (the red path and the blue path). To sum up, with an adequate semantics we can avoid contradictions.

3.3.2 Branching Space-Time and the Time Travel Paradox

In the Priorean semantics branching is a relatively simple issue: for every history the past and the present are uniquely determined and only future alternatives may diverge. Things get more complicated with special relativity: if possible histories are different space-times, how should we understand branching at a specific triple *(e,h,f)*? Three main alternatives are discussed in the literature. One consists in saying that the accessible histories at *(e,h,f)* are all those *h'* which are identical, i.e. verify the same propositions, up to *(e,h,f)* except possibly for what concerns the future light cone of *e*.[14] A second option is instead to consider the larger class of those *h'* that verify the same propositions up to and including the given frame *f*, i.e. that are the same up to the present we are located in. As a third option we may

[14]This is considered, for various reasons, to be the most coherent option by Belnap (1992).

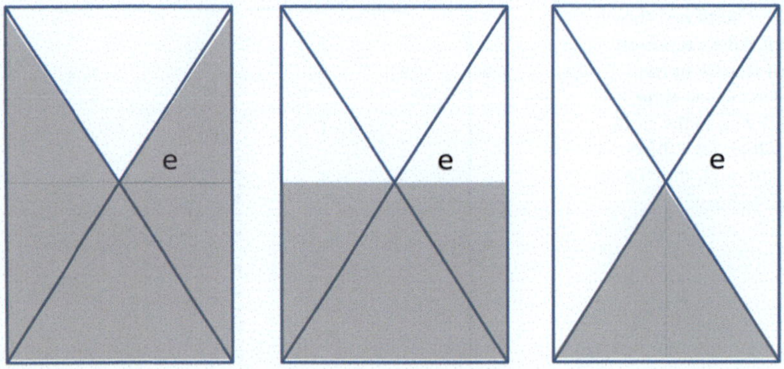

Fig. 3.4 Alternatives for branching

instead allow the even larger class of all the *h'* which verify the same propositions as *(e,h,f)* at least in the past light cone of *e*. These alternatives are illustrated pictorially in Fig. 3.4, from left to right, where the grey areas represent the portions of a history that *h'* must share with *h* to be considered as a possible alternative.

We won't discuss the *pros* and *cons* of these three alternative notions of branching here, but it was important to stress that the question of what should count as an alternative course of events is a very subtle one and should be carefully settled when discussing many puzzles in relativity. It is also important to remark that all these three different conceptions of branching keep the alternative histories within the domain of the "physically possible". In fact, the causal past is not accessible for any of these three options. This is an important point for understanding the time travel paradox.

A simple way to illustrate the time travel paradox is the following. Suppose that it is possible for me to travel faster than the speed of light. It would then in principle be possible for me to reach my causal past and, say, prevent my grandfather from meeting my grandmother, with the consequence that I would end up not being there "here and now". Suppose that *c* stands for the proposition "Carlo exists here" and *g* stands for "Grandfather and Grandmother meet". Where is the paradox? *Prima facie* it seems that our story presupposes that some triple *(e,h,f)* verifies both *c* and $\neg c$ as well as $P^{*}{}_{n}g$[15] and $\neg P^{*}{}_{n}g$. This would entail two contradictions and therefore a paradox in the strong sense. But, is this really so?

The grandfather's story is a *physically impossible* but, according to many, *logically possible* scenario.[16] This means that it should be allowed to rephrase it in a way that does not entail a blatant contradiction. The key lies again in the concept of

---

[15]The truth clause of the operator $P^{*}{}_{n}$ is defined analogously to that of $F^{*}{}_{n}$ (see previous subsection), i.e. $P^{*}{}_{n}\varphi$ is true at *(e,h,f)* if and only if φ is true at *(e',h,f)* for some *e'* such that $e' << e$ and $d(e,e') = n$ w.r.t. the frame *f*.

[16]See Price and Wharton (Chap. 7 of this volume) and Berkovitz (Chap. 8) for a more elaborate discussion of this point (in the framework of retro-causal interpretations of quantum mechanics).

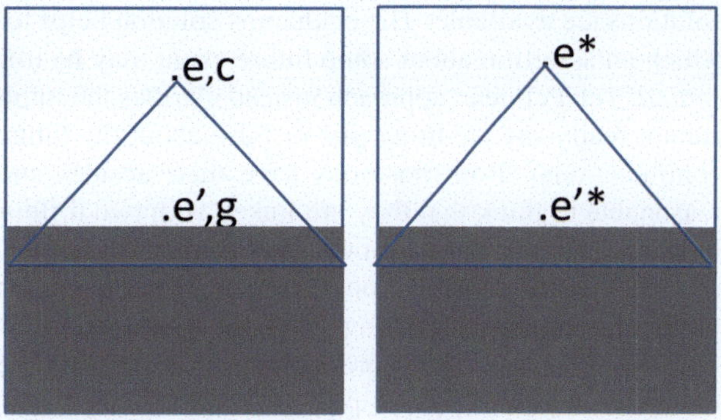

Fig. 3.5 Time travel. The *grey* portions represent the area where the two histories overlap

possibility. Indeed, a way to read our story is to say that I am here, at a given point (e,h,f), and my grandfather has met my grandmother, say at (e',h,f), but I may bring about that my grandfather has not met my grandmother at some $(e^{*},h^{*}f^{*})$ and therefore I am not there, say, at $(e^{*},h^{*}f^{*})$. In formal terms the question is whether it is possible to verify, at some (e,h,f), both $c \wedge P^{*}{}_{n}g$ and $\Diamond(\neg c \wedge \neg P^{*}{}_{n}g)$ according to a proper semantics. If the truth clause of $\Diamond$ is defined according to one of the three previously mentioned accessibility relations this is not possible. To make it possible, we need to further expand the notion of accessibility among histories: the possibility of a time travel would imply that it is possible to reach some alternative history h' that differs from the present one even with respect to a portion of its past light cone (See Fig. 3.5).[17]

This is a non-standard conception of a possible alternative, not contemplated by Belnap (1992) and the subsequent literature and which is also difficult to qualify as "branching", since branching usually tracks the conception of spatio-temporal causality. Whether this notion of accessibility is a reasonable one, i.e. the possibility of travelling in time, is left to physicists to decide. For a logician it is enough to show that the idea may be coherently framed with adequate semantic tools.

3.3.3 Conclusions

We scrutinized a traditional and two modern temporal paradoxes and explained how formal logic can help their understanding and the construction of coherent solutions. It is important to stress how logical analysis helps us understanding such paradoxes. We have seen that for the problem of future contingents at least two

[17]This solution is pretty much the same as the one provided by Lewis (1976). In the same paper, Lewis deems as logically impossible to change one's own past. However, such logical impossibility refers to changing one and the same history by making, e.g., c and $\neg c$ true there. As we have shown, a contradiction may indeed be avoided only by means of two alternative histories. Of course, such an option generates additional problems of trans-world identity and causation which are the object of ongoing philosophical disputes.

solutions are available. The ockhamist solution helps to clarify the precise sense in which an assertion about some future event may be true or false despite not being *settled*. The Peircean approach instead clarifies the intuition according to which we cannot really say anything true or false about the future (except for tautologies or contradictions). Both positions have their upsides and downsides, and it seems reasonable to think that they are linked with two different uses of tenses in natural language. This is the reason why the problem of future contingents is still puzzling and an object of investigation. However, providing a rigorous logical framing of the problems helps to become aware of the consequences of either solution.

Some reflections are in order on the twin and the time travel paradoxes. The first one we deemed as a pseudo-paradox and the second as a genuine one, although both of them are paradoxes only in the weak sense of the word. Deeming the first as a pseudo-paradox depends on the fact that when we agree on special relativity, instead of common-sense, as a basis for our analysis, then the argument is not problematic any more. The time travel paradox is instead more challenging because it asks us to suspend our judgment about a basic assumption of relativity, i.e. that travelling faster than the speed of light is not possible, and to analyze its consequences in a hypothetical scenario. Here logic helps us to see how this scenario can be properly framed and that the formula $(c \wedge P^{*}{}_{n}g) \wedge \Diamond(\neg c \wedge \neg P^{*}{}_{n}g)$ need not to involve a contradiction. Therefore, logical analysis is shown to play an important role for framing thought experiments in physics in a rigorous way.

Acknowledgements The author is grateful to Ilaria Jemos, Iacopo Carusotto and Jeroen Smid for useful theoretical insights and contributions to clarification, and to Andrew McFarland for a thorough language check of the manuscript. This research was supported by the Swedish Research Council.

References

Aquinas, T. 1949. *De veritate. Quaestiones disputatae*. Torino-Roma: Marietti. URL http://books.google.se/booksid=0bRaAAAAYAAJ.

———. 1964. *Summa theologiae: Latin text and English translation, introductions, notes, appendices, and glossaries. Summa theologiae*. London: Blackfriars. URL http://books.google.se/books?id=sc7WAAAAMAAJ.

Aristotle. 1941. De Interpretatione (On interpretation). In *The basic works of Aristotle*, ed. R. McKeon. New York: Random House.

Barjavel, R. 1944. *Le voyageur imprudent* (eng. *Future Times Tree*). Paris: Denoel.

Baudry, L. 1950. *La querelle des futurs contingents (Louvain 1465–1475): Textes inédits*. Paris: Vrin.

Belnap, N. 1992. Branching Space-Time. *Synthese* 92 (3): 385–434.

During, E. 2014. Langevin ou le paradoxe introuvable. *Revue de Métaphysique et de Morale* 84: 513–527.

Einstein, A. 1905. Zur Elektrodynamik bewegter Körper. *Annalen der Physik* 322: 891–921. doi:10.1002/andp.19053221004.

———. 1911. Die Relativitäts Theorie. *Naturforschende Gesellschaft, Zürich,Vierteljahresschrift* 56: 1–14.

Langevin, P. 1911. L'évolution de l'espace et du temps. *Scientia* 10: 31–54. Translated in english as "The Evolution of Space and Time".

Lewis, D. 1976. The paradoxes of time travel. *American Philosophical Quarterly* 13 (2): 145–152.

McCall, S. 1994. *A model of the universe*. Oxford: Oxford University Press.

Prior, A.N. 1957. *Time and modality*. Oxford: Oxford University Press.

———. 1967. *Past, present and future*. Oxford: Oxford University Press.

———. 1968. Tense logic for non-permanent existents. In *Papers on Time and Tense*. Oxford: Clarendon Press.

Reichenbach, H. 1947. *Elements of symbolic logic*. New York: Random House.

Strobach, N. 2007. Fooling around with tenses. *Studies in History and Philosophy of Modern Physics* 38: 653–672.

Weyl, H. 1918. *Raum, Zeit, Materie*. Berlin: Springer.

Chapter 4
Time Variable and Time Scales in Natural Systems and Their Modeling

Annick Lesne

Abstract In this chapter, I discuss the status of time when modeling natural systems. In such an operational perspective, time is defined and measured by a clock, i.e. the comparison with a given physical phenomenon. A central notion is that of time scales. I argue that the representation of time (e.g. discrete vs. continuous) is essentially a modeling choice, depending on the time scale of the observation or description compared to the characteristic times of the system. In integrated approaches developed for investigating systems presenting several levels of organization, it may be fruitful to consider several independent time variables describing evolution at different nested time scales ("multiple scale method"). This would solve some puzzling but recurrent statements, such as an evolution composed of a succession of equilibrium states, and leads to a practical understanding of diverse systems, from combustion and kinetic theory of gases in physics to enzymatic catalysis in biochemistry to adaptive dynamics in ecology. At a more fundamental level, the confrontation of descriptions performed at different scales and within different frameworks, from molecular dynamics to statistical mechanics to thermodynamics, is essential to delineate the limits of each of them, and overall reach a more informative understanding of reality. In particular, such a multi-scale confrontation offers a simple answer to the long-lasting opposition between microscopic reversibility and macroscopic irreversibility of isolated systems (the celebrated Second Principle of the thermodynamics). In biology, time scales are also essential in discussing the robustness of living systems and their evolutionary strategies.

Foreword

As a classical physicist (i.e. not involved in quantum mechanics, cosmology, and general relativity) I do not feel in a position to address the question of the nature of time. I will rather focus on the representation of time when modeling natural

A. Lesne (✉)
Laboratoire de Physique Théorique de la Matière Condensée UMR 7600, CNRS, Université Pierre et Marie Curie, Sorbonne Universités, Paris, France
e-mail: lesne@lptmc.jussieu.fr

C. Bouton, P. Huneman (eds.), *Time of Nature and the Nature of Time*, Boston Studies in the Philosophy and History of Science 326,
DOI 10.1007/978-3-319-53725-2_4

systems and the notion of time scales. In this perspective, time is a relative and flexible notion. I will challenge philosophical discussions with a pragmatic view, as a physicist dealing with *phenomena*, meaning the observable and measurable appearance and not the *nature* of objects and processes.

4.1 Clocks and Time Measurement

The statement attributed to Poincaré (Poincaré 1898; Barbour 2008): "Time is what a clock measures" underlines that the measure of time, and even its everyday definition, rely on the joint observation of two phenomena: that of interest, and a given well-calibrated one, called a clock. In other words, clocks provide a time unit based not on an absolute reference but on a controlled (usually physical) process. Moreover, it indicates that for all practical purposes, the notion of time can be identified with the measure of time. Thus time has been originally defined with respect to the (apparent) motion of sun, thought to be an absolute reference. In parallel to sundials (not always functional nor very accurate), time has been defined using as a reference a process of externally bounded duration: how water flows in a water clock (clepsydra), how sand flows in an hourglass. Time has then been defined with reference to a characteristic time of a system taken as clock. This time unit is shown or assumed to be a constant, invariant as long as the clock is unchanged. The emblematic example is the period of a pendulum, which remains constant when friction damps the pendulum motion, provided the pendulum mass and length remain unchanged. More recently, clocks have been based on more intrinsic characteristic times, thought to provide an absolute unit: the period (inverse frequency) of vibration in a quartz clock, based on the sustained vibrations of an electrically excited crystal of quartz (piezo-electric effect), or the inverse frequency of the photon emitted by an electron when passing from an excited energy level to the ground one, in atomic clocks. In all these cases, time is a phenomenological notion. In classical physics, there is no time without a physical phenomenon, whose course allows measuring and in a way defining time. This operational view, ignoring any preexisting concept that could be formulated independently of the material world, is all the more accepted that it is supplemented with our own intuitive perception, considered as a matter of cognitive science or psychology that does not interfere with common physicists' practice.

4.2 The Concept of Time Scale

Turning to the concept of time scale, a first obvious distinction arises between *epistemic* scales and *intrinsic* scales. The first ones are the scales at which the system is observed or described. The latter are related to the characteristic times of the system, independently of the presence of an observer, for instance: the mean free time between two molecular collisions, the period of vibrations or oscillations, the quantity $1/\nu$ where the frequency ν is related to the energy difference between two atomic levels, the correlation time τ_c of a time-dependent quantity X as defined by the exponential decay of the correlation function $C(t) = < X_t X_0 > - < X_t > < X_0 > \sim A \exp.(-t/\tau_c)$ or the relaxation time τ_r defined by the exponential decay of a time-dependent observable x_t towards its equilibrium value x_0, namely $x_t - x_0 \sim A \exp.(-t/\tau_r)$. Generally a characteristic time is obtained according to a relation: *time = length / velocity* when a characteristic length and a characteristic velocity are known. In living systems, the replication time of the genome, the division time of a cell or the lifetime of an organism complete the previous enumeration (see in particular the Chap. 15 by Nicoglou in the this volume).

4.3 Time Scales: Epistemic vs Systemic Issues

Concurrently with the above two kinds of time scales, we distinguish two issues. A first, *epistemic*, issue is to appreciate how our understanding and description of a system depends on the scale at which it is observed. For instance, as explained in any textbook of statistical mechanics, the impact of air molecules on the wall of a container can be described, at small scale, as a succession of collisions of molecules against the wall, occurring at random times due to the randomness of the thermal motion of the molecules. At a coarser resolution, only the average effect of these collisions is perceptible, corresponding to a time-constant pressure.

A second issue, *intrinsic* to the system of interest, is to understand how the different processes taking place at different time scales are coupled. In *plain physical situations*, microscopic and macroscopic processes are weakly coupled. This means that it is possible to consider separately microscopic elements and macroscopic behaviors. The latter can be obtained by a mere averaging of the microscopic processes (over the system and over intermediary durations): the microscopic fluctuations average out, and what remains is the macroscopic evolution (this is sometimes technically more difficult in practice than in principle). In the above example, the issue is now to describe how the microscopic dynamics ruling the collisions of the molecules on the wall of the container and their kinetic energy will generate at the macroscopic scale the time-variations of pressure and temperature, described autonomously in the framework of classical thermodynamics, with no reference to the microscopic level.

Some situations, termed *"critical phenomena"*, do not obey the rule of separation of scales: details and events at all scales are strongly coupled and microscopic fluctuations have macroscopic repercussions. The hallmarks of such situations are the divergence of the range of temporal and spatial correlations, and singularities in the system's response to perturbations. The latter are typically revealed by a divergence to infinity of the system susceptibilities, meaning that a microscopic (spatially localized or small-amplitude) event is capable of triggering a macroscopic consequence, spreading in the whole system and over long durations

(Laguës and Lesne 2011). For instance, a very localized magnetic perturbation (single spin flip) or the application of a very small magnetic field have observable repercussions at all spatial and temporal scales in a ferromagnet at the critical point, whereas they have no macroscopically observable effect far from the critical point. Although a critical phenomenon occurs in a precisely defined situation (a single specific point in the parameter space), temporally long and spatially extended transients are observed around the critical point. More complicated phenomena, typically slow relaxation dynamics and long-range correlations in an extended domain of the parameter space, may arise in disordered systems in which the control parameter is no longer homogeneous in space (*Griffiths' phases*) (see e.g. Moretti and Muñoz 2013).

While in plain systems, fast microscopic events accumulate and produce collectively a slow macroscopic evolution, the converse may also occur in some systems displaying what has been called *"self-organized criticality"*: a slow accumulation of elementary events relaxes into a sudden and fast collective event, characterized by power-law distributions of the various observables (e.g. the size and duration of the macroscopic events). This concept has been introduced to summarize the complex behavior of avalanches observed in continuously fed sand piles, earthquakes, species extinctions and cascading failures in a network (Bak 1996). Despite all the work and interest it has attracted, the concept of self-organized criticality is still lacking operational power beyond that of an elegant description, and some macroscopic power-law behaviors could have a different origin.

However, the study of criticality and self-organized criticality has underlined the importance of investigating the time scales of the observed phenomenon, in both describing its characteristic features and understanding the underlying mechanisms. For instance, the response of a system may differ according to the frequency f at which its parameters or the applied forces are varied, compared to the (inverse) characteristic times of the system (Yordanov et al. 2011). A system with a reaction time τ does not react if a control parameter changes periodically with a frequency $f >> 2\pi/\tau$; on the contrary, it smoothly follows the changes occuring at a frequency $f << 2\pi/\tau$. An oscillator with natural frequency f_0, displays a huge amplification, called a resonance, if $f \sim f_0$. Scales also matter in any interaction: the scale at which the interaction takes place determines the range of the integration over microscopic structures and processes, which occurs during the interaction. Think e.g. of the interaction between two molecules (integration over the atoms constituting the molecules), the solid friction of a sliding object along a slope (integration of the details of the two contacting surfaces, from the molecular scale to the scale of the object), or the attraction between two planets. This interaction range determines the relevant variables of the simplest effective description. For instance, in first approximation, the interacting planets are described as point particles fully characterized by their mass, and not by their countless components.

Up to now, we have considered scales, which have a clear-cut definition (as seen above). In living systems, the relevant decomposition of the system in nested layers has to follow the different *levels of organization*, rather than scales: one distinguishes an enzyme, a cell, an organ, an organism, even if some organisms (e.g. bacteria) may be smaller than some cells (e.g. human cells). Although these levels somehow follow the scales, their delineation is more qualitative and is associated with a functional or physiological description of the system. Here again, the duality between epistemic decomposition and intrinsic decomposition into nested levels of organization is a delicate issue, which has to be clarified to avoid misinterpretations.

4.4 Relation Between Temporal and Spatial Scales

In the above examples, the systems also display a space dimension, hence not only time scales but also spatial scales are relevant. In physics, fast processes generally correspond to spatially microscopic processes, and slow ones to spatially macroscopic ones. For instance, atomic vibrations are very fast compared to the oscillations of a rope. This intuitive thought is quantitatively supported for linear systems (e.g. wave propagation) by *dispersion relations*, which express the wavelength of the phenomenon as a monotonously increasing (and currently simple) function of its period. (Linear systems are rare, but actually, many systems can be locally linearized, i.e. approximated by a close linear system, exactly as a curve can be locally approximated by its tangent line).

This correspondence between temporal and spatial scales does not hold in living systems. For instance, a spike (elementary activity event) in a neuron involves the transfer of only a few ions in a region of nanometric scale in space, however it extends over about a millisecond in time, far above the usual molecular time scales. Moreover, phenomena like neural coordination or synchronization yield macroscopic events having the same time scale as elementary ones.

4.5 Time Scales, Time Representation and Modeling Choices

The relevant representation of time depends on the observation/description time scale compared to the intrinsic scales of the system. It is actually a modeling choice. Currently three representations of the time variable are encountered:

1. *An integer label qualitatively describing the qualitative ordering of a sequence of events*. An example of such a "qualitative time" is provided by generations, in population dynamics, whatever the duration separating an individual and its offsprings. Another example is a cell, which has a priori its own clock for growth and division. In some cases, e.g. behavioral sequences, only the ordering is kept and the duration of events is eliminated. Due to individual variability and context dependence, what is relevant to understand the behavioral processes is indeed the qualitative succession of events.

2. *An integer variable n measuring a number of time steps.* This amounts to describe the evolution of the system by considering the succession of its states at discrete times $n\Delta t$, where Δt is a fixed unit of reference. If the time step Δt is much smaller than the characteristic time(s) of the system, the discretization is valid with no significant loss of information. If on the contrary the time step is much larger than some characteristic time of the system, the associated processes are not well captured and the discretized dynamics will only reflect properly the slow trends. For instance, a description of the climate dynamics with a step equal or larger than a year will ignore seasonal variations. A more intrinsic (hence more robust) way to discretize the evolution of a system, known as the "Poincaré section method", is possible when the dynamics is close to a periodic regime (limit cycle). One considers an hypersurface of codimension 1 in the state space (e.g. a plane if the state space is of dimension 3), and replaces the study of a trajectory by the study of the sequence of its intersections with the hypersurface; for example, this hypersurface can be defined as the set of points where an observable quantity takes a given value. The time step is now irregular, of duration vaying from one to another, and the discretization amounts to consider a sequence of events. Adaptive time-steps are also encountered in discrete mechanics, where this flexibility is required to preserve both the geometric properties and energy conservation in the use of a discrete variational principle (see Chap. 9 by Ardourel and Barberousse in this volume).
3. *A real time variable t varying continuously.* One should however remember that using a continuous time variable t in a physical model is in fact a mathematical idealization: indeed, the model and associated equations have been derived under some assumptions on the time resolution (the lower time scale of the description), meaning that faster processes have been neglected. Contrary to the mathematical notion, the so-called infinitesimal time increment dt has a physical meaning and should rather be denoted Δt. Similarly the notion of time derivative dx/dt is an idealization (hence an approximation) of a finite rate $\Delta x/\Delta t$.

Also the choice of the framework used to describe the dynamics (i.e. the evolution rules) depends on the relative magnitudes of intrinsic and description scales. They condition for instance the choice of a deterministic or a stochastic dynamic model. The evolution will be described as deterministic if dt is far larger than the correlation time, so that fluctuations average out, or at the opposite (reversible molecular dynamics) if dt is far smaller than the typical time between two successive collisions experienced by a molecule with the other ones (mean free time). In between, fluctuations cannot be ignored and a stochastic model would be more relevant. An important issue is to establish inter-relations between these representations of time and modeling choices (Lesne 2007).

4.6 The Notion of Equilibrium

Seemingly intuitive and commonsense, the notion of equilibrium is in fact complex. An operational definition would be: *a system is in equilibrium if its characteristics do not evolve along time*. In practice, this formulation shows that equilibrium is a relative notion. The definition has to be refined by the prescription of scales. Namely, a system in equilibrum experiences no perceptible changes at a given resolution (lower bound on the time scale) over a given duration (upper bound on the time scale). In plain words, fast processes have evened out and slow ones have not yet begun. Also the resolution on the observable characteristics has to be defined, namely, what is the value of a significant change that can be detected by the observer. The notion of equilibrium is thus deeply related to space and time scales.

The notion of equilibrium involves time in a dual way. Its above definition implicitely refers to the absence of significant dynamic processes; for instance the equilibrium of a (mechanical) scales means that it no longer oscillates unless an additional weight is introduced. However, the very notion of time is related to a rate of change (see above our ways of measuring time); time would be an irrelevant notion for a system in perfect equilibrium. Luckily, this paradox vanishes on its own since there is no such thing as a perfect equilibrium: relaxation becomes infinitely slow near the equilibrium point; surrounding noise unavoidably disturbs the equilibrium; slowly evolving environment modifies what is the equilibrium state; thermal motion implies that equilibrium is an irrelevant notion at molecular scales. In practice, it is safer to speak of "*stationary state*", possibly defined on statistical quantities, and always in reference to prescribed scales and resolutions.

A puzzling statement remains: what is the meaning of the usual definition of an adiabatic evolution in thermodynamics (also encountered under the name of quasi-stationary approximation in numerous studies of complex dynamics) as being "*a succession of equilibrium states*". Answering this question requires a mathematical formulation of time scale separation (Lesne 2006).

4.7 Multiple Time Scales and Quasi-stationary Approximation

A clue to recognize whether a model captures the coexistence of processes at different time scales is the presence of a small parameter ε in the equations of evolution, hinting at the presence of contributions acting on a longer term than the others. Then, the main step is to identify fast and slow variables. Typically, the fast variables describe the microscopic events while the emerging macroscopic trends, corresponding to average or aggregated behaviors, are generally described by slowly varying quantities.

Let us denote x the fast variables reflecting the system micro-evolution, and Y the slowly varying observables accounting for the system macro-evolution. The equations of evolution can be written $[dx/dt = f(x, Y),\ dY/dt = \varepsilon\, g(x, Y)]$ where ε is the small parameter measuring the scale separation between the characteristic times at which x and Y evolve. At a fine temporal resolution, for ε small enough, the quantity x evolves significantly during one time step $\Delta t = 1$, while Y remains constant. This justifies to invoke a *quasi-static approximation* (for Y) in which the second equation of evolution reduces to $dY/dt = 0$ and what remains is an equation for x where Y is a constant parameter – one also speaks of *parametric approximation* (for x).

At a coarser temporal resolution, the time variable becomes $T = \varepsilon t$. The equations of evolution are now written $[\varepsilon\, dx/dT = f(x, Y),\ dY/dT = g(x, Y)]$. During one macro-step $\Delta T = 1$, x completes its evolution at fixed $Y = Y(T)$ and reaches the asymptotic state $x^*[Y(T)]$ satisfying $f(x^*(Y),Y) = 0$. Plugging this asymptotic value turns the equation of evolution of $Y(T)$ into the closed equation $dY/dT = g(x^*(Y), Y)$, where the variation of x^* is *slaved* to Y (Haken 1996). This procedure is termed *quasi-stationary approximation* (for x) since it amounts to replace the full equation of evolution $\varepsilon\, dx/dT = f(x, Y)$ by $f(x,Y) = 0$. Illustrations are given in (Nayfeh 1973; Lesne 2006).

If the evolution is a sustained oscillation, the fast variable x self-averages into a quantity $\underline{x}[Y(T)]$ during one time macro-step $\Delta T = 1$, and the quasi-stationary approximation amounts to replacing x by $\underline{x}[Y(T)]$ in the evolution of Y, yielding a closed equation $dY/dT = g(\underline{x}(Y), Y)$. This reflects the fact that the macroscopic features are not influenced by all the microscopic fluctuations, but are only sensitive to their average and emerging trends.

The quasi-stationary approximation means that the fast processes reach their stationary state before the slower processes begin to evolve, and so on in a nested way from fast small-scale processes to slow macroscopic contributions. In this respect, the slow evolution is described as a succession of equilibria of the fast degrees of freedom, and the above quasi-stationary approximation is the operational way to formulate this idea. A useful spin-off is the derivation of macroscopic evolutions, i.e. closed equations of evolution for a (usually small) number of macroscopic variables, from the knowledge of the microscopic dynamics (Givon et al. 2004; Castiglione et al. 2008).

4.8 The Multiple-Scale Method: Several Independent Times

Let us mention a thought-provoking method to simplify the above decoupling procedure. The idea is to introduce jointly several time variables $t_0 = t$, $t_1 = \varepsilon t$ (slower by a factor of ε since $\Delta t_1 = 1$ for $\Delta t = 1/\varepsilon$), $t_2 = \varepsilon^2 t$ (slower by a factor of ε^2 since $\Delta t_2 = 1$ for $\Delta t = 1/\varepsilon^2$), and to treat them as independent time variables to dissect the processes at different scales. The time derivative becomes

$d/dt = \partial/\partial t_0 + \varepsilon\partial/\partial t_1 + \varepsilon^2\partial/\partial t_2$ and the solution $x(t)$ is searched as a function of the form $X(t_0, t_1, t_2)$. Scale-decoupling approximation here amounts to consider t_0, t_1, and t_2 as formally *independent variables*, with fast variations captured with time variable t_0, slow variations captured with time variable t_1, and slower trends captured with time variable t_2. However, considering independent time variables is only a computational trick, equivalent to (but more concise than) nested quasi-stationary approximations. Ultimately, the solution is obtained under the form $x(t) = X(t, \varepsilon t, \varepsilon^2 t)$ depending on the single original time variable t. This method has been successfully applied to complex evolutions with superimposed time scales, depending on a small parameter ε in a singular way ("singular" insofar as the solution for $\varepsilon = 0$ is qualitatively different from the solution for $\varepsilon > 0$ (whatever small) hence displays a singularity as a function of ε in $\varepsilon = 0$). An example is the treatment of the so-called secular terms in the 3-body problem in astronomy: these terms arise when investigating e.g. the motion of a planet under both the (dominant) influence of the Sun and the (seemingly marginal) influence of other planets. They induce slow variations perceptible only over centuries (hence the name "secular") and cannot be accounted for in a perturbative way starting from the easy case where only the Sun influences the planet motion (Nayfeh 1973; Lesne 2006; Castiglione et al. 2008).

4.9 Time's Arrow

The long debated issue of time's arrow and the apparent inconsistency between the Second Principle of the thermodynamics and the micro-reversibility of molecular dynamics is also a matter of scales. The Second Principle, presented in any textbook of basic thermodynamics, is a macroscopic feature rooted in a statistical argument for a large number of molecules. It is not expected to – and does not – hold for a system of a few molecules (this is rarely underlined in textbooks, which deal with thermodynamic properties, observed in macroscopic systems). There is no such time's arrow in small microscopic systems. For a macroscopic system like a gas in a compartment, the explanation of the time's arrow runs as follows. Considering the expansion of a gas initially confined in a small compartment, the number of reverse trajectories, starting from an expanded gas configuration and leading back to a confined one (i.e. belonging to a very constrained subset of the configuration space), is infinitely small compared to the number of plain trajectories, starting from the expanded gas configuration and leading to another expanded configuration (see e.g. Castiglione et al. 2008). There is thus an (almost) infinitely small probability to observe the spontaneous confinement of an expanded configuration, the smaller the larger the number of molecules in the considered system. Similarly, there is an (almost) infinitely small number of trajectories leading from the confined configuration to another confined configuration, compared to the number of trajectories leading from the confined configuration to an expanded configuration: in practice, at our macroscopic scale, we always observe an expansion of the gas. However, we here see that the Second Principle describes an *improbability* to observe backward evolutions, not their *impossibility*. The reasoning relies on the large size of the system, and fails for a microscopic system composed of a few molecules.

This argument solves the time's arrow issue for the spontaneous evolution of a closed system, that is, the relaxation from a prepared state to an equilibrium state. It does not answer the deeper issue of time's arrow in open non-equilibrium systems (non-equilibrium meaning here "externally driven in a steady state with non-vanishing fluxes"). Fluctuation theorems have established the probability distribution of the time-averaged irreversible entropy production (Kurchan 1998; Cohen and Gallavotti 1999; Evans and Searles 2002) but many questions are still open, motivating on-going studies in the non-equilibrium statistical physics community.

In this context, the question of the "nature of time" is mostly irrelevant for (classical) physicists. They handle time as an effective and operational notion, like the scientists of the nineteenth century handled the notion of temperature: they devised several kinds of thermometers and made efficient thermodynamic machines without understanding the "nature" of temperature.

4.10 Time Scales and Evolution Theory

We have seen that the response of a dynamic system to a perturbation depends on the time scale of this perturbation, compared to characteristic times of the system. This feature is highly relevant not only in the physical world but also in the living world. Typically, living systems evolve robustness to fast transient changes (homeostasis), and adaptation to sustained (directional) changes. Evolutionary strategies are thus to be examined according to the time scale of environmental variation.

More generally, a scale-decoupling argument is invoked in the very notion of genetic equilibrium, assuming that the changes in allelic frequencies from one generation to another are faster than the occurrence of novel mutations, itself faster than the modification of the surrounding ecosystem. If this decoupling applies, the evolution of a species can be seen as a succession of genetic equilibria, each equilibrium corresponding to the local result of natural selection, that is, an optimized distribution of genotypes in the environment present at that time, while the passage from one equilibrium to another is due to mutations, i.e. a change in the set of possible genotypes or/and changes in the environment. Actually, it has been presumed for a long time that populational, ecological and evolutionary processes occur at very different time scales. However, it has recently been realized that ecological and (macro)evolutionary processes can be fast, and scale decoupling is not necessarily valid. If environmental changes or mutations rates are too fast, the above quasi-stationary approximation fails and what is observed for a species is a succession of transient and unpredictable allelic distributions. The decoupling approximation remains most often useful in a first step to identify the different processes and clarify their consequences (see Chap. 14 by Huneman in this volume).

The discussion about quasi-stationary approximation could also enlighten an operational notion of species. I would suggest that considering a species as a quasi-equilibrium and evolution of species as an adiabatic evolution would reconcile the existence of separate species at a given moment and the gradual phenomenon of speciation. Here again nothing relevant could be said without explicitly mentioning the time scale(s) at which processes are discussed.

4.11 Conclusion

In classical physics and natural sciences, time is not a unique absolute concept, but rather a relative and operational one. A physicist has no access to the *nature* of systems and phenomena, in particular the nature of time. The relevant issues are rather time representation and the choice of a proper time variable. In this operational perspective, I have argued that a central notion is that of time scales. It allows clarifying paradoxical issues like the notion of equilibrium, scale separation, decoupling slow and fast processes or even the time's arrow predicted by the Second Principle of the thermodynamics. In a biological context, time scales are especially important in evolutionary studies and the very definition of a species or the fitness of an individual in a population.

References

Bak, P. 1996. *How nature works: The science of self-organized criticality*. New York: Copernicus Press.

Barbour, J. 2008. *The nature of time*. http://www.platonia.com/nature_of_time_essay.pdf.

Castiglione, P., M. Falcioni, A. Lesne, and A. Vulpiani. 2008. *Chaos and coarse-graining in statistical mechanics*. Cambrige: Cambridge University Press.

Cohen, E.G.D., and G. Gallavotti. 1999. Note on two theorems in nonequilibrium statistical mechanics. *Journal of Statistical Physics* 96: 1343–1349.

Evans, D.J., and D.J. Searles. 2002. The fluctuation theorem. *Advances in Physics* 51: 1529–1585.

Givon, D., R. Kupferman, and A. Stuart. 2004. Extracting macroscopic dynamics: Model problems and algorithms. *Nonlinearity* 17: R55–R127.

Haken, H. 1996. Slaving principle revisited. *Physica D* 97: 95–103.

Kurchan, J. 1998. Fluctuation theorem for stochastic dynamics. *Journal of Physics A* 31: 3719–3729.

Laguës, M., and A. Lesne. 2011. *Scale invariance. From phase transitions to turbulence*. Berlin: Springer.

Lesne, A. 2006. Multi-scale approaches. In *Encyclopedia of mathematical physics*, ed. J.P. Françoise, G. Naber, and T.S. Tsun, vol. 3, 465–482. Amsterdam/London: Elsevier.

———. 2007. Discrete vs continuous controversy in physics. *Mathematical Structures in Computational Science* 17: 185–223.

Moretti, P., and M.A. Muñoz. 2013. Griffiths phases and the stretching of criticality in brain networks. *Nature Communications* 4: 2521.

Nayfeh, A.H. 1973. *Perturbation methods*. New York: Wiley.

Poincaré, H. 1898. La mesure du temps. *Revue de Métaphysique et de Morale* 6: 1–13; English translation: The measure of time. In H. Poincaré, *The value of science*, New York: Dover publications, 1958.

Yordanov, P., S. Tyanova, M. Hütt, and A. Lesne. 2011. Asymmetric transitions and time-scale separation in interlinked positive feedback loops. *International Journal of Chaos and Bifurcation* 21: 1895–1905.

Chapter 5
On Time and the Varieties of Science

David Braddon-Mitchell and Kristie Miller

Abstract This paper proffers an account of why interdisciplinary research on, *inter alia,* the nature of time can be fruitful even if the disciplines in question have different explanatory projects. We suggest that the special sciences perform a *subject setting role* for lower-level disciplines such as physics. In essence, they tell us where, amongst a theory of the physical world, we should expect to locate phenomena such as temporality; they tell us what it would take for there to be time. Physical theory tells us whether there is anything like that in the world and what its hidden nature is. Only working in tandem can physics and the special sciences locate and describe the phenomenon that is time.

5.1 Introduction

We are all encouraged to engage in interdisciplinary research and to hold interdisciplinary conferences. The question is: why? What are the kinds of merits that interdisciplinary connection brings to research? Our current focus is, of course, research into the nature of time. This present volume, for example, brings together physicists, biologists, philosophers, geologists, palaeontologists and more, who discuss very different aspects of time and our experience thereof, ranging from Ismael's (Chap. 2) discussion of why we seem to experience time as flowing or passing (even though, according to most physicists it does not do so), through to Huss's (Chap. 10) discussion of 'deep time' and our epistemic access to past events via fossil record. Indeed, we find widely divergent use of the term 'time' throughout this anthology, ranging from Lesne's (Chap. 4) focus on an operational definition of time in terms of the measurement of physical phenomena, through to the physicists' conception of time (embodied in Chaps. 7, (Price and Wharton) 8 (Berkovitz) and 9 (Ardourel and Barberousse)) through to discussion of the concept of time deployed in palaeontology (Huss, Chap. 10 and Tassy, Chap. 12) and to the differences between the paleontological concept and that of the biologist (Ricqlès, Chap. 11).

D. Braddon-Mitchell • K. Miller (✉)
Department of Philosophy, University of Sydney, Sydney, NSW, Australia
e-mail: kristie_miller@yahoo.com

C. Bouton, P. Huneman (eds.), *Time of Nature and the Nature of Time*, Boston Studies in the Philosophy and History of Science 326,
DOI 10.1007/978-3-319-53725-2_5

In some domains the rationale for interdisciplinary research is fairly clear and easy to articulate. Suppose, for example, that you are interested in the spread of the H7N9 bird flu this season. You may want to get epidemiologists, microbiologists, public health information theorists, immunologists, agricultural economists and more into a room. The target for the interaction would be straightforward—we want to minimize the spread and impact of a particular virus. The microbiologists and virologists can tell us about the problem at a cellular level, the epidemiologists can tell us of the likely infection patterns. The immunologists can try to tell us how immune systems will respond, and this is the information that epidemiologists need to know to create their models. The microbiologists and virologists can provide the veterinary biologists with information about the kind of micro-environment that might be relevant to whether the virus will spread amongst birds. The agricultural economists can tell us how economically feasible are the changes to agricultural practice that would minimize the spread of said virus. Public health researchers can tell us how we could change special practice to reduce the spread of the virus, and so on.

In such a case there is reasonably clear focus regarding what information needs to be shared between disciplines: namely just that information that will aid in a particular goal. In the case just discussed that goal is minimizing the spread of the H7N9 virus. Of course exactly what information will aid the goal in question may not always be obvious, so there's reason to pump as much information between disciplines as possible in the hope that there will be serendipitous consequences that will help us fulfil the goal.

At the other end of the scale from our fictional H7N9 project is the current fashion for postmodern exhibitions in galleries and museums. An example is an exhibition at a prominent southern hemisphere art museum entitled "feathers". It included paintings of feathers, feather headdresses, paintings of rooms in which there were items of furniture that included upholstery (which we were presumably meant to suppose contained feathers). It included displays devoted to corrupt politicians, who were feathering their nests. There was no explanatory agenda, perhaps rightly given the context, and it was very unclear that there was much in common between the subject matter of the exhibits barring the fact that each could prompt the use of the English word 'feather'. What one learned from the exhibition was something about the arbitrariness of categorization. One supposes that the point was supposed to be that it was revealing of the arbitrariness of all categorical practice; if so then certainly it revealed it of itself.

And so we come to theorizing about time. Where between these poles does interdisciplinary work about the nature of time fall?

We think that it isn't at the clear and easy end. There's no empirical outcome that rationalizes the research. Sadly we are not in the business of creating more time, or developing time travel, and asking for interdisciplinary contributions whose salience can be measured against that clear objective. Rather, our task is the more nebulous one of gaining understanding of, or explaining, temporal phenomena. The issue of whether the different disciplines are talking about the same or closely connected things thus has independent force.

Let's return briefly to our H7N9 example again. What if we were not explicitly interested in particular outcomes but just interested, ultimately, in something like an account of what H7N9 is—in its nature? You might, at first blush, be tempted to say something like this. At one level only the virologist (from the above list of specializations) is in a position to settle the question, and it is to them we must go for an answer, for H7N9 is, after all, a virus. At another level, though, the other specializations might provide answers, but answers that are at cross-purposes: answers which address different explanatory concerns. Someone might tell you that H7N9 is a major cause on special panic in Hong Kong; others might tell you it is a pandemic amongst poultry in live markets in Vietnam. A pandemic is not a virus. But this is merely apparent disagreement, since what we have are not competing accounts of the very same thing. Notice that such disagreement bears a startling similarity to certain disagreements about time that one sometimes finds at the interdisciplinary coal-face. We recall a time when the claim, made by some anthropologists, that time is circular, was met with an incredulous stare by some physicists and philosophers of science. But there was no real disagreement. Anthropologists claimed, and perhaps rightly so, that some communities of people *represent* time and temporality very differently to others, and that some represent it as being circular or cyclical rather than linear. The anthropologist's interest lay in the way in which certain cultural practices and certain ways of thinking about temporality structured the day-to-day living of societies. This explanatory project was, unsurprisingly, entirely orthogonal to that of the physicist. Two explanatory projects; two merely apparently inconsistent claims. Returning to our example, is there a way in which a single explanatory endeavour—a unified account of H7N9—can be furthered by interdisciplinary cooperation?

Here's a thought. Virology or microbiology can't play, by itself, what we call a subject setting role. Independent of knowing that the target of investigation is the thing that is the cause of the pandemic amongst poultry in live markets in Vietnam, and the thing that is causing special panic in Hong Kong, and so forth, the virologist will have no idea what samples to test, and thus which viruses to sequence. Before virology can begin we first need to identify a phenomenon of illness in birds and humans using, amongst other things, various special sciences. Then we need to establish, via epidemiology, that they likely have a common cause. Then a variety of sciences have to be called in to provide a good account of what samples to take and from where to take them. Only then can the virologist provide an account of what is—described at the level of virology—the target phenomenon. And that account is always subject to correction by other sciences in the event that it turns out not to be an account of the target in question. For perhaps the sequenced virus turns out not to be the cause of most infections, or to be an opportunistic virus that comes along for the ride with the significant, symptom causing, virus. Then although virologists may have told us interesting things about the nature of a particular virus, they will not have told us anything about the nature of the target phenomenon. In this sense, the output of the special sciences at least partially constrains what would count as a good account given in the language of some other science (in this case virology). Nothing that virologists discover will undermine the subject setting role of the other special sciences. Of course, discoveries in virology can lead us to say that there is no H7N9. For it might be that having examined the relevant samples virologists tell us that there is no common cause of both the pandemic amongst poultry and the symptoms amongst humans; there is no single virus responsible for both. Since according to the subject setting special sciences ‘H7N9’ was the name for the target phenomenon that is the common cause of both of these phenomena, and since according to virologists there is nothing that is the common cause of both these phenomena, it follows that there is no H7N9. Thus the interaction between the various disciplines is two-way: the special sciences downwardly *constrain* virology by settling what would count as discovering the H7N9 virus, and virology upwardly *determines* whether there is anything that answers to those constraints.

So the fact that there isn’t a clear and straightforward pragmatic goal for interdisciplinary work on time might not mean that we must look to the postmodern gallery as a model for what we are doing. Even if our interests lie in providing a generalized theoretical account of the nature of time, it might be worth looking for an account of how disciplines can cooperate that involves some disciplines having a kind of subject setting role for others. The subject setting role is important. At present there are physicists who argue that there is no time. That is a startling claim. All the more so because the defence of a particular physical theory about the nature of our world does not, in itself, seem to be in any position to settle whether or not there is time. That’s not a matter that can be determined internal to physics itself, since the corpus of physical theory provides no resources to determine which bits of that theory or sets of theories is rightly called ‘time’ rather than, for instance, which bits of the theory or theories explains the illusion of temporality.

5.2 Folk Roles and the Special Sciences

Here is our proposal: one important way that interdisciplinary work on time is important is that the special sciences play a ‘subject setting’ role with respect to physics in a way which has some structural similarities to the relationship between what some philosophers call ‘folk theories’ and the location of the entities that play the folk roles.

Here’s an indicative sketch of how this story is often told[1]: ordinary common sense gives us ‘roles’ that indicate what it would take for certain kinds of things to exist. These roles are given in ‘folk theories’. There is a theory, for example, that there is something which underlies the bright shafts of light that appear to connect clouds and the earth during storms, and which may explain burnt stumps, bushfires and occasional deaths. There is another theory that there is some substance that

[1]See for instance Frank Jackson (1998); Braddon-Mitchell and Nola (2009) and Braddon-Mitchell (2004).

accounts for most of the liquids that fall from the sky in storms, fills the rivers and lakes, and quenches thirst. These theories tell us, roughly, what it would take for lightning (in the first instance) and water (in the second) to exist. So they play a semantic, subject setting role. They don't, however, tell us the nature of either of these things. It's the empirical discovery that it's a kind of electrical discharge in the first case, and H_2O in the second, that *plays* these roles that allows us to do two things. First, it allows us to be sure that water and lightning exist; for something plays these roles, namely electrical discharge and H_2O. We might instead have discovered that there is no reasonably uniform phenomenon or substance at work and thus discovered that there was, for instance, no water. Second, it allows us to determine the natures of these phenomena and therefore how to locate them in the world. For it might turn out that such phenomena exist even where they do not play the folks roles—suppose there were environments in which H_2O was not clear, potable or liquid and hence does not play the folk roles. Without knowing what it is that plays the folk roles we would be unable to locate that substance as water until we knew water's hidden nature.

One first blush objection to such an outline of the relationship between science—or low-level disciplines like physics—and ordinary thought and talk is that folk theories are being asked to do too much work. Ordinary folk theories are too loose, too likely to be false, and too variable from person to person to make a good starting place for fundamental investigation.

The reply to this is that we aren't concerned with just what people would say or write down. We are instead dealing with a theory that is tacit and perhaps idealized. No doubt this idealization and extraction of tacit theory can proceed in many ways. But one way we think it proceeds is via the transition from folk theory to the empirical special sciences. Some folk theories are a kind of proto-science,[2] and when the different tacit theories of each individual are averaged out and that remaining theory is made consistent with the high level empirical evidence what we get are the high level special sciences. These high level special sciences themselves will still be specifying roles, albeit cleaned up and precisified roles from their folk origins. Thus we propose looking to the role that time plays in the various special sciences and looking for connections between these roles.

Thus the special sciences, insofar as they are successful—explanatory, predictive, and so forth—provide the subject setting role for the phenomenon of time. In a way this is to say that time is just whatever it is that plays some particular functional role, where the functional role is not given by a set of folk platitudes but is given by the role of the "t" term in the successful special sciences. Thus time is, within certain parameters, whatever it is that explains the success of the term "t" in biology, perception, geology, cognitive science, ecology and so forth. We could call the role that is played by "t" in these sciences (severally or jointly depending on context) the t-role. We look to physics to tell us what it is that explains the success of the t-role, but it is the special sciences that constrain what, in the physics,

---

[2]See for instance Stich (1983).

constitutes an answer to the question: what is time? This is because physics will have answered that latter question only when it answers tells us in virtue of what the t-role in the special sciences is explanatory.

So far this is all very abstract. In the following section we will consider what some of the disciplines in the special sciences might bring to the table by way of playing a subject setting role. Then we want to consider the idea of subject setting in its abstract form in order to consider the relationship between the special sciences and fundamental science.

5.3 The Special Sciences and the Role of 't'

So far we have assumed that the various t-roles found in the special sciences are consistent in the following sense: it is possible that there is a unique phenomenon, discoverable by physics, that plays all of those roles. That is, we have assumed that when each of the special sciences settle the subject matter, the subject that they settle on is, by and large, the same subject. Is there, though, any reason to suppose that is true given that the various special sciences have such different explanatory projects?

In order to answer this question we need to examine, in more detail, the disciplines themselves. We cannot hope to canvass all of these. In what follows we consider just a few, beginning with psychological sciences.

5.3.1 Psychology, Cognitive Science and Neuroscience

We make judgements about the order of events into earlier-than and later-than; we judge the duration of events; we judge that events are changing; we judge that events are simultaneous and we judge that events occur in a particular succession and at a particular rate. Call the collective phenomenology involved in thus experiencing events *temporal phenomenology*. How might we begin to try and locate time within the story given to us by physics? We might begin with nothing but physics and our temporal phenomenology and suppose that time is whatever it is in physics that our phenomenology is *about*, or *responsive to*. The problem is that there's no way to find out what that is: there's no bridge between physics, on the one hand, and our phenomenology, on the other. Psychology, we think, can provide that bridge. It can offer us an account of the mechanism of our temporal phenomenology that it is then (in principle) possible to examine to discover to which features of the physical world that mechanism is responsive.[3]

[3]See Cai et al. (2012); Dennett and Kinsbourne (1995); Eagleman (2009); Grondin (2010) and Holcombe (2015).

Psychology's explanatory project is to determine what it is, at the psychological or neural level that explains our temporal phenomenology. Such explanations may do many things: they may tell us what neural states our experience supervenes on (neuroscience) and, at the functional level, what mechanism is responsible for our phenomenology (psychology and cognitive science); they may tell us how, in general, we have come to have neural states of that kind (evolutionary cognitive science and evolutionary biology) and they may also tell us what the local causes of those phenomenologies are (experimental neuroscience and psychology).

Let us say that the usual causes of our phenomenologies are what these phenomenologies *track* or are *responsive to*, in the world. By analogy, one of the things our perceptual system tracks is middle sized objects, since middle sized objects are one of the things in the world that act as causal input to our perceptual mechanisms and their subsequent output. Then one of the subject-setting roles that psychology might play is that it is a desideratum on a theory of time in a lower level discipline that it be a theory of something in the world which contains things ordered in a way which is being tracked by the neural processes which are found to underlie our temporal phenomenology.

Thus suppose we had a completed story about the nature of our world at the most fundamental physical level. In that story we are trying to locate time. Where should we look? We should look for a pervasive structural feature of the world to which our temporal phenomenology is ultimately responsive. To do that we look to see what psychology tells us about the mechanism or mechanisms that are responsible for our temporal phenomenology. Then we look to see what, in the complete physical story, that mechanism tracks. That thing is a candidate to be time.

Unsurprisingly, talk of a mechanism that is responsible for our temporal phenomenology and which, in turn, tracks some feature of the world, is oversimplified. Though much about the cognitive mechanisms that are responsible for our temporal phenomenology is still unknown, it is almost certain that there is no single mechanism at work. It seems likely that there dedicated and autonomous systems that underlie our judgements regarding, and phenomenology of, motion, of change and of duration and ordering. Indeed, it is largely agreed that there are at least two, and perhaps three, systems that underlie our perception of motion.[4] And recently a plethora of new studies and competitor accounts have arisen regarding the mechanisms responsible for order and duration judgements.[5] According to one model, duration is accomplished by means of what has been called a stopwatch: an internal generic timing mechanism that can be applied to all events.[6] An alternative proposal holds that apparent duration of an interval is a function of the number of events or changes within the interval.[7] This is sometimes known as the visual clock, model. Whichever of these, and other, models are right, we know that which areas

---

[4]Nishida et al. (1997).

[5]See Creelman (1962) and Treisman (1963).

[6]See Johnson (2010) and Holcombe (2015) for discussion of recent work.

[7]See Fraisse (1963, 1984); Brown (1995) and Tse et al. (2004).

of the brain are active when duration is encoded depends on the nature of the task and stimulus.[8] This all suggests that there is probably no single mechanism or pattern of activation in the brain that subserves judgements of duration. Matters are equally complicated if we turn to judgements of ordering. It has been suggested that experience may not comprise a single ordered timeline in which all sensations map to some point or interval on the timeline. Instead, it has recently been suggested that we sometimes experience objects or events with undefined temporal relationships.[9] Ordinal processing of times is, we learn, more impoverished than ordinal processing of space. The latter benefits from retinotopic organisation of visual cortices. That is, a mapping of visual input from the retina to neurons within the visual stream. This means a good deal of spatial information can be attained pre-attentively. The brain has no corresponding chronotopy, which might be one reason why evidence suggests that the representation of temporal order is sometimes impoverished. Still, once again, there is as yet little agreement upon exactly what mechanism or mechanisms underlie the phenomenology of temporal order.

So we can expect the task of determining which mechanisms and neural processes underlie our temporal phenomenology to be a complex one that appeals to some quite large number of possibly autonomous systems. Contra what some might pre-theoretically have expected, our phenomenology turns out not to be a single, unified, experience subserved by a single complicated system.

If all of the various different mechanisms that are jointly responsible for our temporal phenomenology are responsive to a single pervasive feature of the world, then that feature of the world is a good candidate to be time. Interestingly, since there are so many different mechanisms responsible for our phenomenology, the (fairly remote) possibility presents itself that some of those mechanisms might be responsive to different features of the world. It would then, we take it, be an open question whether just one of those features of the world is time, and some of our temporal phenomenology is tracking something that is *not* time, or whether none of the features of the world would be time, or whether time would somehow turn out to be some complex mixture of different features in the world.

What possibilities like this tell us is that the subject setting constraints of disciplines like psychology, cognitive science and neuroscience, are important precisifications of some aspects of the folk theory, but are not the whole story. First, they are not the whole story because in the scenario just described nothing in psychology or cognitive science tells us whether we should conclude that time is one of the features of the world being tracked by one of our cognitive mechanisms, while the other features being tracked by our mechanisms are not time, or whether we should conclude that time is multiply realised in the world, or whether we should conclude that there is no time as such, there are just multiple different time-like features of the world each of which is tracked by our phenomenology. Psychology tells us *what* mechanisms are responsible for our temporal phenomenology

[8]Nobre and O'Reilly (2004).

[9]See Holcombe (2015) for more on this.

and psychology tells us what sorts of proximal causes those mechanisms are responsive to. Physics tells us the story of the world at its most fundamental level. Then we try to bring the two stories together. But neither discipline seems well placed to tell us what to do if there is no pervasive structural feature of the world to which *each* of the mechanisms responsible for our temporal phenomenology is responsive.

That is why we think that psychology cannot be the sole subject setter. Suppose each mechanism that is responsible for some aspect of our temporal phenomenology tracks a different feature of the world. How are we to determine which of these various features of the world is time? In part we may be able to answer this by appealing to the subject setting roles of *other* disciplines. For perhaps some, but not all, of these features of the world play the t-role in other disciplines. One is inclined to say that if just one of these features plays the t-role in all the other special sciences, then *that* feature is time. If different features of the world play the t-role in different special sciences then perhaps one is inclined to say that there is no unified phenomenon to which "time" refers but rather, there is a set of features that each have time-like properties and each of which is tracked by some psychological mechanism and plays an explanatory role in some special science. It is then an open question whether one concludes that there is no time, but there is a set of time-like things in the world, or one concludes that there is time, and what it is for there to be time just is for there to exist that set of time-like features of the world.

Which way we answer this last question cannot, it would seem, be determined either by physics, psychology, or any other the other special sciences. It is a question we are faced with once we already know everything there is to know about fundamental physics, on the one hand, and about the constraints placed upon it by the special sciences, on the other hand. We think that if such a scenario were to arise then further higher-level more abstract constraints would be brought to bear and these would arise from our folk concept of time. In particular, they would arise from those aspects of the folk concept not already captured and precisified by the various special sciences. It might be, for instance, that part of the high-level folk concept of time is that time is a unified pervasive structural feature of the world. On the scenario we have just described there is no such feature. So we would conclude that there is no time, there are merely the various time-like features of the world. If the folk concept was somewhat different then we would, instead, conclude that time just is the set of time-like features of the world.

Thus there could be ways things could turn out to be, such that we would need to appeal to the folk theory of time once we had exhausted all of the constraints imposed by the various special sciences. Another scenario in which that might be true would be if there were a unique structural feature of reality that each of the mechanisms that are jointly responsible for our temporal phenomenology track, but where that structural feature is very unlike what we intuitively suppose to be time. Here, again, appealing to the constraints imposed by the other special sciences can be expected to do some work. For if the structural feature tracked by our various psychological mechanisms turned out not to be the structural feature that plays the t-role in the other special sciences then we might conclude that whatever it is that is tracked by those mechanisms is not time. Rather, perhaps time is whatever realises the t-role in the other special sciences and we are subject to some sort of pervasive phenomenological illusion. On the other hand, if what plays the t-role in the other special sciences is exactly what is tracked by the psychological mechanisms that underlie our temporal phenomenology, then it seems more likely we will be inclined to say that time turns out to be radically different than we expected.

Notice that in both these scenarios the constraints arising from the other special sciences are important in determining the right thing to say under various circumstances. Notice also, however, that they are not sufficient to determine what we should say. For it is no part of psychology, or the other special sciences, that if our psychological mechanisms track something that is radically different to how we expect time to be, and if what realises the t-role in the other special sciences is somewhat like we expected time to be, then time turns out to be what realises the t-role in the other special sciences. If that is true, then it is true because of some higher-level constraint: the kind of constraint that would feature in folk theory.

Precisely what folk theory is in these cases is up for grabs; neither of us is committed to the view that it's just what people happen to think at a certain point in time: the kind of thing that could be detected by experimental philosophy.[10] For these are notoriously fragile,[11] especially in the face of surprising empirical discoveries. But there could be more robust facts about what we would be inclined to do in the face of the discovery that the special sciences are not pulling in the same direction. Perhaps, for example, there is a cognitive role for 'time' in action guiding which is univocal even if the phenomena in the world are not, and if so this might pull in the direction of deciding that time exists but is a mixture of naturally disparate phenomena. So exactly what these additional constraints consist in is unclear. What is clear is that a lot of the work of determining what time is, in the world, will be done by the constraints provide by the special sciences and the physical story provided by physics. What is also clear is that some questions can remain unanswered by the combination of those constraints and the physical data, and those questions are ones that require higher-level adjudication that can only be attained by reflection on folk theory.

5.3.2 Special Sciences Sans Subject Setting Role

We think that at least some of the explanatory projects of some of the special sciences regarding temporal phenomena might turn out not to be the kind of projects that play any kind of subject setting role. To see why, consider the examples of evolutionary theory and geology.

[10] See Genone (2012) and Knobe (2007).

[11] Stich (1990).

Very often in interdisciplinary work on time geological time and evolutionary time are mentioned. Both are important concepts: both largely the result of the astonishment with which the century-old discovery that geological and even biological history was many orders of magnitude longer than had been thought. Sometimes debates in evolutionary theory, such as about punctuated equilibria, are taken to be about time: the issue here is whether evolutionary 'time' proceeds smoothly, or speeds up in rapid speciation events and slows down in long periods of relatively slow drift.

There is a real question whether these uses of 'time' are metaphorical. In the case of the deep time of geology the notion of "deep time" can be parsed away with the claim that there has been a surprisingly long time between the present and when certain rocks were formed. Deep time is not a special *kind* of time. In the case of the speeding up and slowing down of time according to some accounts of evolutionary 'time', this notion be replaced with the claim that the amount of evolutionary change that takes place in a fixed amount of time is very variable: sometimes a great amount, sometimes little. No kind of time speeds up or slows down.

Nonetheless once this is done, do these sciences play a subject setting role? It might be thought that they do: time is that thing in which evolutionary and geological processes are embedded, and that might be one of the constraints on lower level theories of time.

Here, we think, matters are a little complicated. Certainly if (as we expect it is) contemporary geology and evolutionary theory are roughly right, then it might count as a constraint on lower level theories that they are theories of, amongst other things, the structure in which the many different geological eras are embedded, and in which different regions of evolutionary history are located. But it's not clear that this t-role has any the improved or clarified status over some of the roles specified by the folk theory. Rather, it is as though these disciplines have embedded in them the *folk* notion of time, without refining that notion in any way. For, in effect, all the t-role articulated above says is that time is the structure in which events are embedded and ordered. But that is, inter alia, what the folk theory says. Of course these theories contain extraordinary discoveries *about* time: how much there is and how surprisingly distributed in it are things of whose past existence we have evidence. But none of this, we suggest, constrains our account of time itself.

Another way to see this is to consider not what the consequences for our account of time would be if what these theories say is true, but, rather, what the consequences would be if they were *false*. In the case of psychology such a discovery could be potentially very important: we can imagine possible, if implausible, scenarios in which psychology discovers that it cannot be the case that the features of the word physics takes to be best candidates for being time are what is being detected or measured by any brain processes. As we noted earlier, this would cause doubts about whether this feature in physics deserves to be called 'time'. But what would the consequences of radical error in evolutionary theory be? If it turns out *per impossibile* that geological processes took only a thousand years then this would be a discovery that would tell us something about the world using the notion of time, but not anything that could change our account of time's own nature. And what if evolution was entirely false? Say the world was built, complete with misleading evidence, by aliens in 1901. And suppose those aliens didn't evolve but were the result of a quantum accident millennia ago, and they have been building worlds ever since—so the galaxy is full of life, but none of it evolved. Improbable and mad beyond belief, of course, but it's a hypothesis that can be entertained, and a hypothesis that would seem to have no effect or constraint on the nature of time at all. So if a theory can be entirely false without affecting time, then it suggests that the theory itself is not part of the subject setting process. To the extent that the theories just discussed mention time, the theoretical apparatus is borrowed from the folk theory (or perhaps defers to other special sciences). Thus if any such theory is falsified without falsifying the folk theory from which the time role was borrowed, the time role remains undisturbed. What we have learned, then, is that in order for a special science to play a subject setting role with respect to time—a role not already played by a higher-level folk theory—that science must take some aspect of the folk theory of time and precisify, clarify, or otherwise build on that theory.

5.4 Different Roles for Different Sciences

When we considered the particular role of psychology in providing constraints on lower level disciplines we noticed that it is possible that each of the various mechanisms jointly responsible for our temporal phenomenology might track a different feature of the world. We wondered, in that context, which of those various features ought to count as being time. That problem is writ large when we consider all of the special sciences and the constraints they jointly bring to bear. For the special sciences might, between them, place entirely different, and inconsistent, constraints on the lower level disciplines. There might be no consistent set of t-roles that includes the t-roles found in each of the special sciences. If that were so, then any phenomenon that would answer to 'time' relative to the constraints of one special science would not count as being time relative to the constraints of another.

This might occur if the original folk theory of time was itself inconsistent, and different special sciences incorporated and made precise different aspects of that theory. If that occurs then, we think, what it is in the world that ought to count as being time cannot entirely be adjudicated from within the special and fundamental sciences. Since the set of t-roles is inconsistent, nothing in the world will answer to those t-roles. Perhaps something will answer to some, but not other, roles. But if there are multiple features of the world each of which answer to some, but not other roles, then we are again left wondering which, if any, of these features is time. One proposal suggests that in this event we return to the folk theory and determine which of the various aspects of the folk theory were most central to that theory—again with the caveat that for this to have normative power those further aspects of folk theory need to be due to deep facts about human cognition, not just transitory weightings on the cluster of t-roles in a certain culture at a certain time. Then if there is a special science, or sciences, that consistently clarifies the more central aspects of folk theory, and if there is some feature of the world that realises that set of roles thus clarified, then that feature of the world is a good deserver to count as being time. Even if this is not the correct proposal, however, it is hard to see how we could determine the right thing to say about time without prescinding from the special sciences to a normatively powerful higher-level theory.

Even supposing there is a consistent set of t-roles it remains an open question whether there is any single phenomenon that realises all, or most, of those roles. If what it takes for there to be time is for there to be a unique realiser of that set of roles then it might turn out that there is no time even if there is something that realises the role of some, but not other, t-variables in some of the special sciences. Or one might be inclined to say that whatever it is that plays a good number of the t-roles counts as time even if the t-roles of some special sciences are not realised at all. Or one might say that there is no unified thing that is time: instead there is time in this special science, and time in that special science, and each of these is realised by some distinctive feature of the complete story of physics. Which of these turns out to be the right story about time will depend, in part, on whether there is a single phenomenon to be found in the physics that realises all or most of the t-roles. But if there is no such phenomenon then the answer to the question ‘what is time’ cannot be answered by attempting to locate time in the physical story of the world by appealing to the constraints of the special sciences. It can only be answered by considering the details of the folk theory of time not found in any of the special sciences. If the folk theory requires only that most of the folk roles (and hence most of the precisified t-roles of the special sciences) need be played, then finding something in physics that plays most of those roles is to find time. If the folk theory requires that all the roles be played, then finding that there is nothing in the physics that plays all the roles is finding that there is no time. If folk theory requires that some central roles be played, but that other roles are less important, then whatever plays those central roles, as delineated by the special sciences, is time, even if it does not play the more peripheral roles.

Despite the fact that we think there are some complicated scenarios in which it would be necessary to appeal to the some version of folk theory to settle the matter of where to locate time in the fundamental theory of the world, we also think that much of this work will be done by the special sciences. We thus think that there is much to say about the interplay between the special sciences and the lower-level disciplines. In what follows we consider just a few of the ways in which the special sciences will interact with and constrain lower level disciplines and what this might reveal to us about the nature of time.

5.4.1 Could Physics Autonomously Abolish Time?

There is an obvious reply to our claim that the special sciences play a constraining, subject-setting role on physics. Physicists might, we think, with good cause go about declaring “there is no time” even if it turns out that there is some unique phenomenon that plays all of the t-roles. It would thus seem as though physics has autonomously abolished time, unconstrained by the special sciences—for the thing *the special sciences* call time might very well still exist. How could that be if the special sciences are the subject setters? Our view is that situations like this are ones in which there is conceptual change; where the term ‘time’ has shifted in meaning within physics until it is no longer constrained by the other sciences. But such a case offers no objection to the claim that the special sciences play the role of settling the general language designation of ‘time’.

Suppose events proceeded as follows. Physicists make the (apparent) discovery that what plays that t-role is some feature of quantum collapse (we do not suggest this is plausible, it is for illustration purposes only). Time is, say the physicists, this process of quantum collapse. For a long while this is the orthodoxy, and indeed, after some long period the various t-roles in the special sciences are largely forgotten within physics. For physicists, time just is the process of quantum collapse. One day an amazing discovery is made: the theory of quantum mechanics, and with it, quantum collapse, is falsified. There is no quantum collapse, and there is nothing even remotely like any such collapse. Orthodoxy is overturned. Not only is this the discovery that a bedrock theory of physics is false, it is also the discovery that there is no time. Or so say many physicists. For ‘time’ for those physicists, has simply come to mean the phenomenon of quantum collapse. Just as, we think, chemists might come to claim “there is no water” if tomorrow we discover that there is no H_2O, likewise these physicists claim that there is no time because there is no quantum collapse. But that, of course, is consistent with there being something that plays the t-roles in the special sciences, just as the discovery that there is no H_2O is consistent with there being something in the world that is clear and potable and hence plays the ‘watery role’. That clear potable substance is likely to still be called ‘water’ in the general vocabulary, if not in the special vocabulary of chemistry. Similarly, we have described a situation in which physicists now take ‘time’ to pick out some particular physical process or structure, which, as it turns out, does not exist. Given this, they are right when they say that there is no time. But the rest of us would equally be right to say that there is time. For in the more general vocabulary of the rest of us, time just the thing, whatever it is, that plays the t-roles. And we have not been told that there is none of that.

This picture leaves room for the following: there could be a radical replacement of theories within physics that would, rightly, lead physicists to claim that there is no time, whilst that same discovery might, rightly, lead the rest of us to claim that time is simply something very different to what physicists had previously told us. This latter will occur when the replacement theories are sufficiently different from the original theories in physics that physicists took to exemplify a theory of time, whilst at the same being such as to vindicate the explanatory role of the t-variable in the special sciences.

The previous scenario is a nice example of the ways in which physics and the special sciences interact. The direction of *constraint* is one-way: from the special sciences to the physics. Even though physicists might announce that there is no time, the special sciences might have legitimate use for the notion. It’s only because of conceptual change in the physics community that their putative discovery counts as a discovery that there is no time. In the ordinary vocabulary it is still the special sciences that play the subject setting role. Physics, on the other hand, is still playing a determining role: it is determining whether there is anything in our world that plays the roles laid out by the special sciences—a question to some degree independent of the question of whether it discovers there is something answering to what it has come to call 'time'.

One might worry, however, that matters are not as straightforward as we have suggested. For if physics tells us that nothing plays the t-roles then it (likely) follows that some or all of the special sciences are false. There is, then, an open question about whether, in the event that the special sciences are false and nothing plays the t-roles, we should conclude that there is no time, or we should conclude that the special sciences were mistaken in their subject setting role: somehow they set the wrong subject. In the former case the special sciences continue to set the subject, it just turns out that nothing in the world answers to the subject that they set; in the latter case it turns out that they do not continue to set the subject and it is then an open question whether or not there is any time. In order to consider this second hypothesis there are two scenarios worth examining.

5.4.2 Timelessness and Unordered Time-Slices

Suppose that physics tells us that our world is like a deck of cards: each card has certain events and objects located on it, but the cards are not in any particular order in the following sense.[12] There is no fact of the matter, for any two cards, whether one card is before, or after, the other card. In this analogy the cards play the role of three-dimensional "instants" of the world: but they are instants without it being the case that one instant occurs before, or after, any other instant.

If our world were this way there is no reason to suppose that the special sciences would not have, and continue to enjoy, the same explanatory and predictive power that they do. Indeed, insofar as palaeontologists tell us that there is some set of instants at which there exist dinosaurs, they are (we may suppose) correct to say so. Insofar as the various theories in the special sciences allow us to predict what occurs at other instants, they are explanatory successful. The use of the term "t" in the various calculations by such disciplines allows them correctly to predict the properties of other instants. Yet strictly speaking the theories of the special sciences are false since their theories are either explicitly or tacitly committed to the existence of temporal orderings and in the scenario under consideration there are no such orderings.

Would this be the discovery that the special sciences play a subject setting role, but there is nothing that answers to the roles that they prescribe must be filled in

[12]This is a simplified version of Barbour's (1994a, b) hypothesis.

order for there to be time, or is this the discovery that the special sciences are mistaken in what they took the subject setting role to be? We see no reason to suppose that this is a case in which physics has undermined the subject setting role of the special sciences. For a start, there will be relatively close descendent theories of the special sciences that preserve their explanatory and predictive power, but not by quantifying over a t term. So much of the theories of the special sciences will, most likely, remain untouched by such a discovery. Given this, it seems that it is primarily those aspects of the special sciences that were precisifications of some aspect of our folk theory that are falsified. If so we have good reason to conclude that the special sciences retain their subject setting role, but there is nothing in the world that answers to that role.

5.4.3 *Simulationism*

Another even more sceptical hypothesis is the simulation hypothesis considered by Bostrom, Chalmers, and Weatherson.[13] According to this hypothesis we are asked to take seriously, due to various interesting arguments that need not detain us here, the claim that the actual world is a computer simulation. This is a kind of scepticism that might seem to have very profound bottom-up influences on the idea of time in the special sciences. You might think this would amount not just to the falsity of detailed claims in the special sciences (there are no geological strata, there never were any trilobites) but also to the possible rejection of the claims about the nature of time in the special sciences (maybe the underlying simulation is static and doesn't have a dimension that corresponds to the special sciences' notion of time). That is, one might wonder whether in such an event one would be inclined to suppose that the special sciences have no subject setting role.

We will proceed by considering two broad kinds of response.

5.4.3.1 Business as Usual

One tempting response to the simulation hypotheses, associated with David Chalmers,[14] is that it leaves everything, including physics, untouched. According to this response, if simulationism is true about the word, then physics is right as far as it goes (and so are the special sciences). It's just that there are more truths about the world than physics mentions. Instead of the ultimate reality being the strings or fundamental particles mentioned in physics, there is a deeper level of reality: the computational substrate that realizes them. But then everything in the special sciences about time is untouched, and is realized by something in physics in

[13]Bostrom (2003, 2005); Chalmers (2005); Weatherson (2003).

[14]Chalmers (2005).

much the way as before. The special sciences never made any claims about the realization, in the physical world, of their claims about where in time to find the origin of the moon or dinosaurs. Simulationism is just one, albeit unlikely, such hypothesis. And equally physics itself perhaps has no 'that's all' clause built into it: that's the job of specifically philosophical theses like physicalism. The physicist means only by 'fundamental' 'our theory doesn't say anything more about the nature of these things beyond what they do in our theory'. So if, say, strings are realized by something odd, interesting, but not in tension with string theory (unless it makes predictions at odds with it) then physics and the special sciences remain true theories, and time turns out to be whatever it is, in the physics, that the special sciences locate by way of the constraints they impose.

5.4.3.2 The Radical Revision Hypothesis

But this isn't the only response you might make: the view that there turns out to be no real time, and perhaps even no real dinosaurs, is defensible. But we think when you understand how this defence proceeds it undermines the idea that there is bottom up subject setting going on.

Here's how the defence might proceed. Perhaps in the simulating universe where the computers are located everything takes very little time. Perhaps it takes no time: the structure of the simulation is static in some way. So if we were to understand how it all worked, we would see it was an illusion in our simulation that dinosaurs are located in time differently from us. So the 't' term in the simulation does not code for time at all (maybe nothing does: maybe one of the spatial dimensions does!). On this hypothesis what the special sciences are saying about time, and perhaps much else, is mistaken. The question, then is whether this is a case in which lower-level disciplines (perhaps lower even than physics) determine that nothing plays the t-roles specified by the special sciences, or whether these lower level disciplines in some way undermine the subject setting role of the special sciences, or otherwise play a subject setting role themselves. In favour of the latter, one might wonder if such a discovery is a discovery that everything we ever thought about time—and what it would take for there to be time—was simply mistaken.

But not so fast. Notice that explicating this scenario required us to imagine a simulating universe a bit like the one we imagine ours to be. If the universe doing the simulating was just a strange mathematical abstraction, for example, then we'd be inclined to default back to the 'business as usual' response. But if we imagine the simulating universe to be very much like the one we imagine our own universe to be, we thereby imagine that there is time in the simulating universe. It is only *then* that we are in a position to say that the various parts of the simulation are not arranged in time in a way that would vindicate the temporal claims of physics or the special sciences.

But how do we locate time in the simulating universe? Only because we imagine things in that universe to be structurally sufficiently similar to the way we thought (perhaps mistakenly) they were in the simulated universe! And that is to say that enough of our understanding of the special sciences at a general level is true of the *simulating universe*, even if not of ours, for us to identify time in it. So the special sciences will be playing a subject setting role for time, even if not for time in our universe.

Thus neither kind of response to simulationism, one of the more radical sceptical hypotheses, undermines the subject setting role of the special sciences.

5.5 Conclusion

So why, then, do we engage in this interdisciplinary work on time of which this book is an example. Many reasons. But we hope we have identified one important one. It's that in answering questions like 'what is time' and 'is there time' no scientific discipline (much less philosophy) alone can do the job. For to identify which features of our fundamental picture of the word we need to look to in order to determine if there is time, and what it is like, we need to compare and contrast the sciences. We need to see whether these features of the world are playing the relevant explanatory and causal roles described in the special sciences. If there is a unique thing that exists and plays those roles and has a certain nature which physics can tell us about, then that's sufficient to answer the question of whether there is time, and what it is. If not, then things get more complicated; but they still involve this interdisciplinary connexion.

Of course not all interdisciplinary meetings will be motivated by the sorts of subject setting considerations we have outlined. Perhaps there are other motivations and useful explanatory projects: and sometimes such meetings will indeed have the flavour of the postmodern gallery. But perhaps we have been a little hard on the postmodern gallery, especially when its exhibits are science and philosophy. For sometimes there are serendipitously useful thoughts that can be had only by strolling its exhibits. We like to think this paper is one of them.

References

Barbour, J. 1994a. The timelessness of quantum gravity: I. the evidence from the classical theory. *Classical and Quantum Gravity* 11 (12): 2853–2873.

———. 1994b. The timelessness of quantum gravity: II. the appearance of dynamics in static configurations. *Classical and Quantum Gravity* 11 (12): 2875–2897.

Bostrom, N. 2003. Are we living in a computer simulation? *Philological Quarterly* 53 (211): 243–255.

———. 2005. The simulation argument: Reply to weatherson. *Philological Quarterly* 55 (218): 90–97.

Braddon-Mitchell, D. 2004. Masters of our meanings. *Philosophical Studies* 118 (1–2): 133–152.

Braddon-Mitchell, D., and R. Nola. 2009. Introducing the Canberra plan. In *Conceptual analysis and philosophical naturalism*, ed. D. Braddon-Mitchell and R. Nola, 1–20. Oxford: Oxford University Press.

Brown, S.W. 1995. Time, change, and motion: The effects of stimulus movement on temporal perception. *Perception & Psychophysics* 57 (1): 105–116.
Cai, M., C. Stetson, and D.M. Eagleman. 2012. A neural model for temporal order judgments and their active recalibration: A common mechanism for space and time? *Frontiers in Psychology* 3 (November): 1–11. doi:10.3389/fpsyg.2012.00470.
Chalmers, D.J. 2005. The matrix as metaphysics. In *Philosophers explore the matrix*, ed. C. Grau, 132–176. Oxford: Oxford University Press.
Creelman, C. 1962. Human discrimination of auditory duration. *The Journal of the Acoustical Society of America* 34: 582–593.
Dennett, D., and M. Kinsbourne. 1995. Time and the observer: The where and when of consciousness in the brain. *Behavioral and Brain Sciences* 15 (1992): 1–35.
Eagleman, D.M. 2009. Brain time. In *What's next: Dispatches from the future of science*, ed. M. Brockman. New York: Vintage Books.
Fraisse, P. 1963. *The psychology of time*. New York: Harper and Row.
———. 1984. Perception and estimation of time. *Annual Review of Psychology* 35: 1–36.
Genone, J. 2012. Theories of reference and experimental philosophy. *Philosophy Compass* 7 (2): 152–163.
Grondin, S. 2010. Timing and time perception: A review of recent behavioral and neuroscience findings. *Attention, Perception & Psychophysics* 72 (3): 561–582. doi:10.3758/APP.
Holcombe, A. 2015. The temporal organisation of perception. In *The oxford handbook of perceptual organisation*, ed. J. Wagemans, 820–839. Oxford: Oxford University Press.
Jackson, F. 1998. *From metaphysics to ethics. A defense of conceptual analysis*. Oxford: Oxford University Press.
Johnson, A. 2010. Modulation of time perception by visual adaptation. In *Attention and time*, ed. A. Johnson and K. Nobre, 187–220. Oxford: Oxford University Press.
Knobe, J. 2007. Experimental philosophy. *Philosophy Compass* 2 (1): 81–92.
Nishida, S., T. Ledgeway, and M. Edwards. 1997. Dual multiple-scale processing for motion in the human visual system. *Vision Research* 37 (19): 2685–2698. doi:10.1016/S0042-6989(97)00092-8.PMID9373668.
Nobre, A.C., and J. O'Reilly. 2004. Time is of the essence. *Trends in Cognitive Sciences* 8 (9): 387–389.
Stich, S.P. 1983. *From folk psychology to cognitive science: The case against belief*. Cambridge, MA: MIT Press.
———. 1990. *The Fragmentation of reason: Preface to a pragmatist theory of cognitive evaluation*. Cambridge, MA: MIT Press.
Treisman, M. 1963. Temporal discrimination and the indifference interval. Implications for a model of the "internal clock". *Psychological Monographs* 77 (13): 1–31.
Tse, P.U., J. Intriligator, J. Rivest, and P. Cavanagh. 2004. Attention and the subjective expansion of time. *Perception & Psychophysics* 66 (7): 1171–1189.
Weatherson, B. 2003. Are you a sim? *Philological Quarterly* 53 (212): 425–431.

Part II
Time Paradoxes in Physics

Chapter 6
Is the Future already Present? The Special Theory of Relativity and the Block Universe View

Christophe Bouton

Abstract It is often claimed that the special theory of relativity (STR) necessitates the block universe view, according to which all the events spread out in space-time are determinate, 'fixed' once and for all. After having briefly recalled the basic notions of STR (principle of relativity, constancy of the speed of light, relativity of simultaneity) (Sect. 6.2), I will distinguish and discuss the different arguments taken from STR in favor of the block universe view: the 'geometric argument' (the spatialization of time) (Sect. 6.3), the 'becoming-argument' (the loss of the becoming) (Sect. 6.4), and the 'determinateness-argument' (the determinateness of the future) (Sect. 6.5). Among these, I will focus in particular on the last argument, which can be summarized as follows: STR entails this puzzling consequence that any future event in a given inertial reference frame is already present or past in another inertial reference frame. Any future event is therefore already determinate. I will show that this argument contains a fallacy, and present several replies to it. My conclusion is that the notion of time in STR does not necessarily imply the block universe view; rather, it gives rise to a new ontology that is more complex but nevertheless consistent.

6.1 Introduction

The special theory of relativity (STR) has managed to convey the illusory nature of the classical conception of time as an absolute flow divided into past, present, and future. One of its most puzzling consequences is the collapse of the notion of an objective present, prevalent in Newtonian physics. This present is independent of a reference frame: a universal now that would be the same for all observers. With the theory of relativity, "'now' loses for the spatially extended world its objective

C. Bouton (✉)
Département de Philosophie/SPH, UFR Humanités, Université Bordeaux Montaigne, Domaine Universitaire, 19 esplanade des Antilles, 33607 Pessac Cedex, France
e-mail: christophe.bouton33@orange.fr

C. Bouton, P. Huneman (eds.), *Time of Nature and the Nature of Time*, Boston Studies in the Philosophy and History of Science 326,
DOI 10.1007/978-3-319-53725-2_6

meaning" (Einstein 1955, p. 149). Einstein pointed out that physics does not establish an unequivocal correspondence between the psychological now and a present state of the world. In his often-quoted letter to the Besso family dated March 21, 1955, he wrote:

> For those of us who believe in physics, this separation between past, present and future is only an illusion, however tenacious. (quoted in von Weizsäcker 2006, p. 288)

Some physicists have long been fascinated by this idea, claiming that the theory of relativity reveals the universe's fundamental atemporal structure. Commenting on STR, Arthur Eddington compared space-time to a "solid block of paper," where there is no place for becoming:

> In a perfectly determinate scheme the past and the future may be regarded as lying mapped out–as much available to present exploration as the distant parts of space. Events do not happen; they are just there, and we come across them. (1990 [1920], p. 51)[1]

Hermann Weyl expressed the same idea in the following passage devoted to the theory of relativity:

> The objective world simply *is*, it does not *happen*. Only the gaze of my consciousness, crawling upward along the life line of my body, does a section of this world come to life as a fleeting image in space which continuously changes in time. (1949, p. 116)

According to Olivier Costa de Beauregard, "relativity [special and general] is a theory in which everything is 'written' and where change is only relative to the perceptual mode of living beings" (1968, p. 430). Another French physicist, Thibault Damour, likens space-time to a musical score that would be created instantaneously in a composer's ingenious brain and that is already written at the moment we hear it (2005, pp. 57–66). The future is not blank, as we may think, but fully determinate. Time, then, is nothing but the subjective experience of a static four-dimensional reality (space-time) in which all of the events are already located. This view, called the 'block universe', seems to be currently the mainstream in the philosophy of time. According to it, past, present and future events equally exist. In the words of Price (1996, p. 3), the block universe is the "view from nowhen", the whole universe considered from a no temporal point of view. It is sometimes based on logical or metaphysical arguments against the "flow of time", in the wake of McTaggart's famous paper on "The Unreality of time" (see McTaggart 1908, and e. g. Smith 2011 and Price 2011). But the block universe view is often connected with the theory of relativity[2]:

---

[1]But Eddington does not exclude the existence of indeterminate events within space-time. In the next page, he gives the example of an "ultimatum," for which "it was possible to happen or not, consistently with the law of nature. Since physics does not attempt to cover indeterminate events of this kind, the distinction of absolute past and future is not directly important for physics" (1990, p. 52).

[2]Besides the quotations above, see Gödel (1949), Grünbaum, (1963, p. 318), Rietdijk (1966), Putnam (1967).

> Relativity theory has led to a revival of Parmenides' ideas, because this theory, formulated within the four-dimensional space-time formalism, has been interpreted in terms of a static four-dimensional block-universe in which there is no room for change or becoming. (Kroes 1985, p. 195)

The block universe view recalls not only Parmenides' ideas, but also that of Plato and his famous myth of the cave. People do not know that the world they are watching is just the temporal reflection of a timeless reality of which they are unaware. Yet if the universe is indeed a timeless block where all physical and psychic events are already there once and for all, how can it be that we have the seemingly vivid experience of time, of a succession of events? In order for such an illusion to appear, the brain would have to change a fixed order, constantly and flawlessly, into a dynamic succession, delivering the mass of information in small successive doses. If past and future do not differ from an ontological point of view, then it would be necessary, for example, to find a neuroscientific function that inhibits or prevents perception of the future as opposed to that of the past (Norton 2010).[3]

Does this mean, then, that we have to accept the block universe view, despite the difficulties it raises, since it is based on a well-established physical theory, the STR? This argument is initially all the more strange given that the concept of the block universe recalls, mutatis mutandis, Laplace's causal determinism; the block universe view would appear to be a revival of it:

> We must consider the present state of the universe as the effect of the earlier state and the cause of the latter state. An intelligence knowing all the forces acting in nature at a given instant, as well as the momentary position of all things in the universe, would be able to comprehend in one single formula the motions of the largest bodies as well as the lightest atoms in the world: to it nothing would be uncertain the future as well as the past would be present to its eyes (Laplace 1951 [1814], p. 4).

This is paradoxical, since the special theory of relativity reduces Laplace's argument to nothing. In STR, "the present state of the universe" has no meaning anymore. Moreover, due to the insurmountable limit of the speed of light in relativistic space-time, Laplace's 'demon' cannot have a complete knowledge of the world at any given moment, since its information is limited to its past light cone.[4] It can then neither retrodict all the past, nor predict all the future.[5] Laplace's determinism and STR do not seem to be happy bedfellows.

To avoid this difficulty, proponents of the block universe view emphasize its difference from causal determinism:

> the world manifold of occurrences, each eternally determi*nate* at its own place and date, may and may not be so determi*ned* in its texture that what occurs at one juncture has its sufficient reason at others (Williams 1951, p. 470).

---

[3]In this volume, Ismael (Chap. 2) provides an interesting solution to this conflict between the psychological time of everyday experience and the static time of the Block Universe.

[4]See Sect. 6.5.1 of this chapter for an explanation of this relativistic notion.

[5]See Popper (2000, pp. 57–61).

The block universe view claims that all the events spread out in space-time are determinate, 'fixed' all at once, whether or not they are determined by causal laws. The expression 'block universe' is therefore misleading (the word 'eternalism' is sometimes preferred[6]) to the extent that William James coined it in 1884 precisely in order to criticize determinism. The deterministic world is like an "iron block," a "block-universe" without possibilities, in which the present state "is compatible with only one totality" (James 1979, p. 118, p. 139, quoted by Dieks 2002, pp. 222–225). Contrary to this initial definition, the block universe view does not involve necessarily a physical determinism.[7] But the conclusion seems to be the same: since all the events are supposed to be fully determinate in space-time, "there is no free will" (Rietdijk 1966, p. 343). What is at stake in the debate on the block universe view is the existence not only of time but also of freedom, conceived of as a choice between several possibilities.[8]

According to Price (1996, pp. 12–13), the theory of the block universe makes three claims:

1. There is no objective present.
2. "There is no more an objective division of the world into the past, the present and the future, than there is an objective division of a region of space into here and there" (Price 1996, p. 12). In other words, there is no ontological difference between past, present and future events, which exist equally (in a tenseless sense of 'exist').
3. There is no objective flow of time.

   The theses 1 and 2 can be defended without the third. This is what Maudlin does (2007, pp. 104–142). For him, there is no objective present. Present, past and future have the same reality. In this sense, he subscribes to the theory of the block universe (2007, p. 109). But he defends the idea of a "passage of time" that leads to changes in the world. It is also possible to accept the thesis 1, without assuming the theses 2 and 3, as I will do in this paper. In other words, the fact that there is no objective present does not imply the ontological equality between past, present and future, nor must we abandon the notion of passage. I will focus on the connection between STR and the block universe view, because STR provides the most commonly used arguments[9] put forward to justify the block

[6]On the different meanings of 'eternalism', see Baron and Miller (2013).

[7]Dieks (2014, pp. 104–105) stressed that the "block determinism" is completely different from the "physical determinism", which deals with the question whether "the equations of motion have unique solutions once boundary and initial conditions have been specified".

[8]See also Petkov (2009, p. 172): "In the Minkowski four-dimensional world, however, there is no free will, since the entire history of every object is realized and given once and for all as the object's worldtube." On the issue of fatalism raised by eternalism, see Miller (2013, pp. 356–358). In this volume (Chap. 8, Sect. 8.5) Berkovitz argues that the block universe view implies the impossibility to change the future but not the impossibility to "*influence* the actual future to be what it is going to be." I will come back to this issue at the end of this chapter.

[9]Even the "strongest arguments", according to Smolin (2013, p. 55), who challenges the block universe view in his last book, by reintroducing the notion of "preferred observer" and "preferred global time" in the framework of the general relativity. See Smolin (2013, chap. 14: "Time Reborn from Relativity", pp. 164–171).

universe view. My question is whether STR necessitates the block universe view. To answer this, I will first establish some of STR's basic notions (Sect. 6.2). Then I will distinguish and discuss the three main arguments taken from the STR in favor of the block universe view: the geometric argument (the spatialization of time) (Sect. 6.3), the becoming-argument (the loss of the becoming) (Sect. 6.4), and the determinateness-argument (the determinateness of the future), which is sometimes called the "Rietdijk-Putnam argument" (Sect. 6.5). In Sect. 6.5, I will point out a fallacy in this third argument, and then present different replies to it: determinateness/indeterminateness can be understood either as intra-referential properties, or as frame independent properties relative to the light cone of the observer. My conclusion is that the notion of time in STR does not imply the block universe view; rather, it gives rise a new ontology that is more complex but nevertheless consistent.

6.2 Basic Notions of STR[10]

6.2.1 The Principle of Relativity

The principle of relativity (in the restricted sense) states that all inertial frames of reference are equivalent for the description of all physical phenomena.

A frame of reference is a Cartesian coordinate system with an origin (reference point), three axes and a clock. Any event can be located in it with four coordinates (x, y, z, t). An inertial frame of reference is a frame of reference in which Newton's first law of motion is valid (law of inertia). The principle of relativity means that the laws of nature are exactly the same in all inertial frames of reference. There is therefore no preferred inertial reference frame, no privileged observer for the description of the physical phenomena. Let us call this principle 'No Privileged Observer'.

6.2.2 The Law of Propagation of Light

Light *in vacuo* is propagated in straight lines with a *constant* velocity ($c =$ approximately 300.000 km/s)

c is frame independent. It has the same value regardless of the motion of the source or the inertial reference frame of the observer.

c is a maximum speed at which all matter and information in the universe can travel.

[10] I refer to Einstein (1905) and (1955).

6.2.3 The Relativity of Simultaneity

The first and the second points entail the *relativity of simultaneity*: two events occurring in two different locations that happen simultaneously in the inertial reference frame K of one observer do not happen simultaneously in the inertial reference frame K' of another observer that is in linear uniform motion relative to the first observer.

The Lorentz transformation specifies the relations between the space-time coordinates (t, x, y, z) of an event in K and the space-time coordinates (t', x', y', z') of the same event in K'. Given that K' is moving relative to K at the speed v,

$$t' = \gamma(t - vx/c^2)$$
$$x' = \gamma(x - vt)$$
$$y' = y$$
$$z' = z.$$

where $\gamma = \dfrac{1}{\sqrt{1 - \dfrac{v^2}{c^2}}}$

Given that two events e_1 and e_2 are simultaneous in K ($\Delta t = t_2\text{-}t_1 = 0$), they are not simultaneous in K′. The Lorenz transformation states:

$$\Delta t' = \gamma\left(\Delta t - \frac{v\Delta x}{c^2}\right)$$

According to this equation, the fact that $\Delta t = 0$ does not imply that $\Delta t' = 0$, except if $\Delta x = 0$ (if e_1 and e_2 occur at the same place).

Einstein pointed out that the relativity of simultaneity entails the loss of the "objective meaning" of the now (Einstein 1955, p. 149). Here 'objective' means 'frame independent'. What is the now? The now is the local now of an observer plus what I would call the *extended now*. It means that the now is the moment of the time where the observer is, and also a property of all the events that are simultaneous with the observer, even if the observer cannot perceive them. Since this distant simultaneity is frame dependent, the extended now is also frame dependent. Distinct observers in motion relative to each other do not have the same extended now, they do not share the same present.

6.2.4 Time Dilatation and Length Contraction

Time lapse between two events and the length between two points are not invariant from one observer to another, but are dependent on the relative speeds of the observers' reference frames (a clock ticks more slowly in motion than at rest, a rod is smaller in motion than at rest).

I will now turn to what appears to me to be the three main arguments taken from STR in order to establish the notion of the block universe.

6.3 The Geometric Argument: The Spatialization of Time

This argument arises in Minkowski's reformulation of STR. In his seminal paper (1909), Minkowski gave a geometrical formulation of STR and coined the notion of 'space-time' in order to emphasize the irreducible connection between space and time. He did not write that the theory of relativity makes an illusion of time. Rather time by itself and space by itself, separated from one another, are an illusion. But the union of the two–the space-time–has an independent reality:

> Henceforth space by itself, and time by itself, are doomed to fade away into mere shadows, and only a kind of union of the two will preserve an independent reality. (Minkowski 1909, p. 104).

How then do we get the idea of a block universe from space-time, which is not explicitly mentioned in Minkowski's paper? The geometric argument consists of the following analogy:

Space-time with four dimensions has a structure analogous to that of three-dimensional Euclidean space. Just as there is no objective 'here' in Euclidean space, there is no objective 'present' in space-time. Just like points in Euclidean space, all events (as 'world-points') are already positioned in space-time with the same reality. Just as the Euclidean space is static, there is no becoming in space-time.

6.3.1 The Analogy Between Euclidean Space and Minkowski's Space-Time

In reference to Minkowski's space-time, Einstein underscored the analogy ('Verwandschaft', 'Analogie' in German) between the four-dimensional space-time of special relativity and three-dimensional Euclidean space. In both cases, there is a continuum:

> Similarly, the world of physical phenomena which was briefly called 'world' by Minkowski is naturally four-dimensional in the space-time sense. It is composed of individual events, each of which is described by four numbers, namely, three space co-ordinate x, y, z, and a time co-ordinate, the time-value t. The 'world' is in this sense also a continuum ; for every event there are many 'neighbouring' events (realized or at least thinkable) as we care to choose, the co-ordinates x_1, y_1, z_1, t_1 of which differ by an indefinitely small amount from those of the event x, y, z, t originally considered. (Einstein 1955, p. 56)

Commenting on Minkowski's formulation of STR, Einstein adds that for the laws of nature that hold in the space-time continuum, "the time coordinate plays exactly the same role as the three space-coordinates" (1955, p. 57). He explains this assumption in the following way. An event with coordinates (x, y, z, t) in a reference frame K is also observed with the coordinates (x', y', z', t') in another reference frame K′ moving uniformly relative to K. What is the relation between (x, y, z, t) and (x', y', z', t')? According to the Lorentz transformation, we get[11]:

$$x^2 + y^2 + z^2 - c^2t^2 = x'^2 + y'^2 + z'^2 - c^2t'^2$$

By substituting the imaginary quantity $\sqrt{-1}ct$ for t, we get a new equation in which the 'imaginary' time coordinate is exactly like the three space coordinates: If $x_1 = x$, $x_2 = y$, $x_3 = z$ and $x_4 = \sqrt{-1}ct$, we get:

$$x_1^2 + x_2^2 + x_3^2 + x_4^2 = x_1'^2 + x_2'^2 + x_3'^2 + x_4'^2$$

The "'time' x_4 enters into natural laws in the same form as the space coordinates x_1, x_2, x_3" (Einstein 1955, p. 122). Einstein deduces from this analogy that Minkowski's space-time is like a "four-dimensional space," where any becoming is absent. In space-time, events do not happen or become, they just exist:

> A four-dimensional continuum described by the 'co-ordinates' x_1,x_2, x_3, x_4, was called 'world' by Minkowski who also termed a point-event a 'world-point'. From a 'happening" in three-dimensional space, physics becomes, as it were, an 'existence' in the four-dimensional 'world'. (Einstein 1955, p. 122)

6.3.2 *The Space-Time Interval*

We can articulate differently the analogy between space and space-time with the notion of a 'space-time interval' that is invariant like a spatial distance. In three-dimensional Euclidean space, the distance between two points is calculated using the Pythagorean Theorem:

$$\Delta l^2 = \Delta x^2 + \Delta y^2 + \Delta z^2$$

In classical physics, if we choose another system of Cartesian coordinates, the distance remains invariant (as the time interval between two events):

$$\Delta l'^2 = \Delta x'^2 + \Delta y'^2 + \Delta z'^2 = \Delta x^2 + \Delta y^2 + \Delta z^2 = \Delta l^2$$

[11]I am summarizing Einstein's demonstration. See Einstein (1955, appendixes I and II) for the details.

It no longer holds true in STR, because of the length contraction and time dilatation (see Sect. 6.2.4). But in Minkowski's space-time, there is a new invariant interval in which space and time are entangled. In a frame of reference, two events are located in x_1, y_1 ,z_1, t_1 and in $x_2 y_2$, z_2, t_2. The space-time interval between these two events corresponds to a physical reality linked to the constant speed of light. It describes the possibility of the connection of two events by one light signal. The square of the space-time interval, introduced by Minkowski, is written as follows:

$$\Delta s^2 = c^2(t_2 - t_1)^2 - (x_2 - x_1)^2 - (y_2 - y_1)^2 - (z_2 - z_1)^2$$

More simply,

$$\Delta s^2 = c^2 \Delta t^2 - \Delta x^2 - \Delta y^2 - \Delta z^2$$

If $\Delta s^2 > 0$ (the interval is 'time-like'), the two events can be connected at least by a light signal and there can exist a causal relationship between them. If $\Delta s^2 = 0$ (the interval is 'light-like'), the two events can be connected *only* by a light signal. If $\Delta s^2 < 0$ (the interval is 'space-like'), the two events cannot be connected by a light signal and therefore, no causal relationship between them can exist. The fundamental point is that the space-time interval between two given events is invariant no matter which inertial reference frame is selected. This is not a surprise, because the space-time interval depends on the speed of light c, which is constant in all the reference frames. Lorentz's transformation shows that:

$$\Delta s'^2 = c^2 \Delta t'^2 - \Delta x'^2 - \Delta y'^2 - \Delta z'^2 = c^2 \Delta t^2 - \Delta x^2 - \Delta y^2 - \Delta z^2 = \Delta s^2$$

Just as the distance between two points in a Euclidean space of three dimensions, the space-time interval between two events, i.e. between two points in space-time, is invariant. Space-time is "pseudo Euclidean," to use Minkowski's expression.

The three theses of the block universe view can be drawn from this analogy between Euclidean space and space-time. If space-time is like a four-dimensional space, there is no objective present in it, no objective distinction between past, present and future (like 'here' and 'there', 'present' is just an indexical notion without any ontological meaning) and no becoming.

6.3.3 *A Reply to the Geometric Argument*

What might be the response to the geometric argument? To begin with, the fact that time is irreducibly linked to space does not mean that it is mixed up with space. Analogy is not identity. Minkowski's space-time does not make time a "fourth dimension of space" and time maintains its specificity (Reichenbach 1957, pp. 110–112, pp. 189–190). In a conference in Paris in 1922, Einstein himself said: “it is certain that in the four-dimensional continuum all dimensions are not equivalent” (quoted in Čapek 1968, p. 435). There are at least two differences:

1. Even in the mathematical symbolism there is a difference between the time coordinate and the three space coordinates (the signs are opposite). We can erase the difference by changing ‘t’ into $\sqrt{-1}ct$. But this imaginary time has no physical meaning.[12]
2. Unlike spatial coordinates, t is irreversible, to the extent that an object can occupy once again the same location in space, but not in time. I will come back in the next section to the notion of irreversibility in STR.

A second response consists of taking up the distinction, pointed out by Lawrence Sklar, between the events and their locations. In STR, possible locations of events that comprise a continuous space-time must be distinguished from the events themselves, which may not fill up all the points of space-time (Sklar 1977, pp. 56–57). This is why Einstein refers to “realized” or “thinkable” events. Space-time is the set of locations of real or possible events. The event is a physical reality (an emission of a light signal, a body falling, a star’s explosion, etc.), distinct from its location, which corresponds to its place and its date, to a point in space-time or a ‘world-point’. Space-time is a framework that defines the whole of *possible* positions of events. It describes, for instance, how future events could be connected to each others, according to which physical constraints, but this does not mean that all events are ‘already there’. Thus, the set of possible positions in a table is not the same thing as the objects that occupy certain positions within this table. Even if all positions were determinate in space-time as they are in space, it would not follow that all events are pre-positioned, as well as the objects lying on a table do not necessarily occupy all the available positions of the table. To say that the universe is the set of events identified with space-time points is to create the idea of the block universe, since all events are then determinate as all points.

To conclude, space-time of STR is a *chrono*geometry, and not a pure geometry. To make space-time into a space is to spatialize time without justification for such a spatialization.

6.4 The Loss of the Becoming

A second argument in favor of the block universe view focuses on the status of the succession in STR. The becoming-argument states:

The relativity of simultaneity entails the impossibility of defining an objective present. The becoming presupposes a succession of presents. The impossibility of

[12]See Eddington (1990, p. 48): “It is not very profitable to speculate on the implication of the mysterious $\sqrt{-1}$, which seems to have the property of turning time into space. It can scarcely be regarded as more than an analytical device.”

the objective present proves the impossibility of the objective becoming. Briefly, the dislocation of simultaneity leads to a dislocation of the becoming.

6.4.1 Gödel's Argument for the Unreality of Time

Let us present this argument with Gödel's famous paper (1949). Gödel proposes to give a new proof of McTaggart's thesis on the unreality of time, "that mysterious and seemingly self-contradictory being,"[13] based on Einstein's theory of relativity (special and general). Concerning the notion of time in STR, which constitutes the first part of the paper, he presents the following argument (1949, pp. 557–558):

- Change is only possible if there is a lapse of time.
- An objective lapse of time assumes that reality comprises infinite layers of now that successively come into existence (a layer of now is a hyperplan[14] that encloses all events that are simultaneously present).
- But since simultaneity is relative to the observer, each observer has his or her own set of nows.
- No series of layers of now can have a prerogative, since each inertial reference frame is valid.
- Therefore, there is no objective change. Change is an illusion or an appearance due to our mode of perception.

The point of the argument is simultaneity's relativity: A now is defined as the observer's now in a given reference frame. As we have seen, the observer's now is not limited to his or her local now (his or her 'here-now'), but designates all events that are simultaneous to this now. This is why this extended now is described as a 'layer' or a hyperplane (it is a "foliation" of space-time). Yet simultaneity between events that are all 'now' is a distant simultaneity, and is thus relative to the observer's reference frame (see above Sect. 6.2.3). Each observer then has his or her own (extended) now. According to the principle 'No Privileged Observer' (see Sect. 6.2.1), no observer has the right to privilege his or her 'now': "Each observer has his own set of "nows," and none of these various systems of layers can claim the prerogative of representing the objective lapse of time" (Gödel 1949, p. 558). For any point in space-time, there are an infinite number of possible and legitimate observers moving uniformly relative to each other, which define an infinite number of presents with their hyperplane of simultaneity at this point. There is a "super-abundance of presents" (Savitt 2000, p. 570). If the becoming (or passage) is defined by "the advance of events (from future to past) through *the* now, then passage is undermined by the fact that there is no longer a unique now to serve as

[13]See Gödel (1949, p. 557). Gödel refers to McTaggart's paper (1908) at the beginning of his article (p. 557, note 1).

[14]A hyperplane of an n-dimentional space is a n-1 dimentional space.

the now" (Savitt 2011, p. 561). There is no objective becoming, because there is no objective present. An event can happen (become present) for an observer, but not for the other.

The becoming-argument shows how the statement 'there is no objective flow of time' (third thesis of the block universe view) can be drawn from the statement 'there is no objective present' (first thesis of the block universe view).

In his book on *Relativity*, Einstein calls upon Gödel's argument against the classical notion of time:

> since there exist in this four-dimensional structure no longer any sections which represent 'now' objectively, the concepts of happening and becoming are indeed not completely suspended, but yet complicated. It appears therefore more natural to think of physical reality as a four-dimensional existence, instead of, as hitherto, the evolution of a three-dimensional existence. (Einstein 1955, p. 150)[15]

Note that Einstein does not conclude, like Gödel, that the becoming is an illusion, but that it is 'complicated'. What is this complication?

6.4.2 Relativistic Becoming

The response to Gödel's argument consists of distinguishing simultaneity and succession. In STR, distant simultaneity is always relative to a frame of reference. Relationships of succession are also relative between two events whose space-time interval is space-like, but absolute between two events whose space-time interval is time-like or light-like, that is, that can be linked by a light signal. In that last case, the order of succession cannot be reversed, because this would imply to exceed the limit of the speed of light, which is impossible (see Sect. 6.2.2). This is a crucial point. It means that if the time lapse measured between two time-like separated events varies from one referential to another (because of the time dilatation, see Sect. 6.2.4), the temporal relationship of anteriority or posteriority nevertheless stays the same. For these events, the time lapse is relative, but the time order is absolute. It is then possible to define an objective succession between events whose space-time interval is time-like. The time lapse between these events is irreducible: It can decrease but not disappear, because they cannot be simultaneous in any reference frame. Moreover, the time order cannot be reversed, and to that extent, the STR confirms the irreversibility of (causal) succession. For example, let's take one particle's world line. The world line of a particle is its path, i.e. the succession of its positions in the four-dimensional space-time (see Fig. 6.1).[16]

Since all events represented in a world line are time-like connected, "the world lines of any kind are irreversible" (Čapek 1968, p. 441). In this case, there is a "*local* process" of events, a becoming that is a "local phenomenon" in space-time, defined

---

[15] This text comes from the Appendix V, added in the second edition of 1952 (first edition : 1917).

[16] For an object with three spatial dimensions, the appropriate notion is a 'world tube'.

Fig. 6.1 The world line of a particule

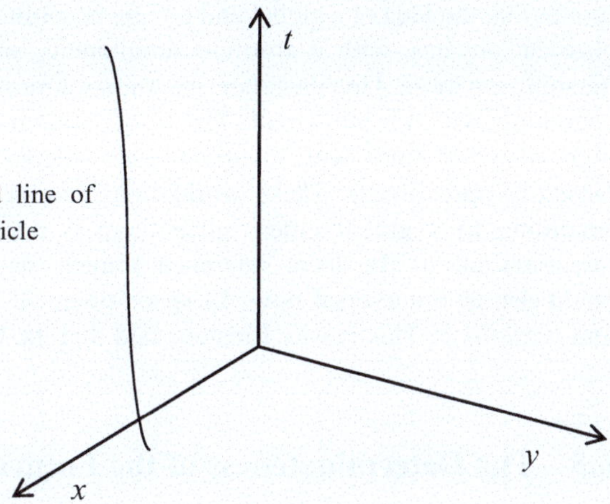

as "the successive occurrence of events" linked to a world line, which does not rely on distant simultaneity nor tense (Savitt 2011, pp. 563–565).

With respect to a world line, it is also possible to define the relativistic notion of 'proper time', which corresponds to the elapsed time of a physical process between two events located on the world line. In Fig. 6.1, the proper time of the particle can be given by the time of a clock that is in immediate proximity of the particle, at rest in relation to it, and accompanies the particle following its world line. Like the time order between two events of the world line, the 'proper time' of a given process between these two events is invariant, it is the same for all observers.[17] However, this does not mean that there is a unique time lapse between two events, as it is the case in classical physics, because "between two events infinitely many connecting worldlines can be drawn, and the time lapse between our events depends on the worldline that is considered." (Dieks 2014, 100) Proper time is frame independent, but world line dependent, whereas the relationship of succession (between time-like separated events) is frame and world line independent.[18] On the one hand, STR

[17]See Arthur (2006, p. 141) : "the quantity of time elapsed for a given process (such as a clock keeping time) is measured by the proper time, a quantity that is calculated by taking the integral along the world line of that process of the quantity: $\tau = \int d\tau$, where $d\tau = \surd\,(c^2dt^2 - dx^2 - dy^2 - dz^2)/c$, where x, y, z and t are the co-ordinates in some given inertial frame, and are considered as functions of the proper time τ. The proper time so calculated is invariant to change of frame: it will come out the same no matter what inertial frame (with co-ordinate values x, y, z and t) is chosen". τ is invariant because $d\tau = ds/c$ and ds (the space-time interval) is invariant (see section 6.3.2, where the square of the space-time interval $\Delta s^2 = c^2\Delta t^2\text{-}\Delta x^2\text{-}\Delta y^2\text{-}\Delta z^2$). If the clock is at rest, its space coordinates don't change: $dx = dy = dz = 0$, so that: $d\tau = \surd\,(c^2dt^2)/c = dt$. In this case, the coordinate time t indicated by the clock is numerically equal to the proper time τ of the particle (Arthur 2008, p. 219).

[18]This can be illustrated by the famous Twin Paradox : the proper time of the twin who stayed on the Earth is greater than the proper time of the twin who travelled on the rocket. But in both cases, the reunion of the twins is located *after* their departure. On this paradox, see Proietti (Chap. 3 in this volume).

proves that the idea of a global and unique becoming, based on the universal time of classical physics, with it absolute simultaneity and its absolute duration, has no meaning anymore. One the other, it assumes a more complex (pluralistic) notion of becoming, which is consistent: The world contains a plurality of objective becomings, of world lines with invariant time orders (succession) and invariant time lapses (proper times). These world lines cannot be ordered among themselves according to a global order, since there is no universal reference frame that integrates all of the local reference frames, or no universal simultaneity that could define a universal now. In space-time, all events are ordered, but *locally* and *partially*.[19] This is why Einstein said that the becoming is "complicated."[20]

6.5 The Determinateness of the Future

The notion of becoming we are dealing with (the "passage of time" or the "flow of time") has different meanings that we have to distinguish. It can be defined by the succession of events from earlier to later (let's call it the B-becoming in reference to McTaggart's B-series). As we have just seen in the last section, the B-becoming keeps an objective meaning in STR, which is local, pluralistic and therefore different from the global notion of becoming in classical physics. Yet there is a second way of understanding the becoming, as the passage of the events from a state where they have not yet happened to a state where they have already happened–from the future to the past through the present (let's call it the A-becoming in reference to McTaggart's A-series). In the framework of STR, this notion of A-becoming seems more difficult to assume: its critique leads us to the third argument in favor of the block universe view, especially to its second thesis of the ontological equality of the past, the present and the future. This argument can be summed up as follows:

Future events are supposed to be indeterminate, whereas past and present events are determinate. But STR states that any future event in a given inertial reference frame is present or past in other inertial reference frames in motion relative to it. There is therefore no ontological difference between future, present, and past, and any future event is already determinate like past and present events.

---

[19]See Dieks (1988, p. 459). Dorato (2006) presents the same kind of argument.

[20]In the second part of his paper, Gödel points out that general theory of relativity allows possible worlds in which closed world lines exist without exceeding the limit of the speed of the light (Gödel, 1949, pp. 559–560). In these worlds, it is possible to travel in any region of the past, and the 'earlier-later' relationship is therefore never objective. But it raises lots of logical and metaphysical paradoxes, like the famous grandfather paradox (see Smeenk and Wüthrich 2011). Commenting Gödel's paper, Einstein concluded: "it will be interesting to weigh whether these [Gödel's cosmological solutions of the gravitation-equations with causal loops] are not to be excluded on physical grounds." (1949, p. 688)

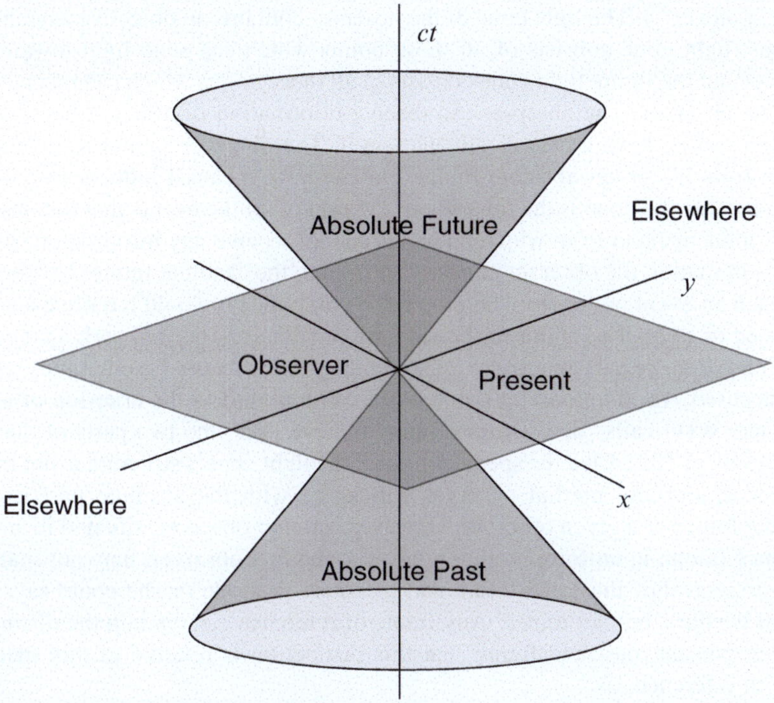

Fig. 6.2 The observer's light cone in space-time

6.5.1 Past, Present and Future in STR

This argument implies that the notion of A-becoming—indeterminate future events becoming determinate (present then past)—has no objective meaning.[21] Before addressing this argument, we have to clarify the meaning of tense in STR. Although the laws of physics do not recognize this difference in their equations, any frame of reference can contain the distinction between present, past and future. This distinction is not arbitrary; it has a physical meaning that appears in the representation of the observer's light cone.

In space-time, any observer is at the point of contact between two cones: the absolute past is the lower cone, the absolute future is the higher cone[22] (see Fig. 6.2). The present of the observer is the hyperplane drawn in dark gray. Contrary to the light cone, it is relative to his or her frame of reference (represented by the

[21]From this point of view, the coming into being of an event (becoming present) is "mind-dependent", it is just "the coming into the present awareness of a sentient organism" (Grünbaum 1963, p. 328).

[22]Absolute means here "frame independent."

axes x, y and ct[23]). The light cone of the observer contains an objective asymmetry. The past light cone consists of all those points which can send light towards the observer and the future light cone consists of all those points which can receive light from the observer. The observer can receive information from his or her absolute past, but cannot have a causal influence on it. The observer cannot receive information from his or her absolute future, but can have a causal influence on it. The region outside the cone is the 'elsewhere', the set of all the events that the observer cannot influence and from which he or she cannot receive any information. As any particle of matter, the observer can only travel into the absolute future. No observer can reach an event in the elsewhere or in the past, since it would require exceeding the speed of light. The "limit to the velocity of signals is our bulwark against that topsy-turvydom of past and future, of which Einstein's theory is sometimes wrongfully accused" (Eddington 1990, p. 57–58). With regard to the criterion of causal influence, Weyl calls "active future" and "passive past" the two parts of the light cone (1949, p. 102). Like the speed of light c, the light cone's structure is not frame-dependent, and any inertial observer will agree with the absolute past and the absolute future of a given observer. However, another observer, situated in another reference frame, in uniform motion relative to the first observer, may not share the same present, thus the same future and the same past. He or she could say: "this present is yours, but not mine." Any frame of reference can contain the distinction between present, past and future, but this distinction is relative to this specified frame of reference.

6.5.2 *From Relativity to Determinateness*

Some philosophers have explicitly used this consequence of STR to justify the future's pre-determination and the block universe view. In his 1967 paper, Putnam begins with this situation: Two observers meet up with a high relative speed (let us call them W_1 and W_2). STR demonstrates that there are future events for one observer that are present or past for the other. Putnam deduces from this that "*future* things (or events) are already real" (1967, p. 242). His argumentation is the following. He recalls that the relation "being simultaneous with" is not transitive across observers in different states of motion.[24] But he defines the relation R "being real for" as *transitive* because of the principle "No Privileged Observer" : if the event X is real for observerW_1, and W_1 is real for observer W_2, thus X is real for W_2. Putnam applies the relation R to the following situation. An event X is future for

[23]By convention, the time coordinate is given by ct where c is the speed of light.

[24]If a and b are simultaneous for O in the inertial frame of reference K, and if b and c are simultaneous for O′ in K′ in uniform linear motion in relation to K, then a and c are not simultaneous for O′ in K′ (since a and b are not simultaneous for O′ in K′). Strictly speaking, the relation "being simultaneous to" is not to be said intransitive: it is transitive only *inside a specified frame of reference*.

observer W_1, and present for observer W_2. X is real for W_2 since all that which is now is real. W_2 is real for W_1, since W_1 and W_2 meet at the same place and the same moment (they are present for each). Thus, according to R, X is also real for W_1, although it is future for her. Putnam deduces two things from this. 1) Contrary to common sense, which holds that only the present is real, it is necessary to states that present, future and past are all real. 2) Contrary to the Aristotelian concept of contingent futures, the future is determinate, and propositions that rely on contingent futures already have a truth value, they are either true or false, even if we don't know in advance what the correct truth value is (Putnam 1967, pp. 243–247).

Two years later, Fitzgerald publishes a paper with the same idea. An observer, Max, is wondering whether it is true that a sea fight will occur tomorrow. Fitzgerald adds to the famous Aristotle's example another observer who zooms past him on a rocket at high velocity. For this observer, the sea fight has already occurred. It seems "that Max should assign a *true* to the rocketeer's statement, and also to his own" (1969, p. 314).

To clarify this argument, I will refer to one of Sklar's thoughts (1977, pp. 272–275). Sklar wants to show that STR refutes "two traditional philosophical claims" about time, which he understands to be the following ontological distinctions:

1. Only those things that are present exist; things that are in the past or the future don't exist. (This is the thesis of presentism.)
2. Whereas present and past are determinate, future is not determinate.

Like Putnam and Fitzgerald, Sklar gives the example of a future event for one observer that is present for another (Fig. 6.3).

In the reference frame K (O, x, t), one observer is situated at event O, and e is an event space-like separated from him or her, i.e. situated in his or her elsewhere, outside his or her light cone, but in the future part of his or her reference frame. STR states that another possible observer exists, who is in uniform linear motion in relation to the first observer and is also situated at O, for whom the event e is present (Imagine that the two observers cross paths in O). In the reference frame K′ (O, x', t') of the second observer, we can see that e is present because it is simultaneous with O. How can the event e not exist for one (be future) and exist for the other (be present)? How can the event be indeterminate for one (be future) and fully determinate for the other (be present)? Sklar deduces from this example that for STR, all events are fully determinate. Past, present and future each have existence and determinateness. From the refutation of presentism (which claims that only the present exists) to the block universe view, there is but one step (note that Sklar does not broach this subject here). More precisely, the conclusion of the argument leads to the second thesis of the block universe view: there is no ontological difference between present, past and future. We get the image of space-time as a 'block', since all the events within it are determinate, 'fixed' once and for all.

At this point, it may be useful to specify the sense of the equivocal notions 'determinate/indeterminate' which are often used in this debate on tense in STR. In this context, they seem to have two complementary meanings:

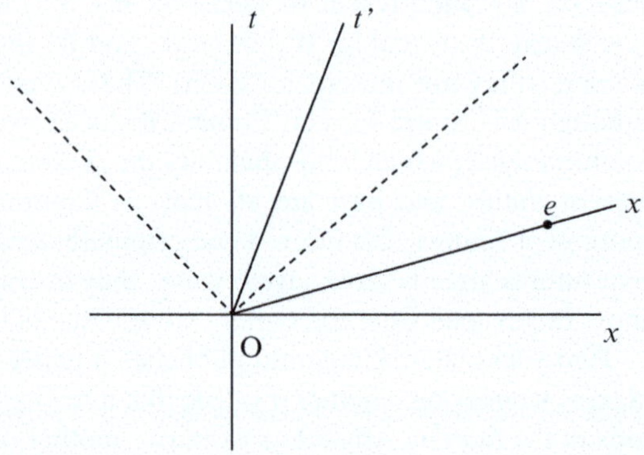

Fig. 6.3 Event *e* is future for one observer in (O, *x*, *t*) and present for another in (O, *x'*, *t'*) (Sklar 1977, p. 274) (I completed the diagram with the axis *t'* and the future light cone (in dotted lines))

- From an ontological point of view (taken by Sklar[25]), an event *e* is determinate if it has occurred or is currently occurring: it is "real", "definite", "fixed", "settled". Stein (1968, 1991) uses the words "determinate" and "definite" with the same sense: an event is determinate or definite if it "has already become".
- From a logical point of view, which is also adopted by Putnam (1967) and Fitzgerald (1969), an event *e* is determinate if the proposition describing *e* is determinate, that is has a truth value (is either true or false).
- Conversely,
- From an ontological point of view, an event *e* is indeterminate if it has not yet occurred.
- From a logical point of view, an event *e* is indeterminate if the proposition describing *e* is indeterminate, that is has no truth value (is neither true nor false).

We can suppose that if *e* is ontologically determinate/indeterminate, *e* is also logically determinate/indeterminate, and vice versa.

The determinateness argument states that the distinction between determinate events (past and present events) and indeterminate events (future events) collapses (there are only determinate events), because for any future event *e* in a reference frame K, there is another observer in K′ in motion relative to K, for whom *e* has already happened.

It seems to me that the determinateness-argument is the strongest argument in favor of the block universe view, because it can resist the previous objection against the becoming-argument. As we have seen, this objection states that there is a plurality of objective B-becomings or world lines in the universe, a plurality of flows of events evolving according different irreversible successive orders. But the determinateness-argument claims that all the events in these world lines are equally determinate, whatever temporal status they may have in a given frame of reference.

[25]See Sklar (1977, p. 272) : "Whereas the present and past, having occurred or currently occurring, can be said to have determinate reality, the events of the future, whatever they *will* be, cannot be said to have any kind of 'determinate' reality".

6.5.3 Absolute Succession and Relative Succession

To confront this issue, we must first set aside one poor solution, which would states that it is a question of several distinct events. In this case, the event would be scattered across as many frames of reference as possible. On the contrary, as Einstein himself stated, it is the *same event* that is located in different reference frames with different temporal statuses.[26] The space-time coordinates are frame relative, not the event.

The first plausible solution is to revisit the distinction between absolute succession (established between time-like and light-like separated events) and relative succession (between space-like separated events). The argument for the future's determinateness only seems to function if the event is taken in the observers' elsewhere. This is the case in Putnam's and Fitzgerald's papers mentioned above. In the opposite perspective, Maxwell (1985) attempts to demonstrate the incompatible relationship between STR and "ontological probabilism".[27] His objective is not to prove determinism, but to point out the imperfection of STR, which supposedly must allow probabilism in order to conform to quantum mechanics. But Maxwell also demonstrates this incompatibility with regard to an event that is beyond the light cone of the observers. In the other case, there is not incompatibility between STR and probabilism. So, if two observers cross paths in an event E_1, and if a second event E_2 is located in E_1's future light cone, "E_2 may well be ontologically indefinite" for both observers (Maxwell 1985, p. 25).

This raises the following question. Let us look at Sklar's space-time diagram with a second event e_2 located in the future light cone of the observers situated at event O (see Fig. 6.4).

The event e, which is outside O's future light cone (represented in dotted lines), is future is in the reference frame (O, x, t) and present in (O, x', t'). If we consider another event e_2 in O's future light cone, e_2 is future in the two reference frames (O, x, t) and (O, x', t'). e_2 is located after O in both reference frames and the succession between O and e_2 is absolute. But is e_2 future in all the frames of reference? As we will see in the next subsection, this is not necessarily the case and here again, the before/after relationship and the distinction between past, present and future must be carefully distinguished from one another.

[26]See Einstein (1955, p. 32): "Relative to K$'$, *the same event* would be fixed in respect of space and time by corresponding values x', y', z', t', which of course are not identical with x, y, z, t" (my emphasis).

[27]See Maxwell (1985, p. 23) : "Probabilism, as understood here, is the thesis that the universe is such that, at any instant, there is only one past but many alternative possible futures-the fundamental laws of the universe being probabilistic and not deterministic".

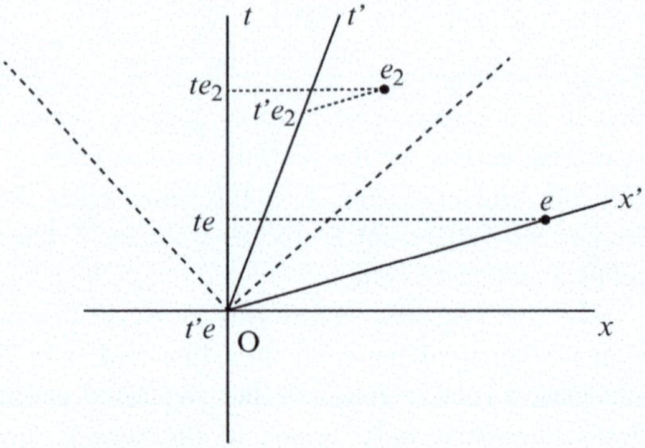

Fig. 6.4 Sklar's diagram with a second event e_2 in the future light cone of the observers

6.5.4 *Back to Determinism?*

In an article written in the same period as Putnam's paper (1967), Rietdijk (1966) aims to demonstrate that STR proves determinism. In the literature on this topic, one speaks sometimes of the "Rietdijk–Putnam argument" (see for instance Petkov 2006). But this expression overlooks the fact that even if the two papers lead to the same conclusion (the connection between STR and determinism), their demonstrations, however, are quite different. Instead of making the two observers cross paths in the same point of space-time, Rietdijk considers two observers at two different places: an observerW_2 in uniform linear motion approaches another observer W_1 (see Fig. 6.5).

There is an inertial frame of reference S_1 (O_1, X_1, T_1) with an observerW_1 originally situated in O_1 (event *A*) and another inertial frame S_2 (O_2, X_2, T_2) in motion in relation to S_1 (S_1 is represented in red, S_2 in blue). The observer W_2 is originally situated in O_2 (event *B*). At $T_1 = 0$, there is an event *P* (e.g. the emission of a light signal), *situated in W_1's light cone,* which is future in S_1, and present for W_2 in S_2. Note that *A* and *B* are simultaneous in S_1, *B* and *P* are simultaneous in S_2, but *A* and *P* are successive in S_1 and S_2. In S_1, *A* is present and *P* is future. In S_2, *A* is past and *P* is present. The succession between *A* and *P* is the same in both reference frames, but the temporal properties of these events (past, present or future) change from one reference frame to another. Although *P* is absolutely situated after *A*, the fact that *P* is future in S_1 does not mean that *P* is future in S_2. Rietdijk concludes from this that "*P* is completely determined at the moment T_1=0 of observer W_1" (1966, p. 341). In effect, given that *P* is present for W_2, it is a wholly determined event. Even if *P* is future for W_1, it cannot be conceived of as undetermined, otherwise it could not be a present event for W_2. The consequence is that according to Rietdijk, STR proves determinism in a specific sense.

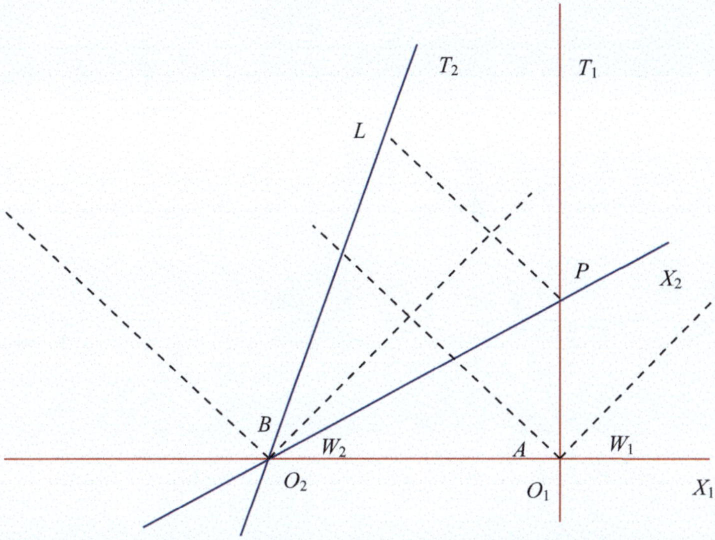

Fig. 6.5 The observer W_2 moves with a constant velocity in the direction of W_1 (Rietdijk 1966, p. 341) (I simplified Rietdijk's diagram in order to make it more readable)

Rietdijk redefines determinism—in a different way than Laplace's determinism[28]—as follows: An event P is (pre)determined if for any observer W_1 that has P in his or her absolute future, there exists a possible observer W_2 who can prove at a certain moment T_p that W_1 could not have influenced P at any moment where P was future for W_1 (Rietdijk, 1966, p. 342). Now, if we call "event L" the reception of the light signal from P by the observer W_2 (assumed that P is the emission of a light signal), then, in Fig. 6.5, "the moment T_p of the definition will occur, e.g., when W_2 passes event L" (1966, p. 343). At this moment, W_2 will receive the information that P was already present, determined in S_2, and, therefore, W_2 will know that from the moment $T_1 = 0$, W_1 was not able to 'influence' P in S_1, although P was future for W_1.

One could reply that W_1 was able to influence P before $T_1 = 0$, i.e. before P became present for W_2, when P was more remote in the future of W_1. It is the reason why Rietdijk adds this indication in his paper: "By taking the distance AB arbitrarily long we can make AP correspond with an arbitrarily long time, that is, P is pre-determined 'from time immemorial'" (1966, p. 343). If we increase the distance AB between the two observers (see the distance A'B' in Fig. 6.6), that is, if the new observer W_3 (with the inertial reference frame (O_3, X_3, T_3) represented in

[28]Contrary to Laplace's determinism, that only has meaning in a Newtonian time, Rietdijk's pre-determinism is defined without reference to natural laws or causation. Strictly speaking, we should say that for Rietdijk, the event P is not determined but determinate, according to Williams' distinction mentioned earlier in the introduction of this chapter. The word "determined" refers to causal law, which is not the case here.

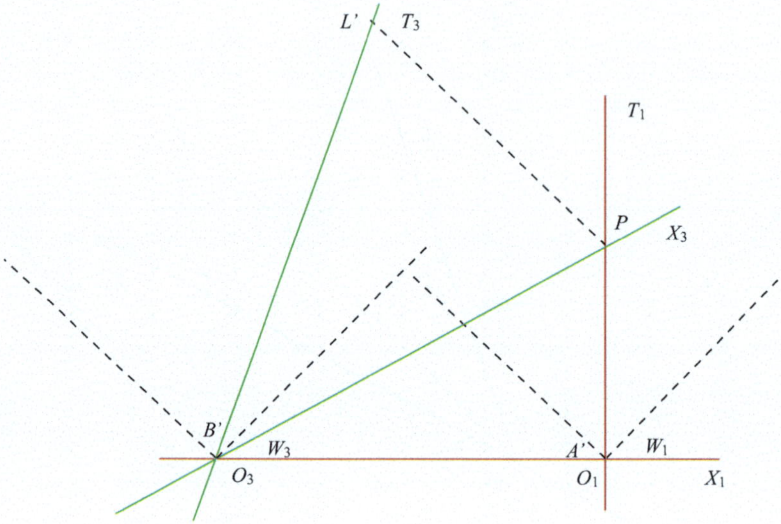

Fig. 6.6 Rietdijk's diagram with a greater distance between the two observers W_1 and W_3 (W_3 approaches W_1 in uniform linear motion)

green) is more distant from the first observer W_1, the event P is more remote in the future of W_1, but nevertheless present for W_3. This means, according to Rietdijk's definition of determinism, that P is pre-determined not only at the moment $T_1 = 0$, but at any moment when P is future for W_1, however remote it may be in this future, since it is always present for another distant observer. The more distant in the future is the event for the first observer, the more distant in the space is the second observer for whom the event is present. Therefore, *for any future event relative to a given inertial reference frame, there are always other possible inertial reference frames moving toward it, relative to which this future event is already present or even past.*[29] Unlike Putnam's argument, this holds even if the event in question is situated in the first observer'sabsolute future (the forward light cone). Being in the absolute future of an event doesn't mean being absolutely future.

6.5.5 A Paradoxical Determinism

To deal with the determinateness-argument, I will consider three possible strategies. We can raise a first objection about this kind of pre-determinism. Looking at Rietdijk's example more closely (in Fig. 6.5), neither W_2 nor W_1 can predict P, in the ordinary sense of predicting, i.e., given all informations available presently, one

[29]Note that we can use the same argument with another reference frames relative to which P is past. See Rietdijk (1976, p. 599).

can know in advance what is the content of a future event. In effect, at $T_1 = 0$, P is future for observer W_1 and present for observer W_2, but W_2 can absolutely not give this information to W_1, since W_2 does not yet have any information about P: P is unobservable, because it is simultaneous with his or her present, it is situated in W_2's elsewhere (see above Sect. 6.5.1). When W_2 receives information about P in L, P is already past (P is located before L in W_2's reference frame). Then, it still requires time for this information to reach W_1, insofar as W_1 will find out that event P was pre-determined, when P is already past for him too. Rietdijk recognizes this limit: "signals predicting the future for W_1 would, of course, always be too late, because of the limiting effect of the velocity of light" (1966, p. 341). If the travel of the signal were instantaneous, the prediction would be possible. But because of the limit of the speed of light, the prediction of P by W_2 or W_1 (informed by W_2) could be made only after P took place in their reference frames. Usually, a prediction is supposed to be made *before* the future event occurs (i.e. becomes present). But in this case, the prediction can be made only *after* the occurrence of the event. The observer W_2 can just know retrospectively that the event *was* already determined for W_1 when it was future in S_1. In other words, the determinism Rietdijk puts forward is paradoxically a pre-determinism in which, strictly speaking, any *pre*diction is impossible—a retro-determinism.

6.5.6 A Fallacy in the Determinateness-Argument

My second objection is that the determinateness-argument based on STR in favor of the block universe view contains a fallacy that has not been noticed. Let us return to the core of this argument (Fig. 6.5). At the moment event P is present in the reference frame S_2, it is future in the reference frame S_1. The same event P is future in S_1 for observer W_1 and present in S_2 for observer W_2. If we say that the future is indeterminate and the present determinate, we come across a contradiction: P is both indeterminate and determinate. To remove this contradiction, we claim that P is in reality determinate for W_1 as for W_2, and that the difference between future and present is purely psychological. The impossibility of objective present implies the determinateness of the future, which has therefore the same ontological status as the past and the present (thesis 2 of the block universe view).

At first sight, this reasoning would only have meaning if P were considered *in the same frame of reference*. Yet precisely because there are two reference frames, all we can say is that *in* S_1, P is future and indeterminate and *in* S_2, P is present and determinate. There isn't any contradiction in this assumption. Between the two observers in motion relative to each other, difference is not disagreement. There would be a contradiction if within *one frame of reference*, an event were at the same time future and indeterminate for one observer and present and determinate for another. In this case, we could logically conclude that only one of the two observers, probably the second, has the correct information.

There is something fallacious in the determinateness-argument, since it plays out on two levels. *On the one hand*, it emphasizes the principle of relativity: there is 'No

Privileged Observer', no objective present: the event is therefore future in S_1 *and* present in S_2; *on the other*, it conflicts with the principle of relativity, because it privileges implicitly the viewpoint of the observer for whom the event is present and determinate. W_2 is right and W_1 is not, as if the faster observer (W_2) was the winner, the best observer! Why could we not say after all, that the event is in reality indeterminate, and just apparently determinate for W_2?

But according to the principle of relativity, no observer is allowed to privilege his or her own frame of reference. Let us draw all the consequences of this point. The relativity of simultaneity in STR means that there is no universal simultaneity beyond the different reference frames; there is no objective or global present. Simultaneity is an *intra-referential* property, it is always a triadic relation between at least two events and a frame of reference. The same could be true for the properties of being determinate or being indeterminate. They hold only in a specified frame of reference (indeterminate/determinate is always indeterminate/ determinate *in relation to* a specified reference frame). Arguing against Putnam (1967), the relation R 'being real for', like 'being simultaneous with' and 'being determinate', is transitive only in a specified reference frame, and not across the observers in motion, precisely because no observer can affirm he or she is more right than another, without violating a basic principle of STR (the principle of relativity, according to which there is 'No Privileged Observer').[30] If we accept that the event P is present in S_2 and future in S_1 without contradiction and privileged observer, we also have to accept the consequence that this event is determinate in S_2 and indeterminate in S_1 without contradiction and privileged observer.

This relativistic ontology is a way between the acceptance of the block universe a view and the rejection of STR. This way is "surprising, no doubt, but by no means inconsistent or patently absurd" (Sklar 1977, p. 275). It is all the more surprising that considering humans' weak speed relative to the speed of light, human time is roughly Newtonian and apparently absolute. It is the reason why STR, with its relative present, seems to us to be odd or fantastic. Yet the relativistic ontology faces serious objections: Does it entail a relativization of existence? Is reality fragmented in all reference frames? The relativistic ontology seems to lead to Fine's fragmentalism (2006), according to which the reality is not "of a piece", but rather composed of a myriad of different fragments that cannot be joined together in a single coherent whole. Each fragment contains tensed facts (like "2016 is the present year", "2015 is the last past year"), which are coherent with each other in this fragment, but not with other tense facts belonging to other fragments. Fine's interpretation of STR states that each fragment corresponds to a particular frame of reference, so that "each standpoint will then give raise to its own reality" (Fine 2006, p. 412). The consequence of this realism about tense is the "giving up of the idea that there is a single coherent reality" (Fine 2006, p. 413). This conclusion seems to me problematic, because it gives up Einstein's idea that it

[30] See Sklar (1985, p. 291) who rejects Putnam's claim that the principle of 'No Privileged Observer' leads to affirm the transitivity of 'reality for' beyond the frames of reference.

is the *same* event (then the *same* reality) which is described by different observers. Moreover, the different frames of reference in STR are not *incompatible* parts of a larger fragmented reality, because the apparent contradictions between tensed facts (like ‘event *e* is future in S_1 and present in S_2’) can be perfectly explained and calculated by the Lorenz transformations.[31]

Dealing with the relativistic ontology does not require abandoning the idea that there is a single reality. There is a single reality with the same events in all the reference frames, then for all observers, but these events are determinate for some, indeterminate for others. To make this reply consistent, we must distinguish the existence, which is frame independent, from the determinateness, which is frame dependent.[32] Existence is the position in space-time in accordance with the laws of physics. Granted that a future event like ‘the sea-fight tomorrow’ is indeterminate in a given reference frame (it has not yet occurred), it still exists as a possible event located in space-time.

6.5.7 Local Present Versus Distant Present

According to this relativistic ontology, the same event can be indeterminate (future) for some observers, and determinate (present or past) for other observers. It may happen in some reference frames, and has happened is other reference frames. To avoid this puzzling situation, there is another option that has been pioneered by Stein (1968 and 1991). This third strategy against the determinateness-argument consists in rejecting the notion of extended present. In his comments on Rietdijk’s article, Stein (1968, p. 16) notes that the event *B*, where the second observer W_2 is situated, is outside the light cones of the events *A* and *P*. Henceforth, we cannot say that for any observer W_2 located in *B* either that *P* is present or that *A* is past, nor that *A* and *P* have already become, since no information can come to W_2 that concerns *A* and *P*, which are situated in W_2’s elsewhere. In contrast, *A* has already become for *P*, and can influence it. More generally, for an event at a space-time point *a*, “those events, and only those, have *already become* (real or determinate), which occur at points in the topological closure of the past of *a* [i. e. in the past light cone of *a*]” (Stein 1968, p. 14). This point renders Rietdijk’s argumentation in favor of determinism impossible.

In a later paper entitled “On Relativity Theory and Openness of the Future” (1991), Stein gives another formulation of his argument against determinism in STR. He defines the relation *R* in Minkowski space-time as the following: *Rab*

[31]This point is made by Hofweber and Lange (2016).

[32]If existence is reduced to determinateness, it is relative to the different reference frames and we face Gödel’s objection (1949, p. 558, note 5) that “existence is by its nature something absolute”. See also Mozersky (2000, p. 232), who argues that all events are determinate, because he identifies, like many others, existence and determinateness.

signifies "*b* is definite as of *a*" (or "for *a*, *b* has already become"). *R* is reflexive (*a* is definite as of *a*), transitive (if *b* is definite as of *a*, and *c* is definite as of *b*, then *c* is definite as of *a*), but never symmetric (if *a* is definite as for *b*, *b* is not definite as of *a*). Then Stein shows that the relation *R* holds in Minkowski space-time only if the vector from *a* to *b* is a past-pointing vector, i.e. if *b* is in the past light cone of *a*.[33] This implies that all the events located outside the past light cone of the event *a* are not "definite", they can't have "already become". To that extent, the future light cone of the event *a* (its 'absolute future') and its elsewhere are indeterminate. Stein concludes that STR is "perfectly compatible" with Maxwell's "ontological probabilism"[34] (1991, p. 164). In other terms, STR is compatible with the idea that all the space-time region outside the past light cone of an event *a* contains several ontologically real alternative possibilities as of *a*.[35] In this view, events are indeterminate not only because they have not yet occurred, but also in the sense that they may or may not happen. Stein's argument allows us to conceive of a local A-becoming dependent on the world line of one observer, for whom some indeterminate possible events, located in his or her future light cone, become progressively determinate by entering his or her local present and then his or her past light cone, while others possible events are not actualized (they "might have been").

This solution sounds very effective, to the extent that it destroys all the versions of the determinateness-argument, but it raises at least two difficulties. The first problem with Stein's position is that it renounces the idea of a spatially extended present relative to a reference frame, based on the idea of distant simultaneity. The present is limited to the local present of an event that is here and now: "in Einstein-Minkowski space-time, *an event's present is constituted by itself alone*. In this theory, therefore, the present tense can *never* be applied correctly to 'foreign' objects" (Stein 1968, p. 15); "in the theory of relativity, the only reasonable notion of "*present to a spatial-time point*" is that of the mere-identity relation: present to a given point is that point alone" (Stein, 1991, p. 159). From Stein's point of view, W_2's present (in Fig. 6.5) is not a hyperplane that groups together all simultaneous events with W_2 in his or her reference frame, but rather it is a unique point: event *B* where W_2 is located. In this case, the observer's reality is limited to the observer's

[33]Stein's argument has been further developed by Clifton and Hogarth (1995), who show that Stein's proof can be extended to cover a more general kind of objective becoming that is understood to occur from the standpoint of events on the particular worldlines followed by observers.

[34]See above note 27.

[35]I follow the interpretation of Stein by Pooley (2013, p. 353): "So, to say that all and only events in the past light cone have become, as of some space-time point *p*, is to say that, while there is a unique matter of fact concerning what has occurred in all regions to the past of *p*, there is (as of *p*) a plurality of possibilities open for regions of space-time to *p*'s absolute future *and in its elsewhere*. This suggests that a first step towards a relativistic version of the open-future view of passage should be a relativistic generalization of branching-time models to structures that encodes this pattern of relational indeterminacy". In the rest of his paper, Pooley studies the theories of 'branching space-time' of McCall (1994) and more specifically of Benalp (1992). On branching space-time, see also Proietti (Chap. 3 in this volume).

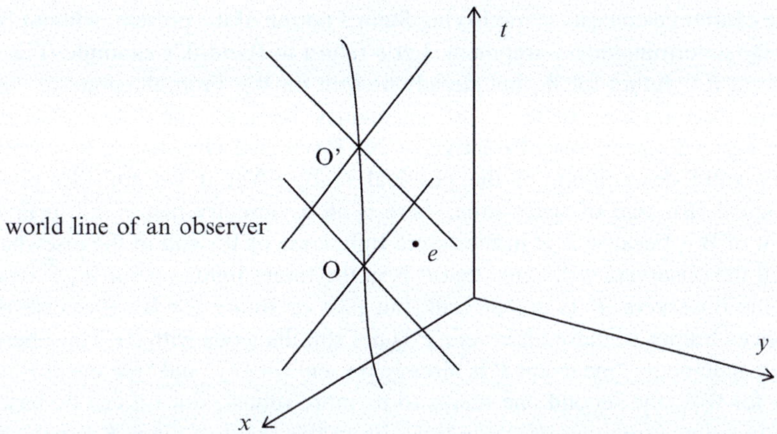

Fig. 6.7 The event e is located in the observer's elsewhere when the observer is at O, and in the observer's past light cone when the observer is at O′

local present with his or her past light cone, which leads to a "peculiarly extreme (but pluralistic!) form of solipsism" assumed by Stein (1968, p. 18). This "uniqueness assumption" (for each observer, there is a unique present: the here-now) seems at odds with the ordinary concept of time (Callender 2000, p. 594), and it is not sure that it fits with STR. Einstein (1905) did want to show that distant simultaneity was frame relative, but his purpose was not to reduce this notion to nothing. Even if there is no experience of the present of remote events, he explained how it is possible to define a distant simultaneity in an inertial reference frame with a set of synchronized clocks.[36]

Stein's rejection of distant simultaneity and extended present raises a second difficulty.[37] Some events can be in the observer's elsewhere at a certain moment, but according to the evolution of the observer's world line, these events may enter later in his or her backward light cone. They become past without ever having been present! (see Fig. 6.7).

If the concept of spatially extended present, based on distant simultaneity and relative to a reference frame, has to be preserved, Rietdijk can say that for observer W_2, P is present, since it is situated on the axis (O_2X_2) that defines W_2's (extended) present in the diagram of the Fig. 6.5. Still, the temporal status of the event is relative to the observer's reference frame. We are back to the determinateness-argument: Does the fact that the event is future for observer W_1 but present for observer W_2 then imply a form of determinism? Does the challenge of Stein's uniqueness assumption entail the block universe view[38]?

[36]On the question (which I can't address in this paper), whether simultaneity in STR is a mere convention or a physical fact, see Dieks's presentation (2014, pp. 107–112).

[37]See Callender (2000, pp. 593–594).

[38]This is Callender's conclusion (2000, pp. 592–593).

The challenge consists in modifying Stein's notion of the present, without facing again the determinateness-argument. Let's return to Rietdijk's example (Fig. 6.5). The event P is future for W_1 but already present for W_2. From the point of view of the reference frames, there is no asymmetry between these two statements: according to the principle of relativity ('No Privileged Observer'), they both are equally valid descriptions of the physical reality. Yet if we consider now the ontological structure of space-time, there is an asymmetry that is not in favor of W_2, but of W_1, because P is in the future light cone of W_1 and in the elsewhere of W_2. All the observers will agree that P is in the future light cone of W_1,[39] whereas for some observers, P is not present, but past or future for W_2 (because in the references frames of these observers, P is not simultaneous with B). Thus, between the two statements "the event P is already present for W_2" and "the event P is still future for W_1", the second one seems to be more robust, since it can be based on frame invariant structures of space-time. From this point of view, P is not already present (for W_2), rather it is still future (for W_1).

What is therefore the ontological status of P? For W_1, P is indeterminate because it is in her or his future light cone. And for W_2? We could consider that even if P is present in the reference frame of W_2, it is still indeterminate for W_2 too, to the extent that it is in her or his elsewhere. P will be determinate for W_1 and W_2, if it enters the local present of W_1 and then the past light cone of W_2. Contrary to Stein's approach, this view can assume the distinction between the local present, the 'here-now' of an observer, and the extended or distant present of this observer defined by a hyper-plane of simultaneity relative to a given reference frame. The local present is absolute (frame invariant) and determinate, whereas the distant present is not only relative (frame dependent), but also indeterminate, it is hypothetically present so to speak: an event located in the distant present of an observer may or may not happen; one can say that it happened, it has been present, *only if* it enters the observer's past light cone. The distant present is always a perfect present (it *has been* present). The same goes for future and past events located in the elsewhere of the observer. It is only if they come in the past light cone of the observer that he or she can establish that they have been future or past in his or her reference frame.

So we have now three interpretations of Rietdijk's thought experiment:

- Block universe view (it is Rietdijk's conclusion): P is determinate for W_1 as for W_2
- Relativistic ontology I, with frame-relative determinateness: P is indeterminate for W_1 and determinate for W_2
- Relativistic ontology II, with frame-independent determinateness: P is indeterminate for W_2 as for W_1

In this second version of the relativistic ontology, determinateness and indeterminateness are not intra-referential properties (like in Sect. 6.5.6), but frame

[39]Since A and P are time-like separated, the relation of succession between them (the fact that P is after A) is absolute (see above Sect. 6.4.2).

independent properties relative to the light cone of the observer. This solution allows us to keep the notion of distant or extended present (in a deflationary sense), so that all the events that are in the past light cone of an observer have been present (locally or remotely) before being past. However, since the events belonging to the distant present of the observer are located in the elsewhere of the observer, they are indeterminate, they always have to be confirmed by the observer's past light cone. In Fig. 6.5, P is future and indeterminate for W_1, it may or may not happen. If P doesn't happen for W_1, it will never enter the past light cone of W_2. If P happens for W_1, it may enter later the past light cone of W_2 evolving along his or her world line, for example when W_2 is in L. In this case, when P was future for W_1, it was not present locally (then determinate) for W_2, but remotely or hypothetically present (then indeterminate), so that one cannot conclude, like Rietdijk, that P was 'already' determinate for W_2. According to this view, Rietdijk's example does not mean that the event P is determinate for W_1 as for W_2, but on the contrary, that the event P is indeterminate for W_2 as for W_1. The principle of relativity only states that P is future for W_1 and present for W_2: these are two correct descriptions of the reality. But the ontological consequences of this situation (the determinateness/indeterminateness of P) must be drawn from the position of P in space-time relative to the observers' light cones. According to this relativistic ontology II, the determinateness/indeterminateness of an event cannot be defined *simpliciter*, it is relative to a specified event or to the position of an observer in space-time. This does not mean that the observer could arbitrarily choose his or her own reality, because the observer cannot choose his or her position in space-time and his or her light cone.[40]

When it comes to the ontological status of an event P relative to an observer, there are three possibilities. (1) If P is in the local present of the observer (his or her 'here-now'), or in his or her past light cone, P is determinate in the logical and the ontological meaning of 'determinate'. (2) If P is in the absolute future of the observer, P is indeterminate, not because it doesn't exist, but rather because it *may* exist. Events in the future light cone of the observer have not yet occurred, they may or may not happen (ontological indeterminateness), and the future-tensed propositions on these events have no determinate truth value (logical indeterminateness). They can have various degrees of probability, but this point is beyond the scope of this chap. (3) If P is located in the elsewhere of the observer, there is an alternative: either we say with Stein that P is indeterminate without any temporal properties, and we lose the notion of extended present (in a relativistic sense). Or we say that the events in the elsewhere are determinate once and for all, and space-time is therefore a block universe, because any event (even in the future light cone of the observer) is in the elsewhere of another event.[41] There is a third way—that I am

[40]See Clifton and Hogarth (1995, pp. 383–384) : "this situation is hardly more problematic than the corresponding assertion in Newtonian spacetime – that what is real for an observer depends upon where she is located in time. Indeed, location in *spacetime* would seem to be the most appropriate relativistic generalization of that Newtonian idea."

[41]See Weingard (1972) who uses this reasoning to claim "that *all* events in special relativistic space-time are real" (p. 121).

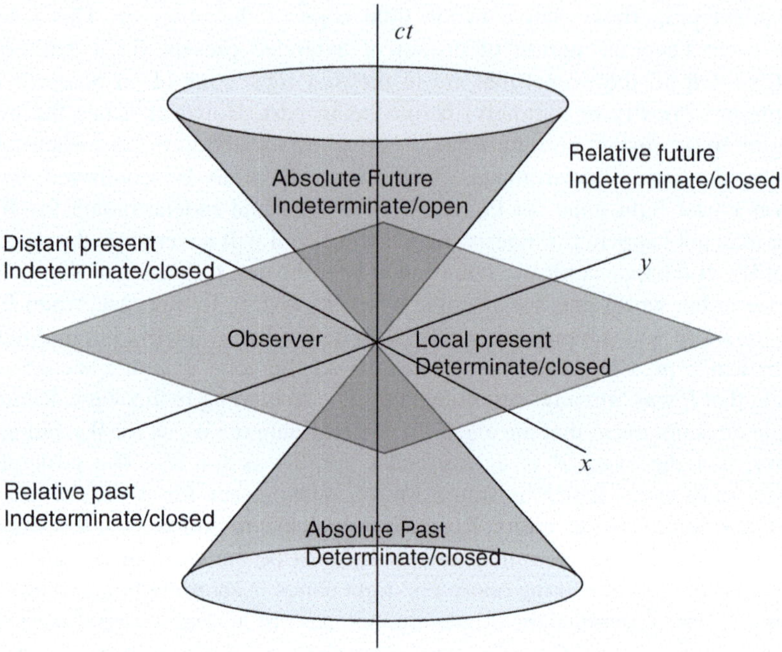

Fig. 6.8 The relativistic ontology (II) of the observer's light cone in space-time

following in this subsection—according to which an event *P* located in the elsewhere of an observer is indeterminate, but it has a temporal status relative to the frame of reference of the observer: *P* is present, past or future, as a function of its position relative to the observer's extended present. If it is simultaneous with the observer's present,[42] it is present; if it is earlier than it, it is past; if it is later, it is future. In the elsewhere, present, past and future events are indeterminate, they may or may not happen until they enter the past light cone of the observer. We could speak of distant present, relative past and relative future, to describe the temporal status of the events that are located in the elsewhere of the observer.

To sum up (see Fig. 6.8), for any observer located in space-time with a reference frame, we have to distinguish the local present and the absolute past that are determinate, the elsewhere (divided in distant present, relative past and relative future) that is indeterminate, and the absolute future that is also indeterminate. What is the difference between the elsewhere and the absolute future? The answer is well known: it is the causal influence. From a practical point of view, this means that the future light cone of the observer (what Wheyl called the "active future") is "open", to the extent that he or she can have a causal influence on it. The observer can bring about certain events and prevent other events to happen (in that case, he or

[42]That is, if *P* is in the hyperplane of present of the observer.

she can "change" the future). On the contrary, all the regions outside the future light cone of the observer are not "open", they are "closed", since the observer cannot have any causal influence on it.

6.6 Conclusion

Does STR imply the block universe view? Let us return to the three theses that usually define the block universe:

1. There is no objective present.
2. Past, present and future events equally exist.
3. There is no objective flow of time.

The fact that there is no objective present for STR does not imply that present, past and future are on a par with each other. On the contrary, STR is compatible with the indeterminateness of the future, understood either as an intra-referential property (relativistic ontology I), or as a frame independent property relative to the light cone of the observer (relativistic ontology II). Moreover, STR allows objective flows of events, which are located on world lines. In each world line, the temporal ordering is always preserved. We can conclude that the block universe view—as it is defined in this chapter—is a questionable interpretation of STR, and not its necessary consequence.

Acknowledgments I wish to thank audiences in Paris (« Philosophie & Physique » seminar, University of Paris Diderot), Sydney (Centre for time, University of Sydney) and Bordeaux (Centre SPH, University Bordeaux Montaigne) for helpful comments and discussions. In particular, I wish to thank Elie During, Philippe Huneman, Kristie Miller, Huw Price, Carlo Rovelli and Blai Pié i Valls for their critical remarks, which helped improving the final version of the paper. An earlier draft was written while I was visiting scholar at the University of Sydney with a grant of the Institut Universitaire de France (IUF). I thank all the members of the philosophy department for their marvelous hospitality.

References

Arthur, R.T.W. 2006. Minkowski space-time and the dimensions of the present. In *The ontology of space-time*, ed. D. Dieks, 129–155. Amsterdam: Elsevier.

———. 2008. Time lapse and the degeneracy of time: Gödel, proper time and becoming in relativity theory. In *The ontology of space-time*, ed. D. Dieks, vol. 2, 202–227. Amsterdam: Elsevier.

Baron, S., and K. Miller. 2013. Characterising eternalism. In *New papers on the present*, ed. G. Torrengo, K. Miller, and R. Ciuini, 31–62. München: Philosophia Verlag.

Benalp, N. 1992. Branching space-time. *Synthese* 92: 385–434.

Callender, C. 2000. Shedding light on time. *Philosophy of Science*, *67*, Supplement, 587–599.

Čapek, M. 1968. Time in relativity theory : arguments for a philosophy of becoming. In *The voices of time*, ed. J.T. Fraser, 434–454. London: Allen Lane The Penguin Press.

Clifton, R., and M. Hogarth. 1995. The definability of objective becoming in Minkowski space-time. *Synthese* 103: 355–387.
Costa de Beauregard, O. 1968. Time in relativity theory : Argument for a philosophy of being. In *The voices of time*, ed. J.T. Fraser, 417–433. London: Allen Lane The Penguin Press.
Damour, Th. 2005. *Si Einstein m'était conté*. Paris: Le Cherche midi.
Dieks, D. 1988. Special relativity and the flow of time. *Philosophy of Science* 55 (3): 456–460.
———. 2002. Does chance make a difference ? The philosophical significance of indeterminism. In *Between chance and choice. Interdisciplinary perspectives on determinism*, ed. H. Altmanspacher and R. Bishop, 209–236. Imprint Academic: Exeter/Charlottesville.
———. 2014. Time in special relativity. In *Springer handbook of space-time*, ed. A. Ashtekar and V. Petkov, 91–113. Berlin: Springer.
Dorato, M. 2006. The irrelevance of the presentist/eternalist debate for the ontology of Minkowski space-time. In *The ontology of space-time*, ed. D. Dieks, 93–109. Amsterdam: Elsevier.
Eddington, A. 1990 [1920]. *Space, time and gravitation. An outline of the general relativity theory*. Cambridge: Cambridge University Press.
Einstein, A. 1905. Zur Elektrodynamik bewegter Körper. *Annalen der Physik* 17: 891–921.
———. 1949. Reply to criticisms. In *Albert Einstein : Philosopher-Scientist*, ed. P.A. Schilpp, 665–688. La Salle, Illinois: Open Court.
———. 1955. *Relativity. The special and the general relativity*. Trans. by. R. W. Lawson. London: Methuen & Co. Ltd.
Fine, K. 2006. The reality of tense. *Synthese* 150: 399–414.
Fitzgerald, P. 1969. The truth about tomorrow's sea fight. *The Journal of Philosophy* 66 (11): 307–329.
Gödel, K. 1949. A remark about the relationship between relativity theory and idealistic philosophy. In *Albert Einstein : Philosopher-Scientist*, ed. P.A. Schilpp, 557–562. La Salle, Illinois: Open Court.
Grünbaum, A. 1963. *Philosophical problems of space and time*. New York: Alfred A. Knopf.
Hofweber, T., and M. Lange. 2016. Fine's fragmentalist interpretation of special relativity. *Noûs*. doi:10.1111/nous.12150.
James, W. 1979. *The will to believe*. Cambridge, MA: Harvard University Press.
Kroes, P. 1985. *Time : Its structure and role in physical theories*. Dordrecht: D. Reidel Publishing Company.
Laplace, P. S. 1951 [1814]. *A philosophical essay on probabilities*. Transl. by F. Truscott and F. Emory. New York: Dover Publications.
McCall, S. 1994. *A model of the universe. Space-time, probability, and decision*. Oxford: Clarendon Press.
Maudlin, T. 2007. *The metaphysics within physics*. Oxford: Oxford University Press.
Maxwell, N. 1985. Are probabilism and special relativity incompatible ? *Philosophy of Science* 52 (1): 23–43.
McTaggart, J.M.E. 1908. The unreality of time. *Mind* 17: 457–474.
Miller, K. 2013. Presentism, eternalism, and the growing block. In *A companion to the philosophy of time*, ed. H. Dyke and A. Bardon, 345–364. Chichester: Wiley-Blackwell.
Minkowski, H. 1909. Raum und Zeit. *Physikalische Zeitschrift* 10: 104–111.
Mozersky, J. 2000. Time, tense and special relativity. *International Studies in the Philosophy of Science* 14 (3): 221–236.
Norton, J.D. 2010. Time really passes. *Humana Mente : Journal of Philosophical Studies* 13: 23–34.
Petkov, V. 2006. Is there an alternative to the block universe view? In *The ontology of space-time*, ed. D. Dieks, 207–228. Amsterdam: Elsevier.
———. 2009. *Relativity and the nature of space-time*. 2nd ed. Berlin, Heidelberg: Springer.
Pooley, O. 2013. Relativity, the open future, and the passage of time. *Proceedings of the Aristotelian Society* 113: 321–363.
Popper, K. 2000. *The open universe: An argument for indeterminism*. London: Routledge.

Price, H. 1996. *Time's arrow and archimedes' point*. New York/Oxford: Oxford University Press.
———. 2011. The flow of time. In *The Oxford handbook of philosophy of time*, ed. C. Callender, 277–311. Oxford: Oxford University Press.
Putnam, H. 1967. Time and physical geometry. *The Journal of Philosophy* 64 (8): 240–247.
Reichenbach, H. 1957 [1927]. *The philosophy of space and time*. New York: Dover Publications Inc..
Rietdijk, C.W. 1966. A rigorous proof of determinism derived from the special theory of relativity. *Philosophy of Science* 33 (4): 341–344.
———. 1976. Special relativity and determinism. *Philosophy of Science* 43 (4): 598–609.
Savitt, S. 2000. There's no time like the Present (in Minkowski Space-time). *Philosophy of Science* 67 (2000): 563–574.
———. 2011. Time in the special theory of relativity. In *The Oxford handbook of philosophy of time*, ed. C. Callender, 546–570. Oxford: Oxford University Press.
Sklar, L. 1977. *Space, time, and space-time*. Berkeley/Los Angeles: University of California Press.
———. 1985. *Philosophy and space-time in physics*. Berkeley/Los Angeles: University of California Press.
Smeenk, Ch., and C. Wüthrich. 2011. Time travel and time machine. In *The Oxford handbook of philosophy of time*, ed. C. Callender, 577–630. Oxford: Oxford University Press.
Smith, J. J. N. 2011. Inconsistency in the A-Theory. *Philosophical Studies* 156: 231–247.
Smolin, L. 2013. *Time reborn: From the crisis in physics to the future of the universe*. Boston, New York: Houghton Mifflin Harcourt.
Stein, H. 1968. On Einstein-Minkowski space-time. *The Journal of Philosophy* 65 (1): 5–23.
———. 1991. On relativity theory and openness of the future. *Philosophy of Science* 58 (2): 147–167.
Von Weizsäcker, C.F. 2006. *The structure of physics*. Dordrecht: Springer.
Weingard, R. 1972. Relativity and the reality of past and future events. *The British Journal for the Philosophy of Science* 23 (2): 119–121.
Weyl, H. 1949. *Philosophy of mathematics and natural science*. Princeton: Princeton University Press.
Williams, D.C. 1951. The myth of passage. *Journal of Philosophy* 48: 457–472.

Chapter 7
Dispelling the Quantum Spooks: A Clue That Einstein Missed?

Huw Price and Ken Wharton

Abstract It is well-known that Bell's Theorem and other No Hidden Variable theorems have a "retrocausal loophole", because they assume that the values of pre-existing hidden variables are independent of future measurement settings. (This is often referred to, misleadingly, as the assumption of "free will".) However, it seems to have gone unnoticed until recently that a violation of this assumption is a straightforward consequence of time-symmetry, given an understanding of the quantization of light that would have seemed natural to Einstein after 1905. The new argument shows precisely why quantization makes a difference, and why time-symmetry alone does not imply retrocausality, in the classical context. It is true that later developments in quantum theory provide a way to avoid retrocausality, without violating time-symmetry; but this escape route relies on the "ontic" conception of the wave function that Einstein rejected. Had this new argument been noticed much sooner, then, it seems likely that retrocausality would have been regarded as the default option for hidden variables theories (a fact that would then have seemed confirmed by Bell's Theorem and the No Hidden Variable theorems). This paper presents these ideas at a level intended to be accessible to general readers.

7.1 Einstein's Dream

Late in his life, Einstein told Max Born that he couldn't take quantum mechanics seriously, "because it cannot be reconciled with the idea that physics should represent a reality in time and space, free from spooky actions at a distance." (Einstein, Letter to Max Born, 3 March 1947, in Born 1971, p. 178). Most physicists

---

H. Price (✉)
Trinity College, Cambridge CB2 1TQ, UK
e-mail: hp331@cam.ac.uk

K. Wharton
Department of Physics and Astronomy, San José State University, San José, CA, USA
e-mail: kenneth.wharton@sjsu.edu

C. Bouton, P. Huneman (eds.), *Time of Nature and the Nature of Time*, Boston Studies in the Philosophy and History of Science 326,
DOI 10.1007/978-3-319-53725-2_7

think of this as the sad lament of an old giant, beached on the wrong side of history – unable to accept the revolution in physics that he himself had done so much to foment, decades before, with his 1905 discovery that light is absorbed in discrete "quanta".

According to this orthodox view, John Bell put the final nail in the coffin of the dream of a spook-free successor to quantum theory, a decade after Einstein's death. Bell's Theorem seemed to show that any theory of the kind Einstein hoped for – an addition to quantum theory, describing an underlying "reality in space and time" – is bound to involve the kind of "action at a distance", or "nonlocality", to which Einstein was objecting.

So Einstein's spooks have long seemed well-entrenched in the quantum world – and this despite the fact that they still seem to threaten Einstein's other great discovery from 1905, special relativity. As David Albert and Rivka Galchen put it in a recent piece in *Scientific American*, writing about the intuition of "locality": "Quantum mechanics has upended many an intuition, but none deeper than this one. And this particular upending carries with it a threat, as yet unresolved, to special relativity – a foundation stone of our twenty-first-century physics." (2009, p. 32)

But could this accepted wisdom be due for a shake-up? Could Einstein have the last laugh after all? Intriguingly, it turns out there's a new reason for taking seriously a little-explored loophole in Bell's Theorem.[1] Even more intriguingly, it's a reason that Einstein himself could have spotted as early as 1905, since it is a simple consequence of the quantization of light, together with another assumption that he certainly accepted at that time.

The loophole stems from the fact that Bell's argument assumes that our measurement choices cannot influence the *past* behaviour of the systems we choose to measure. This may seem quite uncontroversial. In the familiar world of our experience, after all, causation doesn't work "backwards". But a few physicists have challenged this assumption, proposing theories in which causation can run backwards in the quantum world. This idea – "retrocausality", as it is often called – has been enjoying a small renaissance. (See Box 7.1.)

Box 7.1: The Retrocausal Mini-Renaissance
Recent interest in retrocausality in QM comes from several directions:

1. Supporters of Yakir Aharonov's **Time Symmetric Quantum Mechanics** (see popular account here: bit.ly/tsqm2010) have claimed that because this theory is retrocausal, it provides a superior account of the results of so-called "weak measurements".

(continued)

[1]The same loophole exists in the other so-called "No Hidden Variable theorems", that seek to prove that there can't be a deeper reality underlying quantum theory, of the kind Einstein hoped for. For simplicity we'll focus on Bell's Theorem, but the spook-dispelling technique we're talking about works equally well in the other cases.

Box 7.1 (continued)

2. Some authors have suggested that there is evidence of retro-causality in recent versions (e.g. Ma et al. 2012) of John Wheeler's **Delayed Choice Experiment**.
3. The recent **Pusey-Barrett-Rudolph (PBR) Theorem** (Pusey et al. 2012) puts tough new restrictions on interpretations of QM that regard the wave function, as Einstein preferred, merely as a representation of our *knowledge* of a physical system. Commentators have noted that retrocausality may provide the most appealing escapehatch – see Matt Leifer's summary: tinyurl.com/LeiferPBR.
4. Recent work by Ruth Kastner [tinyurl.com/SciAmTI] has revived interest in Cramer's **Transactional Interpretation**, an early retrocausal proposal inspired by work by Wheeler and Feynman.

Until very recently (Price 2012), however, no one seems to have noticed that there is a simple argument that could have put retrocausality at centre-stage *well before* the development of quantum theory. As we explain below, the argument shows that retrocausality follows directly from the quantization of light, so long as fundamental physics is *time-symmetric* (meaning that any physical process allowed in one time direction is equally allowed in reverse).

Many ordinary physical processes, such as cream mixing in coffee, don't appear to be time-symmetric. (The cream mixes in, but never mixes out!) But these are processes involving large numbers of microscopic constituents, and the constituents seem to behave with complete time-symmetry. Cream "unmixing" out of coffee is physically possible, just very improbable. Einstein himself defended the view that "irreversibility depends exclusively upon reasons of probability" (Ritz and Einstein 1909, p. 324), and is not evidence of a fundamental time-asymmetry.

As we shall explain, the new argument shows that quantization makes a crucial difference. Time-symmetry alone doesn't guarantee that causation ever works backwards, but quantization gives us a new kind of influence, which – assuming time-symmetry – must work *in both temporal directions*. This new kind of influence is so subtle that it can evade spooky nonlocality, without giving us an even more spooky ability to send signals into the past. One of the striking things about the apparent action at a distance in quantum mechanics (QM) is that it, too, is subtle in just this way: there's no way to use it to build a "Bell Telephone", allowing superluminal communications. The argument hints how this subtlety might arise, as a consequence of quantization, from an underlying reality that smoothly links everything together, via pathways permitted by relativity.

What would Einstein have thought, had he noticed this argument? We can only speculate, but one thing is clear. If physics had already noticed that quantization *implies* retrocausality (given fundamental time-symmetry, which was widely accepted), then it wouldn't have been possible for later physicists simply to ignore this option, as often happens today. It is true that the later development of quantum theory provides a way to avoid retrocausality, without violating time-asymmetry. But as we'll see, this requires the very bit of quantum theory that Einstein disliked the most – the strange, nonlocal wave function, that Einstein told Born he couldn't take seriously.

By Einstein's lights, then, the new argument seems important. If it had been on the table in the early years of quantum theory, the famous debates in the 1920s and 1930s about the meaning of the new theory couldn't possibly have ignored the hypothesis that a quantum world of the kind Einstein hoped for would need to be retrocausal.

Similarly, the significance of Bell's work in the 1960s would have seemed quite different. Bell's work shows that *if there's no retrocausality*, then QM is nonlocal, in apparent tension with special relativity.[2] Bell knew of the retrocausal loophole, of course, but was disinclined to explore it. He once said that when he tried to think about backward causation he "lapsed quickly into fatalism" (Bell 1988). But if retrocausality had been on the table for half a century, it would presumably have been natural for Bell and others to take this option more seriously. Indeed, Bell's Theorem itself might then have been interpreted as a *second* strong clue that the quantum world is retrocausal – the first clue (the one that Einstein missed) being the argument that the quantization of light implies retrocausality, given time-symmetry.

In this counterfactual history, physicists in the 1960s – two generations ago – would already have known a remarkable fact about light quantization and special relativity, Einstein's two greatest discoveries from 1905. They would have known that both seem to point in the direction of retrocausality.[3] What they would have made of this remarkable fact, we cannot yet say, but we think it's time we tried to find out.

7.2 Polarization, from Classical to Quantum

The new argument emerges most simply from thinking about a property of light called "polarization". (It can also be formulated using other kinds of properties, such as spin, but the polarization case is the one that Einstein could have used in 1905.) So let's start with the basics of polarization.

Imagine a wave travelling down a string. Looking down the string's axis you would see the string moving in some perpendicular direction; horizontal, vertical,

[2]Bell (1986) himself certainly thought there was such a tension – as he puts it at one point, "the cheapest resolution [of the puzzle of nonlocality] is something like going back to relativity as it was before Einstein, when people like Lorentz and Poincaré thought that there was an aether."

[3]Quantization does so via the argument below. Relativity does so via Bell's Theorem, since the retrocausal loophole provides a way to escape the apparent tension between Bell's result and relativity.

or at some other angle. This angle is strongly analogous to the "polarization angle" of an electromagnetic wave in Classical Electromagnetism (CEM). For a linearly polarized CEM wave, the electric field is oscillating at some polarization angle, perpendicular to the direction the wave is travelling.

Polarized sunglasses contain filters that allow light with one polarization angle to pass through (vertical, 0°), while absorbing the other component (horizontal, 90°). For light with some other polarization angle (say, 45°), such a polarizing filter naturally divides the electric field into horizontal and vertical components, absorbing the former and passing the latter. (The exact two components depend on the angle of the filter, which you can confirm by watching the blue sky gradually change color as you tip your head while wearing polarized sunglasses.)

But there are also optical devices that *reflect* one polarization component while passing the other; an example of such a "polarizing cube" is depicted in Fig. 7.1. This cube has a controllable angle σ_R that splits the incoming light into two components. The portion polarized at angle σ_R passes into the R1 path, and the portion polarized at $\sigma_R + 90°$ reflects into the R0 path.

A striking feature of CEM is that the equations are all time-symmetric, meaning that if you can run any solution to the equations in reverse, you'll have another solution. Imagining the time-reverse of Fig. 7.1 is the easiest way to see that these polarizing cubes can also combine two inputs into a single output beam, as in Fig. 7.2. Different input combinations will produce different output polarizations τ. Putting all the input beam on L1 will force τ to be equal to the angle σ_L; an L0 input will force $\tau = \sigma_L + 90°$; and an appropriate mix of L1 and L0 can produce any polarization angle in between these extremes.

One of Einstein's great discoveries in 1905 – he thought of it as the most important of all – is that CEM is not always correct. It breaks down when the energy in the electromagnetic field is very small – i.e., at the levels of Einstein's light quanta, or photons, as we now call them. Einstein himself does not seem to have thought about the implications of this "quantization" for polarization, and the classic early discussion (to which Einstein later refers approvingly) is due to Paul Dirac in 1930 (1930). In terms of our example, the crucial point is that in the case of a single photon, the light passing through cube in Fig. 7.1 will *not* split into two paths. Instead, when measured, the photon is always found to be on R1 or on R0, not on both. (The probabilities of these two outcomes are weighted such that in the many-photon limit the CEM prediction is recovered, on average.)

This change may seem innocuous, but it has profound implications. At the very least, it shows that CEM fails and has to change. But how? Einstein wrestled with the problem of how to combine quantization with the classical theory, even as he moved on to what he thought of as the *simpler* problem of general relativity! As for Dirac, he later simply dodged the issue, by asserting that it wasn't the business of physics to look for a story about the underlying reality: "Questions about what decides whether the photon [goes one way or the other] and how it changes its direction of polarization when it does [so] cannot be investigated by experiment and should be regarded as outside the domain of science." (Dirac 1930, p. 6)

Fig. 7.1 A polarizing cube

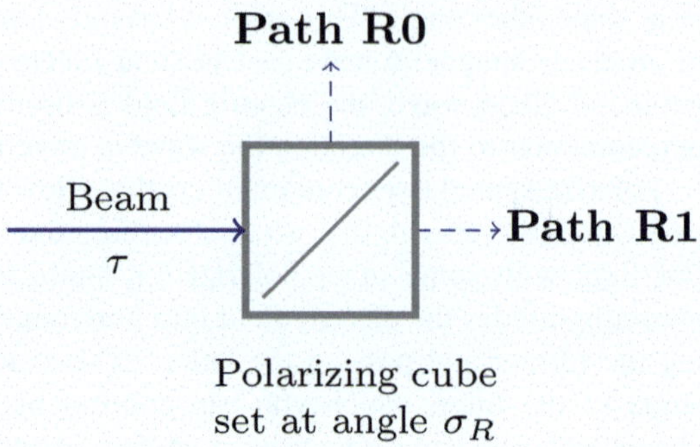

Fig. 7.2 The polarizing cube in reverse

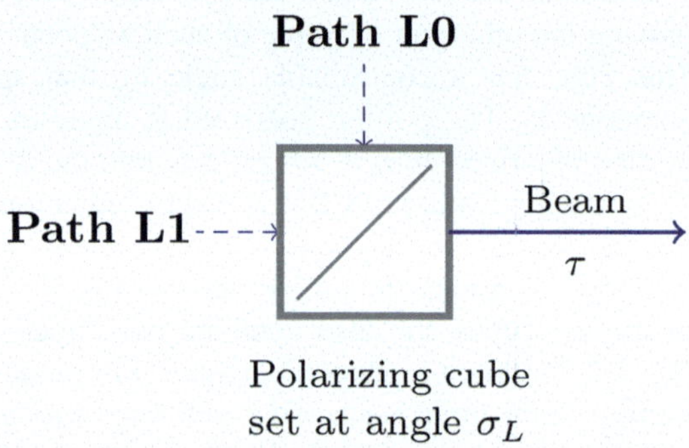

To the extent that quantum theory does provide us with a picture, the orthodox view is to let CEM run its normal course until the photon passes through the cube and encounters some detectors.[4] Then, at that very last moment, the measurement process "collapses" all the light onto only one detector, either on R0 or R1, just as it arrives.

Einstein disliked this "collapse" – it is what he had in mind when complaining about "spooky action at a distance" (since the R0 and R1 detectors may be far apart). He thought that the so-called collapse was just the usual process of updating of our knowledge after acquiring new information, and didn't involve any real change in the physical system. And he hoped for a theory that would fill the gaps in our incomplete knowledge (via "hidden variables"), providing a description of how light quanta actually behave. On such a view, presumably, the photon will be regarded as "making a choice" as it passes through the cube – though not necessarily a choice determined by any pre-existing property.[5]

[4]Technically, the orthodox view uses an equation that evolves a quantum state instead of electromagnetic fields, but the sentiment is the same.

[5]As Einstein says, "Dirac ... rightly points out that it would probably be difficult, for example, to give a theoretical description of a photon such as would give enough information to enable one to decide whether it will pass [one way or the other through] a polarizer placed (obliquely) in its way." (Einstein 1954, p. 270)

Neither Einstein nor Dirac seems to have considered the question of the time-symmetry of these simple polarization experiments. On the face of it, however, both a collapse picture and the "choosy photon" picture turn out to be incompatible with the principle that the world is time-symmetric at a fundamental level. To see what that means, and why it matters, let's turn to time-symmetry.

7.3 Time-symmetry

How can it be that the laws of fundamental physics are time-symmetric? After all, as we already noted, many ordinary processes seem to have a clear temporal preference – like cream mixing in coffee, they happen one way but not in reverse. But by the end of the nineteenth century, many physicists were coming to the view that these cases are not fundamental. They reflect the statistical behaviour of huge numbers of microscopic constituents of ordinary stuff, and the fact that for some reason we still don't fully understand, stuff was much more organized – had very low *entropy*, as physicists say – some time in the past.[6] But this explanation seems fully compatible with time-symmetry at the fundamental level.

This didn't mean that fundamental physics *has* to be time-symmetric. In principle, there might be some deep time-asymmetry, waiting to be discovered. But by this stage, a century ago, the tide was running firmly in the direction of fundamental symmetry. And certainly Einstein read it this way. Although he recognised that time-asymmetric fundamental laws were possible in principle, he thought them unlikely. Writing to Michele Besso in August 1918, he says, "Judging from all I know, I believe in the reversibility of elementary events. All temporal bias seems to be based on 'order'." (Einstein 1998, p. 630)

Since then, the evidence has changed in one important respect. We now know that there is a subtle time-asymmetry in particle physics. But there is a more general symmetry, called CPT-symmetry, that is widely regarded as fundamental (and is equivalent to time-symmetry for the cases we are discussing here). So for practical purposes, the consensus remains the same: time-symmetry (or strictly speaking CPT-symmetry) is something fundamental.

Still, this attitude to time-symmetry "in principle" often co-exists with a distinctly time-asymmetric working model of the quantum world, as we can see by thinking about how CEM is typically modified to explain Einstein's photons. It turns out that both the "collapse" view and the "choosy photon" view described above are time-asymmetric.

To see why, we need to know that the crucial test for a time-symmetric theory is not that everything that happens should look exactly the same in reverse. For example, a rock travelling from Earth to Mars looks different in reverse – it now goes from Mars to Earth – even though the underlying mechanics is thoroughly

[6]See Sean Carroll's blog [bit.ly/SCeternity] for an accessible introduction, and Price (1996) for more.

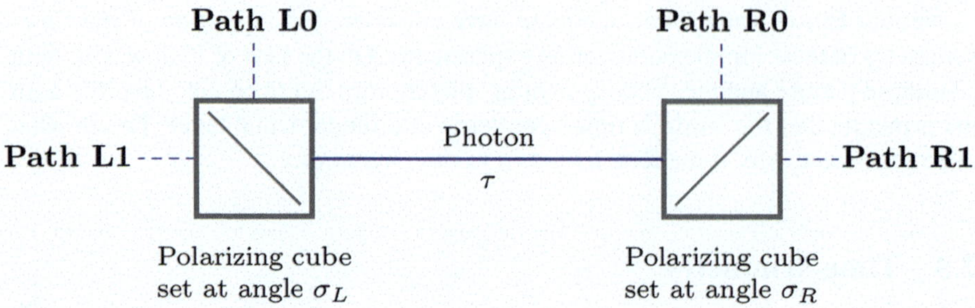

Fig. 7.3 Combining two cubes

time-symmetric. The test is that if the theory allows one process then it also *allows* the reverse process, so that we can't tell by looking at a video of the process whether it is being played forwards or backwards. (If one process isn't allowed, then it is easy to tell: if the video shows the illegal process, it must be being played backwards.)

With this in mind, think of a video of the apparatus shown in Fig. 7.3, for a case in which a photon came in from the L1 path and was then detected on the R1 path.[7] This video shows the photon entering through a cube set to σ_L, exiting through a cube set to σ_R, with a polarization $\tau = \sigma_L$ in between.

Reversing the video shows a photon entering the apparatus through a cube set to σ_R, exiting through a cube set to σ_L, on the same paths as before, and *still with a polarization* $\tau = \sigma_L$ *in between*. And this isn't allowed by the theory (except in the special case where $\sigma_L = \sigma_R$), according to either of these models. On the contrary, if the photon enters through a cube set to σ_R, using Path 1, the intermediate polarization should be $\tau = \sigma_R$, not $\tau = \sigma_L$. In these models, then, time-reversing a legal process often leads to an illegal process.

This time-asymmetry is entirely a result of quantization. In the classical case, the light can always be split between the two paths, at one end of the experiment or both. This ensures that the reversed version of any process allowed by CEM in the apparatus in Fig. 7.3 is also a legal process. In the classical case, therefore, we can't tell which way a video is being played.

Somehow, then, combining discrete inputs and outputs with CEM destroys time-symmetry (at least if one tries to preserve standard CEM between the two cubes, as in the collapse view and the choosy photon view). In the next section we'll explore why this happened, and the surprising consequences of insisting that time-symmetry be saved.

---

[7] We can't literally make videos of individual photons, but we can make computer-generated videos showing what our theories say about photons, and they work just as well, for applying this test.

7.4 Control – Classical and Quantum

Does time-symmetry mean that the future controls the past just as much as the past controls the future? That can't be true in general, because CEM is perfectly time-symmetric, and yet it is easy to see that we can use the apparatus in Fig. 7.3 to send signals (and hence, if we wish, control things) from past to future – i.e., left to right – but not in the other direction. If we feed in a classical light beam on path L1, then by controlling the angle σ_L we can control τ, and hence encode a signal in the light beam. By measuring τ on the right hand side, we can then decode the message, or control some piece of machinery. But we can't do that from right to left, of course, even though the equations of CEM are perfectly time-symmetric. So where does the difference come from?

The answer is that there's still a big difference between the left and the right of the experiment, the past and the future, despite the fact that the laws of CEM are time-symmetric. On the left side, we normally control whether the light comes from L0 or L1, and as we'll see in a moment, this is crucial to our ability to use the device to signal to the future. On the right side, we don't control how much of the light goes out via R0 and R1, and it's really this difference that explains why we can't signal to the past.

We can check this diagnosis by imagining a version of the full experiment in which we *don't* control the inputs on the left, so that our control, or lack of it, is symmetric on the two sides. To help us imagine this, let's introduce a Demon (from the distinguished family whose ancestors have been so helpful at other points in the history of physics!) Let's give the Demon exactly the control on the left that Nature has (and we lack) on the right. The Demon therefore controls the light beam intensities on the two channels L0 and L1. It knows what setting σ_L we have chosen for the angle of the left cube, and shares with us the goal of having a single light beam emerge from the cube in the direction shown. The question is, who now controls the polarization τ? Is it us, or the Demon, or is the control shared in some way?

It turns out that the Demon now has full control. The Demon can make τ take any value, by choosing appropriate intensities for the two input beams on L0 and L1. So the Demon could use the device to signal to the future, by controlling the polarization – but we can't! (See Box 7.2, for a simple analogy – we are in the same position as Kirk, and the Demon is Scotty, with full control of the direction of the ship, so long as she can use two thrusters.)

Box 7.2: How Discreteness Makes a Difference

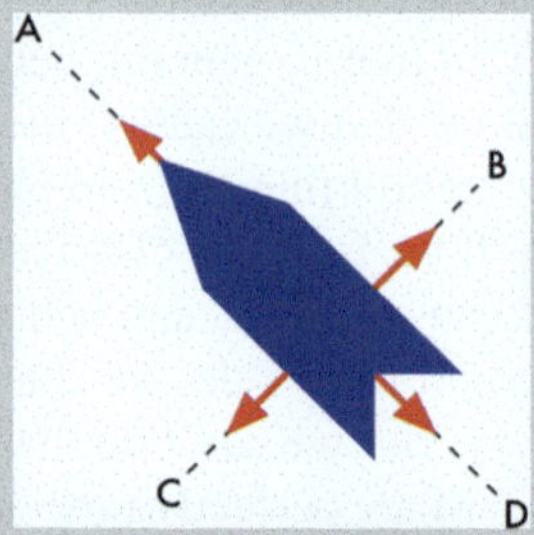

If Scotty has variable control of forward–reverse and left–right thrusters (shown as red arrows), then it doesn't matter which way Kirk points the *Enterprise* (e.g., towards A) – Scotty can still make the ship travel in whatever direction she chooses, by putting an appropriate portion of the power on two different thrusters at 90° to each other. But if Scotty can only use one thruster, then the *Enterprise* has to travel *either* in the orientation that Kirk chooses (along the line AD, in one direction or other), *or* at 90° to that direction (along the line BC, in one direction or other). So Kirk now has at least *some* control of the ship's direction, rather than none at all! This is exactly the difference that quantization makes, in the case of polarization. (For precision, we specify (i) that this *Enterprise* lives in Flatland, and hence travels in only two dimensions, and (ii) that Scotty cannot vary the power of the thrusters during use – any thruster in use burns for the same time, at its pre-set intensity.)

We can now see why in the ordinary case, we can't use the device in Fig. 7.3 to signal to the past, *even though the underlying physics is time-symmetric*. The problem is simply that we don't control the outputs, the amount of light that leaves the apparatus on the two beams R0 and R1. Nature controls that, not us. If we make a change in σ_R, Nature makes a corresponding change in the amount of light on the two beams R0 and R1, and nothing changes in the past.

But now let's see what happens if we "quantize" our Demon, by insisting that it input a light beam on a single channel – either L0 or L1, *but not both at the same time*. It turns out that this makes a big difference to what we can control on the left side of the apparatus – though still not quite enough to enable us to signal to the future!

With this new restriction on the Demon, choosing σ_L for the orientation of the left cube means that there are only two possibilities for τ. Either $\tau = \sigma_L$, in the case in which the Demon uses channel L1, or $\tau = \sigma_L + 90°$, in the case in which the Demon uses channel L0. This means that while we don't control τ completely, we do have a lot more control than before we quantized our Demon. We can now limit τ to one of just two possibilities, 90° different from one another. (See Box 7.2 again

– this is like the case in which Scotty can only use one thruster, giving Kirk some control over where the *Enterprise* goes.)

Summing up, the quantization restriction on the Demon gives us *some* control from left to right, just by controlling the angle σ_L, of a kind we don't have without quantization. But now symmetry shows that if Nature is quantized in the same sense – if Nature has to put the output beam on R0 or R1 *but not both* – then we have the same new kind of control from right to left, just by controlling the angle σ_R. If the underlying physics is time-symmetric, and if the outputs on the right are quantized, controlling σ_R gives us some control over the polarization of the photon, immediately *before* it reaches the right hand cube!

To visualise this possibility, we need to modify our diagram a little. In place of the original τ, controlled by σ_L, let's now write τ_L. And let's introduce a new label, τ_R, for the property that must be controlled by σ_R, if time-symmetry is to be preserved. This gives us Fig. 7.4 – we place τ_L and τ_R at left and right, respectively, next to the piece of the apparatus by whose setting they are (partially) controlled. (We'll come back to the question what might happen in the middle – for the moment the "?" stands for "unknown physics"!)

Summing up, we saw that the quantization restriction on the Demon gave us some control over τ_L by manipulating σ_L. Time-symmetry implies that an analogous quantization restriction on Nature gives us control over τ_R by manipulating σ_R. Again, we can't control τ_R completely, but we can limit it to two possibilities: either $\tau_R = \sigma_R$ or $\tau_R = \sigma_R + 90°$. (Note that this argument doesn't depend on the details of QM in any way – we would have this new kind of control in CEM, if Nature cooperated in this way.)

7.4.1 *In Einstein's Shoes*

Now think about this argument from Einstein's perspective, in 1905, after he's discovered that light is quantized, and before he's heard about quantum mechanics – it lies twenty years in the future, after all! He knows that there's a minimum possible amount of light that can pass through the apparatus, what we now call a single photon.

For a single photon, it seems, the quantization restriction must hold. The photon must travel on one path or other, *at both ends of the experiment*. And that means that it is subject to this new kind of control – the control that results from quantization – *in both directions*. (We are relying on time-symmetry here, of course – if the laws governing the photon are allowed to be time-asymmetric, one sort of control, left to right, or right to left, can vanish.)

This simple argument shows that if we accept assumptions that would have seemed natural to Einstein after 1905, we do have a subtle kind of influence over the past, when we choose such things as the orientation of a polarizing cube. This is precisely the possibility called "retrocausality" in contemporary discussions of QM. As we noted above, most physicists think that they are justified in ignoring

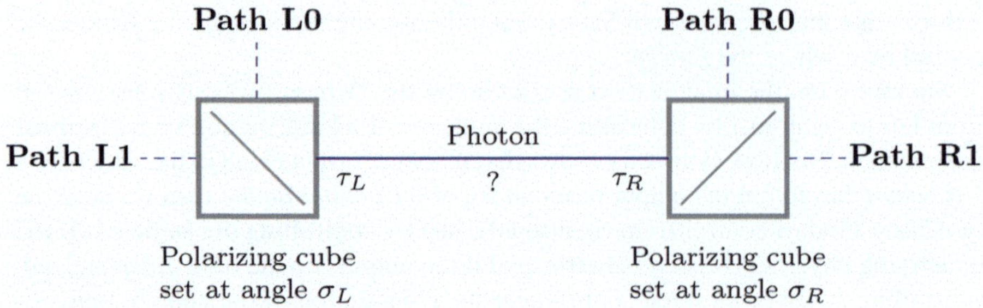

Fig. 7.4 Time-symmetric polarization

it, and using "No retrocausality" as an uncontroversial assumption in various arguments intended to put further nails in the coffin of Einstein's view of the quantum world. And yet it turns out to have been hiding just out of sight, all these years!

7.4.2 Spooks Against Retrocausality

Defenders of orthodox QM, the view that Einstein came to dislike, are not going to be impressed by this argument. They'll point out that quantum theory allows us to avoid one or both of the two crucial assumptions, of time-symmetry and discrete (or quantized) outcomes. Both escape routes depend on the quantum wave function – the strange object whose properties Einstein disliked so much.

On some views, the wave function undergoes time-asymmetric "collapse", and so the underlying physics is not regarded as time-asymmetric. On other views, the wave function itself divides between the two output channels, R0 and R1. By varying the "proportion" of the wave function on each channel, Nature then has the flexibility she needs to respond to changes we make the angle σ_R, preventing us from controlling the past. Either way, these versions of QM can escape the argument, and avoid retrocausality.

But this escape from retrocausality rides on the back of the wave function – the spooky, nonlocal object, not respectably resident in time and space, that Einstein deplored in his letter to Born. From Einstein's point of view, then, the argument seems to present us with a choice between spooks and retrocausality. We cannot say for certain which option he would have chosen – "There is nothing one would not consider when one is in a predicament!", as he put it at one point, writing about the problem of reconciling CEM with quantization (Letter to A. Sommerfeld, 19 January 1910 in Einstein 1995, p. 147) – but the choice makes it clear that anyone who simply *assumes* "No retrocausality" in arguing with Einstein is tying their famous opponent's arms behind his back. In the light of the argument above, retrocausality is the natural way of avoiding the spooks. To rule it out before the game starts is to deprive Einstein of what may be his star player.

7.4.3 Control Without Signalling

At this point, Einstein's contemporary opponents are likely to object that retrocausality is even worse than the so-called quantum spooks. After all, doesn't it lead to the famous paradoxes? Couldn't we use it to advise our young grandmother to avoid her unhappy marriage to grandfather, for example, and hence ensure that we had never been born?

This is where the subtle nature of the new kind of control allowed by quantization is important. Let's go back to the case of the quantized Demon, who controls the inputs on the left side of Fig. 7.4, but can only use one channel at a time: all the input must be on either L0 or L1, and not on both at the same time. We saw that we do have some control over the polarization τ_L in this case, but not complete control. The Demon always has the option of varying the polarization we aim for by 90°. But this means that we can't control the polarization enough to send a signal – intuitively, whatever signal we try to send, the Demon always has the option of turning it into exactly the opposite signal, by adding a factor of 90°.[8]

So the kind of control introduced by quantization is too subtle to allow us to signal left to right, in the experiment in Fig. 7.4. Hence, by symmetry, it is also too subtle to allow us to signal from right to left, or future to past. We couldn't use it to send a message to our grandmother, and history seems safe from paradox. Despite this, we can still have some control over hidden variables in the past, of the kind needed for retrocausality to resolve the challenge of Bell's Theorem in Einstein's favor.

7.5 Where Next?

So discussions of the meaning of quantum theory might have been very different, if Einstein had noticed how quantization suggests retrocausality. But would it have made any difference in the end? Is there any practical way to rescue Einstein's dream of a spook-free quantum world, if we do allow it to be retrocausal?

As we noted earlier (see Box 7.1), there are several retrocausal proposals on the table. But some, like the Aharonov-Vaidman Two State proposal, or the earlier Transactional Interpretation, try to build their retrocausal models with the same kind of elements that Einstein objected to – wave functions not properly located in space and time.

If we want to stay close to the spirit of Einstein's program, then, we'll need to start somewhere else. One such proposal aims to extract a realistic model of

[8]More formally, this is a consequence of the so-called No Signalling Theorem. A protocol that allowed us to signal in this case would also allow signalling between two arms of an analogous entanglement experiment with polarizers, in violation of the No Signalling Theorem – for the correlations in the two cases are exactly the same (Evans, Price, Wharton, 2013).

quantum reality from the techniques developed by another of the twentieth century's quantum giants, Richard Feynman. Feynman's "path integral" computes quantum probabilities by considering all the possible ways a quantum system might get from one event to another, and assigning a weight, or "amplitude" to each possible "path".[9]

The path integral is usually regarded simply as a calculational device. Most people, including Feynman himself, have thought that it doesn't make sense to interpret it as telling us that the system *really* follows one particular history, even if we don't know which. Attractive as this thought seems in principle, the probabilities of the various paths just don't add up in the right way. But there are some clues that this might be made to work, using non-standard versions of the path integral.[10]

Applied to the case we have been discussing, this approach suggests that electromagnetic fields are not strictly constrained by the equations of CEM (a possibility that Einstein certainly contemplated). Instead, the path integral considers *all* possible field histories, even apparently crazy cases where a photon's polarization *rotates* from τ_L to τ_R in empty space. In fact, only by considering such non-classical cases can the path integral approach make accurate predictions. Such a scenario, involving a real polarization rotation between the cubes, is exactly the sort of process that could restore time-symmetry to the case of single photons.

Any approach that does succeed in interpreting the Feynman path integral realistically – i.e., that makes sense of the idea that the system actually follows just one possible history – is likely to be retrocausal (Wharton 2016). Why? Simply because we have control over some aspects of the "endpoint", when we choose to make a particular measurement (e.g., when we choose σ_R in Fig. 7.4). In general, our choice makes a difference to the possible paths that lead to this endpoint, and if one of those paths must be real, then we are affecting reality, to some extent.

In putting future and past on an equal footing, this kind of approach is different in spirit from (and quite possibly formally incompatible with) a more familiar style of physics: one in which the past continually generates the future, like a computer running through the steps in an algorithm. However, our usual preference for computer-style physics may simply reflect an anthropocentric bias. It is a good model for creatures like us, who acquire knowledge sequentially, past to future, and hence find it useful to update their predictions in the same way. But there is no guarantee that the principles on which the universe is constructed are of the sort that happens to be useful to creatures in our particular situation.

Physics has certainly overcome such biases before – the Earth isn't the center of the universe, our sun is just one of many, there is no preferred frame of reference. Now, perhaps there's one further anthropocentric attitude that needs to go: the idea that the universe is as "in the dark" about the future as we are ourselves.

---

[9]These "paths" are not necessarily localized trajectories of particles, but could be the entire "histories", between the two events in question, of something spread out in space, such as an electromagnetic field. (This extension from the particle-path integral to a field-history integral is used in quantum field theory.)

[10]As in a few recent efforts Wharton (2013, 2016), Kent (2013); see also Sorkin (2007).

It is too early to be sure that the spooks are going to be driven out of the quantum world; too early to be entirely confident that retrocausality will rescue Einstein's dream of a relativity-friendly quantum reality, living in space and time. What is clear, we think, is that there are some intriguing hints that point in that direction. Most intriguing of all, there's a simple new argument that suggests that reality *must* be retrocausal, by Einstein's lights, if we don't go out of our way to avoid it. It is too early to award Einstein the last laugh, certainly; but too soon to dismiss the dreams of the old giant.

Note Added in Proof The argument discussed in this paper has recently been extended by Leifer and Pusey (2016), who claim a significant generalization.

Acknowledgements The authors would like to thank Blai Píe i Valls for detailed and useful comments.

References

Albert, D., and R. Galchen. 2009. A quantum threat to special relativity. *Scientific American* 300: 32–39.

Bell, J.S. 1986. Interview. In *The ghost in the atom*, ed. P.C.W. Davies and J.R. Brown, 48–49. Cambridge: Cambridge University Press.

———. 1988. Letter to H. Price, 8 June 1988, quoted in Huw Price, *Time's arrow and Archimedes' point*. Oxford: Oxford University Press, 1996, 241.

Born, M., ed. 1971. *The Born-Einstein letters: friendship, politics and physics in uncertain times*. New York: Macmillan.

Dirac, P. 1930. *The principles of quantum mechanics*. Oxford: Oxford University Press.

Einstein, A. 1954. Maxwell's influence on the idea of physical reality. In *ideas and opinions*. New York : Bonanza Books, 266–270 (originally published in *James Clerk Maxwell: A Commemorative Volume*. Cambridge: Cambridge University Press, 1931).

———. 1995. *The collected papers of Albert Einstein*. vol. 5: The Swiss Years: Correspondence, 1902–1914. Trans. A. Beck. Princeton: Princeton University Press.

———. 1998. *The collected papers of Albert Einstein*. Vol. 8: The Berlin Years: Correspondence, 1914–1918. Trans. A. M. Hentschel. Princeton: Princeton University Press.

Evans, P., Price, H., and K. Wharton 2013. New slant on the EPR-Bell experiment. *British Journal for the Philosophy of Science* 64(2): 297–324. arXiv:1001.5057.

Kent, A. 2013. Path integrals and reality. arXiv:1305.6565.

Leifer, M.S., and M.F. Pusey. 2016. Is a time symmetric interpretation of quantum theory possible without retrocausality? arXiv: 1607.07871.

Ma, X.-S. et al. 2012. Experimental delayed-choice entanglement swapping. *Nature Physics* 8: 479–484. arXiv:1203.4834

Price, H. 1996. *Time's arrow and archimedes' point*. Oxford: Oxford University Press.

———. 2012. Does time-symmetry imply retrocausality? How the quantum world says "maybe". *Studies in History and Philosophy of Modern Physics* 43: 75–83. arXiv:1002.0906.

Pusey, M.F. et al. 2012. On the reality of the quantum state. *Nature Physics* 8: 476–479. arXiv:1111.3328.

Ritz, W., and A. Einstein. 1909. Zum gegenwärtigen Stand des Strahlungsproblems [On the Current State of the Radiation Problem]. *Physikalische Zeitschrift* 10: 323–324.

Sorkin, R.D. 2007. Quantum dynamics without the wave function. *Journal of Physics* A 40: 3207–3222. arXiv:quant-ph/0610204.

Wharton, K. 2013. Lagrangian-only quantum theory. arXiv:1301.7012.

———. 2016. Towards a realistic parsing of the Feynman path integral. *Quanta* 5: 1–11. arXiv:1512.00740.

Chapter 8
On Time, Causation and Explanation in the Causally Symmetric Bohmian Model of Quantum Mechanics

Joseph Berkovitz

Abstract Quantum mechanics portrays the universe as involving non-local influences that are difficult to reconcile with relativity theory. By postulating backward causation, retro-causal interpretations of quantum mechanics could circumvent such influences and, accordingly, increase the prospects of reconciling these theories. The postulation of backward causation poses various challenges for retro-causal interpretations of quantum mechanics and for the existing conceptual frameworks for analyzing counterfactual dependence, causation and causal explanation, which are important for studying these interpretations. In this chapter, we consider the nature of time, causation and explanation in a local, deterministic retro-causal interpretation of quantum mechanics that is inspired by Bohmian mechanics. This interpretation, the 'causally symmetric Bohmian model', offers a deterministic, local 'hidden-variables' model of the Einstein-Podolsky-Rosen/Bohm experiment that presents a new challenge for Reichenbach's principle of the common cause. In this model, the common cause – the 'complete' state of the particles at the emission from the source – screens off the correlation between its effects – the distant measurement outcomes – but nevertheless fails to explain it.

8.1 Introduction: The Background and the Plan of the Chapter

The arrow of time and the time-asymmetry of causation are closely related. Yet, the exact relation between them is a matter of ongoing discussion and controversy. Some authors maintain that causal asymmetry is related to temporal asymmetry by definition. They define causation in terms of temporal asymmetry, so that causes

J. Berkovitz (✉)
Institute for the History and Philosophy of Science and Technology, University of Toronto, 91 Charles St. West, Toronto M5S 1K7, ON, Canada
e-mail: joseph.berkovitz@utoronto.ca

C. Bouton, P. Huneman (eds.), *Time of Nature and the Nature of Time*, Boston Studies in the Philosophy and History of Science 326,
DOI 10.1007/978-3-319-53725-2_8

precede their effects by definition. David Hume's characterization of causation in terms of constant conjunction is a famous example:

> We may define a cause to be *an object, followed by another, and where all the objects similar to the first are followed by objects similar to the second* (Hume 1777, Section VII, Part II).

Hume also relates the time-asymmetry of causation to two other asymmetries: time-asymmetry of counterfactual dependence and a corresponding asymmetry in our experience. Thus, he adds:

> Or in other words *where, if the first object had not been, the second never had existed.* The appearance of a cause always conveys the mind, by a customary transition, to the idea of the effect. Of this also we have experience. We may, therefore, suitably to this experience, form another definition of cause, and call it, *an object followed by another, and whose appearance always conveys the thought to that other*. (*ibid.*)

Other authors argue that temporal asymmetry supervenes upon causal asymmetry (Reichenbach 1956, Mellor 1981, 1998),[1] though there is a controversy as to whether the causal asymmetry is primitive (Mellor 1981, 1998) or supervenes on the thermodynamic asymmetry (Reichenbach 1956).[2] Reichenbach (1956) proposes that the direction of time could be reduced to the 'fork' asymmetry in macroscopic phenomena – namely, the fact that, in such phenomena, all the open v-shaped causal forks seem to be in the same direction (see Fig. 8.1a) – and that this asymmetry could be reduced to the thermodynamic asymmetry. By a cause, Reichenbach meant a probabilistic cause, i.e. a cause that raises the probability of its effects (for more details see Sect. 8.9).[3]

Accounts of causation, like Hume's, that define the direction of causation to be the same as the direction of time exclude the possibility of backward causation, i.e. causation from the present to the past or from the future to the present. While our experience seems to suggest that backward causation does not exist, it is commonly thought that this kind of causation is conceptually, metaphysically and, perhaps, physically possible. Indeed, in the physics literature, there have been proposals to interpret classical electromagnetism and quantum mechanics (henceforth, QM) as involving backward causation. Wheeler and Feynman (1945, 1949) introduce the absorber theory of electromagnetism as a time-symmetric alternative to conventional electromagnetism, which, unlike the latter, imposes no

[1]Cover (1997, p. 306) thinks that it is "not unreasonable to read Kant's Second Analogy as expressing a causal theory of time."

[2]For discussions of the relations between the direction of time, the time-asymmetry of causation and the thermodynamic asymmetry, see for example Price (1996), Albert (2000), Kutach (2001, 2002, 2007), Loewer (2007), Price and Weslake (2009) and Frisch (2013).

[3]To simplify things, in what follows in Sects. 8.1–8.8 we shall not discuss the exact characterization of probabilistic causation, though the approximate contours of the concept of causation we have in mind will become clearer as we go. In Sect. 8.9, we shall characterize Reichenbach's concept of probabilistic causation more precisely and discuss it in the context of a deterministic, retro-causal interpretation of QM.

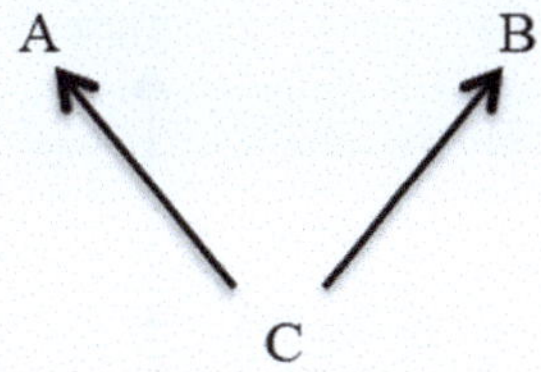

Fig. 8.1a V-shaped causal fork open toward the future, constituted by a common cause, *C*, and its joint effects, *A* and *B*. *Arrows* denote causal connections

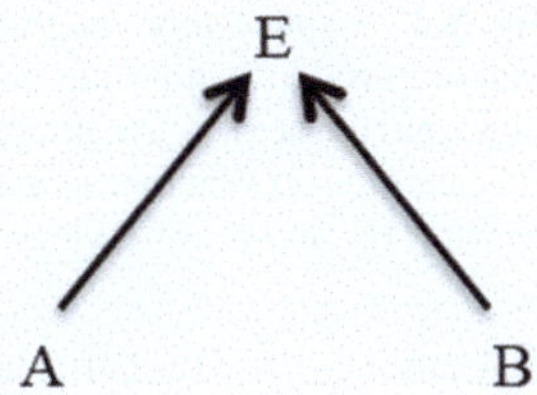

Fig. 8.1b V-shaped causal fork open toward the past, constituted by two causes, *A* and *B*, and their joint effect, *E*

ad hoc time direction on electromagnetic processes (for a discussion of this theory, see Crammer 1983). And there have been various retro-causal interpretations of quantum mechanics[4] (henceforth, RCIQM) that postulate the existence of backward causation in an attempt to circumvent the non-local influences that the mainstream and more conventional interpretations of QM postulate, and accordingly increase the prospects of reconciling QM with relativity theory.

In what follows, our focus will be on RCIQM. We shall analyze the causal structures that models of RCIQM predict in Bohm's (1951) version of the Einstein-Podolsky-Rosen experiment (henceforth, the EPR/B experiment) and some variants of it (see Sect. 8.2). In the EPR/B experiment, pairs of particles are emitted from a source and when they are far away from each other undergo measurements. The distant (space-like) measurement outcomes are curiously correlated, and these correlations suggest the existence of non-local influences between the outcomes (see Fig. 8.3), which are difficult to reconcile with relativity theory. The RCIQM we shall discuss provide local common-cause models of this experiment: the measurement outcomes are the effects of a common cause – the 'complete' pair-state at the emission from the source – which occurs in the intersection of their backward light cones (see Fig. 8.4).

We shall consider the nature of the backward causation that these interpretations postulate and the closed causal loops they predict, as well as the predictive and explanatory challenges that these causal patterns raise. Here, and henceforth, by a causal loop we mean a sequence of events (or facts) that loops back in such a way

---

[4]See, for example, Costa de Beauregard (1953, 1977, 1979, 1985), Cramer (1980, 1983, 1986, 1988), Sutherland (1983, 1998, 2008), Price (1984, 1994, 1996, 2008, 2012), Reznik and Aharonov (1995), Miller (1996, 2008), Berkovitz (2002a, 2008, 2011), Gruss (2000), Aharonov and Gruss (2005), Aharonov and Tollaksen (2007), Kastner (2012), and Price and Wharton (Chap. 7 in this volume).

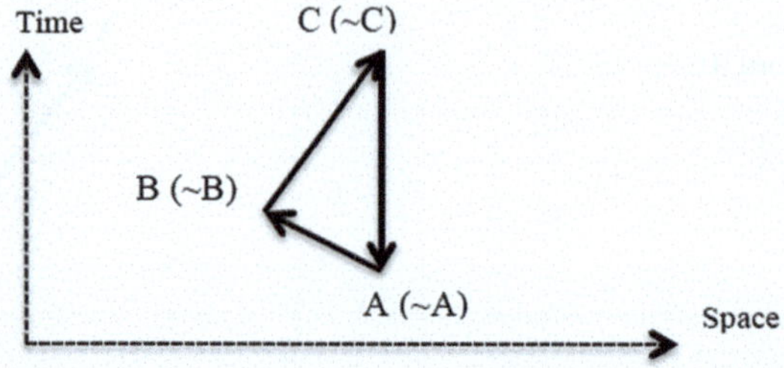

Fig. 8.2 A space-time diagram of a causal loop that involves backward causation, where *arrows* denote causal influences. In this loop, *A* is a forward cause of *B*, *B* is a forward cause of *C* and *C* is a backward cause of *A*. An alternative possible loop is one in which ~*A* causes ~*B*, ~*B* causes ~*C* and ~*C* causes ~*A*. (Whether event causation could exist between events which are absences of other 'events' is a matter of controversy (see, for example, Schaffer 2003/2016))

that each event (fact) in the loop is an indirect cause of itself (Horwich 1995, p. 259) (for an example of such a loop, see Fig. 8.2).[5] We shall assume for the sake of consideration that backward causation is metaphysically possible and, moreover, that our universe might comprise such causation if any RCIQM were an appropriate interpretation of QM. This assumption excludes theories of time and theories of causation that prohibit backward causation.

In the philosophical literature, there are various arguments for the impossibility of backward causation. Further, as we shall see, the backward causation that RCIQM postulate gives rise to causal loops, and there are arguments for the impossibility of such loops. There are also arguments for the inconsistency of RCIQM. We shall argue that all these arguments are based on disputable premises, and that backward causation, causal loops and RCIQM could be consistent. We shall then analyze the nature of time, causation, probability and explanation in the context of RCIQM.

In the next section, we present the main ideas of RCIQM. In order to prepare the ground for the study of the backward causation that RCIQM postulate and the causal loops that they predict, in Sect. 8.3 we discuss the challenges that analyses of backward causation and causal loops encounter. In Sect. 8.4, we review the main arguments for the impossibility of backward causation, causal loops and RCIQM, and argue that they have all been challenged. In Sect. 8.5, we present the block universe, arguably the dominant ontological framework for understanding space and time in contemporary physics and metaphysics, and consider the representation of backward causation and causal loops in this framework. In Sect. 8.6, we discuss indeterministic RCIQM and argue that the causal loops that these interpretations

[5]In the philosophical literature, the nature of causation is controversial. It is common to think of causation as a relation, though different accounts explicate causation in terms of different relations, and there is a controversy about whether the relata are events or facts. There are also theories of causation that explicate causation in terms of processes (see Sect. 8.5). For a review of the metaphysics of causation, see Schaffer (2003/2016) and references therein; and for a discussion of whether causation is a relation, see Hausman (1998, Sect. 2.3).

predict pose serious challenges for their predictive and explanatory power. The question arises as to whether deterministic RCIQM fare better. Sutherland (2008) proposes a deterministic RCIQM that is based on Bohmian mechanics, 'the causally symmetric Bohmian model' (henceforth, CSBM). In Sect. 8.7 we briefly review Bohmian mechanics (henceforth, BM), and in Sect. 8.8 we present and analyze CSBM.

In Sect. 8.9, we consider Reichenbach's Principle of the Common Cause (henceforth, PCC). There have been various objections to PCC. However, as we shall see, none of them apply to the common-cause models that CSBM predicts. In Sect. 8.10, we argue that CSBM presents a new challenge for PCC. In the common-cause model that this interpretation predicts for the EPR/B experiment, the common-cause screens off the correlation between its effects: the probability of each of the effects given the common cause is independent of the other effect. Thus, it is natural to expect that the correlation between the effects be explained by the common cause. But, as we shall see, this is not the case. In this model, the correlation between the effects has no causal explanation: it is a matter of a brute fact.

That CSBM postulates causally unexplained correlations might not come as a surprise to some students of causal loops. Indeed, it has been argued that causal loops involve causally inexplicable correlations (Horwich 1995). Yet, this argument has been contested (see Sect. 8.10), and in any case the causally unexplained correlations that CSBM postulates are of a different kind.

In Sect. 8.11 we conclude by considering whether the backward causation that CSBM postulates is compatible with causal theories of the direction of time, and whether the failure of PCC in the common-cause models that CSBM predicts is due to backward causation.

8.2 The Main Idea of Retro-Causal Interpretations of Quantum Mechanics

RCIQM postulate backward causation. The state of a measured system before the measurement is influenced by the state of the measurement apparatus during the measurement or the outcome of the measurement (or some corresponding event/state that carries the same relevant information).[6] These backward influences are a radical deviation from the mainstream metaphysics of science and they pose challenges for the predictive and explanatory power of RCIQM (Berkovitz 2001, 2002a, 2008, 2011). So why bother with such interpretations of QM? RCIQM are supposed to overcome one of the main foundational problems that the mainstream and more conventional interpretations of QM encounter. The mainstream and more conventional interpretations portray the quantum realm as non-local (Bell 1987,

[6]For an example of such corresponding events/states, which are not effects of the measurement outcomes, see Sects. 8.8 and 8.10.

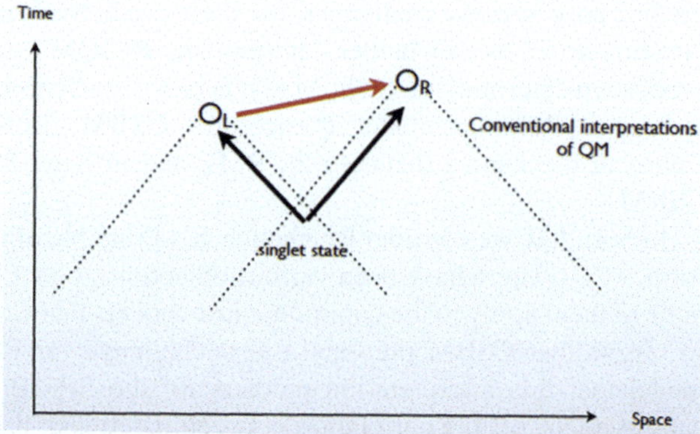

Fig. 8.3 A Space-time diagram of non-local influences in the EPR/B experiment in 'conventional' interpretations of QM. The left-wing measurement outcome influences the right-wing measurement outcome. O_L (O_R) denotes the left- (right-) wing measurement outcome

Redhead 1987, Butterfield 1992, Berkovitz 2007/2016), and it is very difficult to reconcile this non-locality with relativity theory (Maudlin 1994/2011, Berkovitz 2007/2016, Sect. 10). By postulating backward causation, RCIQM could avoid this kind of non-locality and consequently increase the prospects of reconciling QM with relativity theory. Further, orthodox QM encounters the infamous measurement problem (Myrvold 2016, Sect. 4), and some RCIQM, including CSBM, also address this problem.

The main idea of RCIQM could be illustrated by a simple example. Consider the EPR/B experiment. Pairs of particles are emitted from a source in opposite directions in the singlet state for spin.[7] When the particles are far away from each other, they encounter apparatuses that can be set to measure spin properties along various directions. Each of the measurements occurs outside the backward light cone of the other measurement, so that there could not be any ('direct') subluminal or luminal influences between them (see Fig. 8.3). According to orthodox QM, the quantum realm is indeterministic and the outcome of each of the distant measurements is a matter of sheer chance. Yet, the measurement outcomes are curiously correlated: the probability of spin 'up' along the direction n in one wing of the experiment depends on whether the measurement outcome in the other wing is spin 'up' or spin 'down' along the direction m. This correlation suggests the existence of a curious influence between the distant (space-like separated) measurement outcomes, and indeed the more conventional interpretations of QM predict the existence of such non-local influence (see Fig. 8.3). In the experiment depicted in Fig. 8.3, pair-state at the emission and the setting of the left-wing measurement apparatus jointly determine the outcome or the probability of the outcome of the left-wing measurement, and the pair-state at the emission, the setting of the right-wing measurement apparatus and the outcome of the left-wing measurement jointly determine the

[7]In Dirac's notation, the singlet state for spin is expressed as follows: $|\psi\rangle = \frac{1}{\sqrt{2}}\left(|n+\rangle_1|n-\rangle_2 - |n-\rangle_1|n+\rangle_2\right)$, where the indexes '1' and '2' denote the first and the second particle respectively, and $|n+\rangle_1$ $\left(|n+\rangle_2\right)$ and $|n-\rangle_1$ $\left(|n-\rangle_2\right)$ are the states of the first (second) particle having respectively spin 'up' and spin 'down' along the direction n.

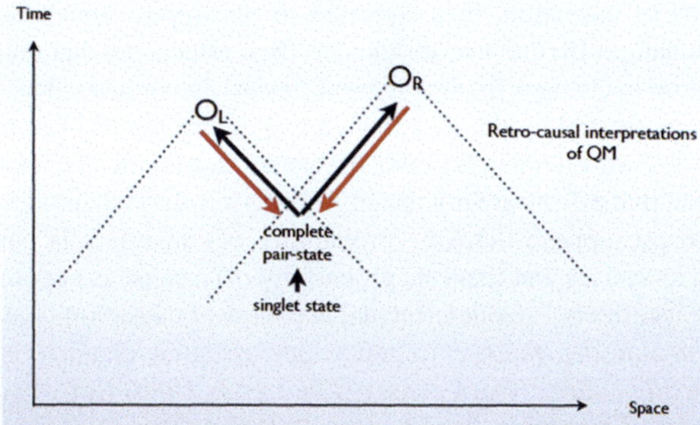

Fig. 8.4 A space-time diagram of the causal influences in local retro-causal interpretations of QM. The probability of each of the measurement outcomes is determined by the 'complete' pair-state at the emission from the source (and the setting of the local measurement apparatus), and this state is influenced by the settings of the measurement apparatuses or the measurement outcomes (or corresponding events that carry the same relevant information), depending on the model

outcome or the probability of the outcome of the right-wing measurement.[8] This kind of non-local influences between the measurement events is difficult to reconcile with relativity theory.

RCIQM provide quite different models of the EPR/B experiment (see Fig. 8.4). These interpretations of QM are, in effect, 'hidden-variables' theories, and accordingly in the models they prescribe for the EPR/B experiment the state of the particle pair is more 'complete' than its QM wavefunction. In these models, the probability of each of the measurement outcomes is determined by the complete pair-state at the emission from the source (and the setting of the local measurement apparatus), and this state is influenced by the settings of the measurement apparatuses or the measurement outcomes, depending on the model. In Sect. 8.8, we shall discuss the nature of the complete pair-state in the CSBM model of the EPR/B experiment.

8.3 Backward Causation and Causal Loops: The Complications

The study of backward causation poses serious challenges for the existing conceptual frameworks for analyzing causation. Let us consider, for example, two of the main asymmetries that are presupposed in analyses of causation: the asymmetry of manipulation and the asymmetry of counterfactual dependence.

[8]Some hidden-variables interpretations of QM, like Bohmian mechanics, postulate the existence of 'parameter dependence', i.e. the dependence of the probability of the distant measurement outcome on the setting of the nearby measurement apparatus in the EPR/B experiment. But, as Berkovitz (1998, Sections 2.3–2.4, 2007/2016, Section 8.4) argues, Bohmian mechanics involves 'specific' outcome dependence, and parameter dependence is the result of this outcome dependence. For the characterization of 'specific measurement outcomes' in this experiment, see Fnt. 16.

In discussions of causation, it is common to presuppose that manipulations of causes result in changes in (the probabilities of) their effects but that manipulations of effects do not result in changes in (the probabilities of) their causes. This *asymmetry of manipulation* is motivated by the idea that causes are means of bringing about their effects – whenever an effect is an end, its causes automatically supply means of achieving it[9] – but that effects are not means of bringing about their causes. Relatedly, it is common to presuppose that (the probability of) an effect is counterfactually dependent upon its causes, but that (the probability of) a cause is not counterfactually dependent upon its effects.[10] Counterfactual accounts of causation take this *asymmetry of counterfactual dependence* to reflect an essential characteristic of causal relations (Hume 1777, Sect. VII, Lewis 1973a, Lewis 1986) (for a review of counterfactual accounts of causation, see Menzies 2001/2014).

It is difficult to account for these asymmetries in the context of backward causation. First, the standard understanding of the counterfactuals that are involved in causation seems to exclude backward causation. The most influential analysis of such counterfactuals is due to Lewis (1973b, 1986, Chaps 16–18). In this analysis, the explication of causal dependence should not be in terms of 'backtracking' counterfactuals – namely, counterfactuals that state that if things had been different at a certain time, some things would have been different at earlier times.[11] Accordingly, Lewis' account of causation, and any other account that explicates causal dependence in terms of non-backtracking counterfactuals, is inapplicable to backward causation. Thus, if causation is to be explicated in terms of counterfactuals, we need to look for a different analysis of counterfactuals. Since non-backtracking counterfactuals are also central to analyses of manipulation, a similar challenge exists for attempts to account for the asymmetry of manipulation in the context of backward causation. Berkovitz (1998, Sect. 2.6) proposes a sketch of an account of causation in which causal dependence is analyzed in terms of backtracking counterfactuals, and this account is applicable to backward causation.

It is noteworthy that Lewis (1979) motivated the *time asymmetry of counterfactual dependence* – according to which future events might be counterfactually dependent on past events, but past events could not be counterfactually dependent on future events – by the *asymmetry of overdetermination*.[12] Lewis framed his reasoning in terms of determinants. A *determinant* of an event is any set of

[9]See Mellor (1995, p. 79) and Menzies and Price (1993, p. 187).

[10]More precisely, it is common to assume that (the probability of) an effect is counterfactually dependent on its cause, or that there is a chain of counterfactual dependencies that connect the effect to the cause (see, for example, Lewis 1986, Chap. 21).

[11]In this analysis, in resolving the vagueness of 'non-backtracking' counterfactuals, we typically hold the past fixed until the time in which the antecedent of the counterfactual is supposed to obtain.

[12]The time asymmetry of counterfactual dependence is the same as the asymmetry of counterfactual dependence in forward causation. But in a universe with backward causation the asymmetry of counterfactual dependence may obtain while the time asymmetry of counterfactual dependence is violated.

conditions that are jointly sufficient, given the laws of nature, for the event's occurrence. Thus, determinants of an event may be its causes or its effects (i.e. 'traces of its occurrence'). Lewis held that it is contingently true that events typically have very few earlier determinants but very many later determinants. As Price and Welsake (2009) comment, Lewis believed originally that the asymmetry of overdetermination is not statistical, in which case it should be distinct from the thermodynamic asymmetry. Elga (2000) argues, though, that the dynamical properties of thermodynamic irreversible processes show that in many ordinary cases Lewis' deterministic asymmetry of overdetermination fails.

In Sects. 8.6, 8.8 and 8.10, we shall argue that backward causation could give rise to causal loops, and in such loops the asymmetry of manipulation, the asymmetry of counterfactual dependence, and the time asymmetry of counterfactual dependence do not obtain. Yet, even if we abandoned the idea of explicating causation in terms of these asymmetries, counterfactual dependence and manipulation would still be important concepts for the analysis of backward causation and causal loops. Thus, we would still face the challenge of developing accounts of counterfactual dependence and manipulation that are applicable in these contexts.

Backward causation raises some additional challenges. In a universe in which backward causation exists, events might be caused by both forward and backward causes or be parts of causal loops, and accordingly be much more constrained. In such circumstances, the relations between events and the behavior of physical systems are expected to be very different from linear causation. Yet, while the heavy constraints on events in causal loops pose serious challenges for the study of such loops, they also provide a key for analyzing the causal loops that RCIQM predict; for, as we shall see in Sects. 8.6 and 8.8, these constraints help identify the range of possible causal loops (Berkovitz 2001, 2008, 2011). But first we need to counter the arguments for the impossibility of backward causation, causal loops and RCIQM.

8.4 Arguments for the Impossibility of Backward Causation, Causal Loops and Retro-causal Interpretations of Quantum Mechanics

In the literature, there are various arguments for the impossibility of backward causation. Some arguments are based on accounts of causation that exclude its possibility by definition. As we have seen above, such accounts are out of favor. Other arguments are based on accounts of causation that do not exclude backward causation by definition. For example, if causation requires that the cause and the effect both exist (as it is commonly assumed), backward causation cannot exist according to tensed theories of time. In these theories, the future is unreal, and accordingly backward causation would require that effects occur without causes. In what follows, we shall assume the 'block universe' framework of space and time (see Sect. 8.5), and this ontological framework does not exclude the possibility of backward causation.

Another line of argument for the impossibility of backward causation is the so-called 'bilking argument' (Flew 1954, Black 1956, Dummett 1964, Faye 2001/2015). The Bilking argument attempts to demonstrate that backward causation is impossible or, at least, that a belief in its existence can never be justified. The main idea is that if an effect *E* occurred before its cause *C*, one could observe *E* and then try to prevent *C* and thus 'bilk' the alleged backward causation from *C* to *E*. If *C* could be prevented, then it cannot be a cause of *E*. And if it is never possible to prevent *C* when *E* occurs, then there is no ground to claim that *C* is a backward cause of *E*.

Bilking arguments have been challenged (see, for instance, Dummett 1964, Horwich 1987, Tooley 1997, pp. 48–52). Moreover, the common view is that backward causation is conceptually, metaphysically, and perhaps also physically possible. Yet, in RCIQM backward causation may lead to causal loops and there are various arguments for the impossibility of such loops. The most popular arguments attempt to show that if causal loops were possible, paradoxes would follow. For example, it is argued that if a person could travel to the past, she could kill her grandfather even before her mother is conceived, thus making her birth impossible (for a causal analysis of such a scenario see Sect. 8.10). Another argument for the inconsistency of causal loops, put forward by Mellor (1995, Chap. 17.3, 1998, Chap. 12), is based on a probabilistic account of causation, where causation is explicated in terms of the objective single-case probabilities that an effect has with its cause and would have had without it.[13] Mellor argues that these probabilities are supposed to constrain the long-run frequencies of the effect with the cause and without it, but in causal loops they fail to do so (for an analysis of this argument, see Berkovitz 2001).

Other arguments are more modest. Rather than striving to establish the impossibility of causal loops, they aim to demonstrate that such loops would involve anomalies, such as improbable or inexplicable coincidences. For instance, no matter how many times a time traveller attempts to kill her grandfather before her mother is conceived, she is destined to fail because of some commonplace reasons, like slipping on banana skins. But, it is argued, the coincidences between the time traveller's attempts and such commonplace causes of the attempts' failures would be causally inexplicable because the coincidences would be due to neither 'direct' causal connections nor common causes (Horwich 1987, Chap. 7, 1995).

In reply to the above impossibility/improbability/inexplicability arguments, one may point out that causal loops impose heavy consistency constraints on the events that occur in them, whereas the above arguments overlook these constraints (Lewis 1986, Chap. 18, Horwich 1987, Chap. 7, 1995, Smith 1997, Berkovitz 2001, 2002a, 2008, 2011, Dowe 2003). Once we take account of these constraints, we find that

[13]For such interpretations of probability, see Mellor (1971, 1995), Giere (1973a, b), Lewis (1986, Chap. 19), Humphreys (1989), Popper (1990), Suarez (2013) and Berkovitz (2015).

causal loops are neither impossible nor necessarily improbable, though they may appear paradoxical, counterintuitive, or inexplicable (see Sects. 8.6, 8.8 and 8.10 for some of the curious characteristics of causal loops). Another reply, which we shall consider in Sect. 8.10, is that the apparent improbable or inexplicable correlations between events in causal loops do have causal explanations.

Finally, there have also been arguments for the inconsistency of RCIQM. In particular, Maudlin (1994/2011) argues that these interpretations of QM are destined to be inconsistent. The argument is, in effect, that RCIQM predict causal loops in which the frequencies of measurement outcomes are bound to deviate from their objective single-case probabilities. However, as Berkovitz (2002a, Sects. 5.2–5.4, 2008, Sect. 8) shows, Maudlin's line of reasoning is based on premises that are natural in linear causation but untenable in the context of causal loops.

8.5 The Block-Universe, Backward Causation and Causal Loops

In current physics and metaphysics, the 'block universe' is the dominant ontological framework for thinking about space and time, and in what follows we shall assume it. We now turn to present this framework and then consider the representation of backward causation and causal loops in it.

In the block-universe framework, the universe is depicted as a four-dimensional block, where three dimensions represent space and the fourth dimension represents time. Events may be thought of as the properties or states of space-time regions in the block, or properties or states of things in space-time regions. All past, present, and future events exist (but not at the same time). There is no ontological difference between past, present and future events, and the division between them is relational to a standpoint. While the four-dimensional events in a block never change, changes are accounted for in terms of the patterns of and the relations between such events. For example, the motion of an object is characterized in terms of the object being at different locations at different times.

It is frequently claimed that the block-universe framework dictates determinism and excludes chancy events and free will. The idea is that if all future events exist, the future is predetermined. That is, if future events exist and it is a fact about the future that an event E occurs, then E is bound to occur, and no past, present, or future events could cause E not to occur. Thus, it is argued that the future is not open to different possibilities, as indeterminism requires, and that free will is impossible and fatalism is unavoidable.

This reasoning is based on a failure to distinguish between 'the impossibility to *change* the actual future from what it is going to be' and the 'impossibility to *influence* the actual future to be what it is going to be'. It is impossible to change the actual future from what it is going to be, but it is possible to influence it to be what it is going to be. The events in the four-dimensional block that represents our universe reflect the actual past, present, and future of our universe. However, our universe could have been different. Had things been different, a different block, i.e. a block with different events, would have represented our universe. In particular, had we taken different decisions, our future would have been different. That is, the unrealized possibilities can account for our capacity to make free choices that partly shape the actual future to be what it is going to be. The unrealized possibilities are also a key for explicating causation, indeterminism and probabilities.

Consider causation. In the literature, there are various accounts of causation. A popular account is Lewis's (1986) counterfactual theory of causation. Lewis' theory fits in the block-universe framework. In fact, this is the framework that Lewis had in mind. In Lewis's theory of *deterministic* causation, an actual event *E* in a universe *w* is said to be *causally dependent* on a distinct, actual event *C* in *w* just in case *E* does not occur in the most similar universes in which *C* does not occur. In Lewis's theory of indeterministic causation, an actual event *E* in *w* is *causally dependent* on a distinct, actual event *C* in *w* just in case the objective single-case probability of *E* in *w* is higher than it would have been in the most similar universes in which *C* does not occur.[14] Other accounts of causation also fit in the block-universe framework. In particular, process theories of causation (Dowe 1996/2007), which explicate causation in terms of physical processes, such as processes that transport 'marks' (Salmon 1998) or conserved quantities (Aronson 1971, Fair 1979, Dowe 1992, 1995), fit well within this framework. Yet, pace the arguments of some propronents of these theories, it is doubtful that explications of causal processes could circumvent the need to appeal to counterfactual reasoning.

Next, consider indeterminism. Like in a deterministic universe, in an indeterministic universe the actual future cannot be different from what it is going to be. But unlike deterministic universe, the past is compatible with different futures, and things could have been different: the past could have been followed by a different future. Thus, in the block-universe framework, indeterminism is accounted for by multiple blocks, each representing a possible universe that has the same past and laws but a different future.

Objective single-case probabilities of events are also explicated in terms of possibilities. Suppose, for example, that the objective single-case probability of 'heads' in a given coin toss is 0.5. This probability, which is a property of the actual circumstances, is partially explicated in terms of long-run frequencies of 'heads' and 'tails' in the same kind of tosses (i.e., tosses in the same kind of circumstances) in the universes that are the most similar to our universe. In particular, in a long series of independent tosses of the same kind in the most similar universes, the frequency of 'heads' will almost certainly be 0.5.

[14]Two comments: (i) Lewis talks about 'worlds' but really mean universes. (ii) In fact, the above definition is a slight modification of Lewis' account, as Lewis (1986, pp. 176–177) requires that the objective single-case probability that the effect has with the cause be much higher than it would have been without the cause; where 'much higher' means by a large factor, though not necessarily by a large difference.

In the block-universe framework, there is no reason to exclude backward causation as a matter of principle. If backward causation existed, causal loops might exist. We may think of causal loops as sets of events that are related to each other in a cyclical way (see Fig. 8.2). In consistent loops, each event in the loop's set (together with the relevant circumstances) determines the next event in the set or its probability according to the laws governing the situation. That is, in consistent loops, the relevant laws do not lead to inconsistency, i.e. to events that are incompatible with the loop's set.

In the context of backward causation and causal loops, we may still think of causal dependence as implying some kind of counterfactual dependence. Indeed, in Lewis' theory the counterfactuals are non-backtracking, and as we have seen in Sect. 8.3 such counterfactuals exclude the possibility of backward causation. But, first, a theory of causation could be based on backtracking counterfactuals (Berkovitz 1998, Sect. 2.6). Second, while the question of the analysis of counterfactual dependence in the context of backward causation is complicated, we could apply some counterfactual reasoning to analyze the range of the possible causal loops in a given circumstance even without having a general account of counterfactual dependence for backward causation and causal loops. For example, in circumstance S in which A is a forward cause of B, B is a forward cause of C and C is a backward cause of A, the following counterfactuals obtain: if A occurred in S, the causal loop $A \Rightarrow B \Rightarrow C \Rightarrow A$ would occur, where '$\Rightarrow$' denote causal dependence; and if $\sim A$ had occurred in S, the loop $\sim A \Rightarrow \sim B \Rightarrow \sim C \Rightarrow \sim A$ would have occurred (see Fig. 8.2).[15]

Finally, process and manipulation accounts of causation could also be useful in analyses of backward causation and causal loops, though, as we have seen above, their application in this context is not straightforward and requires further study.

8.6 On Prediction and Explanation in Indeterministic Retro-Causal Interpretations of Quantum Mechanics

There are three main kinds of indeterministic RCIQM. The distinction between them could be illustrated by reference to the kind of backward causation they postulate: (I) RCIQM in which the setting of a measurement apparatus during a measurement influences the 'complete' state of the measured system before the measurement; (II) RCIQM in which the measurement outcome influences the complete state of the measured system before the measurement; and (III) RCIQM in which events that correspond to the measurement outcome (but not the outcome itself) influence the complete state of the measured system before the

[15]Here for the sake of simplicity, we suppose that S is compatible with both loops.

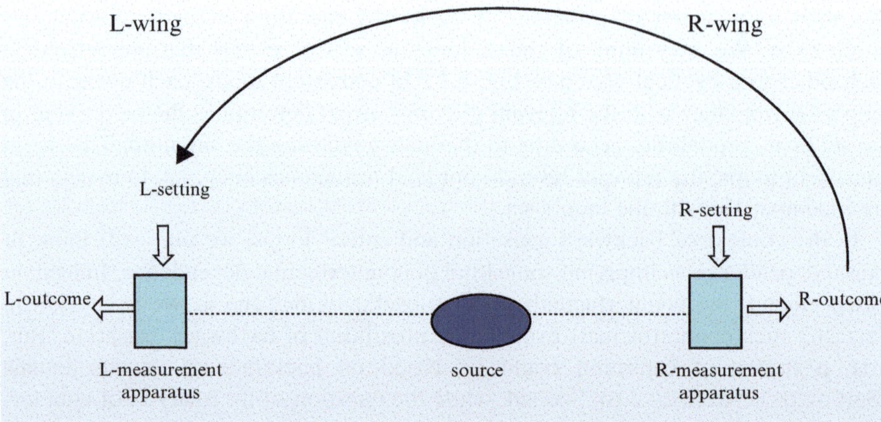

Fig. 8.5 The 'contingent EPR/B experiment', where the R-outcome determines the setting of the L-apparatus by a subluminal signal. For a space-time diagram of the causal connections in this experiment, see Fig. 8.6

measurement. In this section we shall focus on indeterministic RCIQM of the first and second kind, and in Sects. 8.8 and 8.10 we shall discuss a deterministic RCIQM of the third kind.

In the EPR/B experiment, the first kind of indeterministic RCIQM postulate that the settings of the measurement apparatuses during the measurements influence the complete pair-state at the emission from the source, and the second kind of indeterministic RCIQM postulate that the measurement outcomes influence the complete pair-state at the emission.[16] Both kinds of RCIQM predict the existence of causal loops. An example of a set up in which such loops are predicted is the following EPR/B-like experiment, henceforth the 'contingent EPR/B experiment' (see Fig. 8.5).

In the contingent EPR/B experiment, the measurement at the right wing of the experiment (henceforth, the *R*-measurement) occurs before the measurement at the left wing of the experiment (henceforth, the *L*-measurement). The *R*-apparatus is set up to measure spin along the direction *r*. The outcome of the *R*-measurement determines, by a subluminal signal, the setting of the *L*-apparatus: if the *R*-outcome is spin 'down' along the direction *r*, the L-apparatus is set up to measure spin along

[16]In the EPR/B experiment, the measurements are of spin quantities. Unless said otherwise, by 'spin measurement outcomes', we shall mean 'specific' outcomes, e.g. spin 'up' in the direction *m* (rather than the 'non-specific' outcome spin 'up', which appears in various discussions of quantum non-locality). In RCIQM of the second kind, models of the EPR/B experiments postulate that specific measurement outcomes influence the complete pair-state at the emission, and so the backward influences of the measurement outcomes also embody information about the apparatus settings.

the direction *l*, which is the same as the direction *r*; and if the *R*-outcome is spin 'up' along the direction *r*, the *L*-apparatus is set up to measure spin along a different direction, *l**. Figure 8.6 provides a space-time diagram of the causal connections in this experiment. RCIQM of the first kind predict the existence of Loop 1 of Fig. 8.7, and RCIQM of the second kind predict the existence of Loop 2 of Fig. 8.7. In Loop

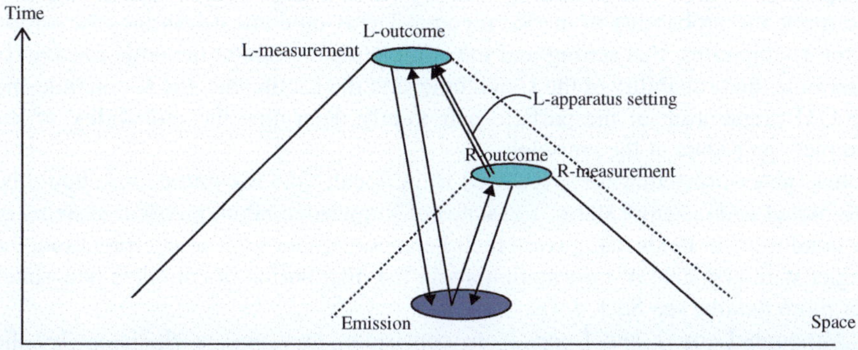

Fig. 8.6 A space-time diagram of the causal connections in the contingent EPR/B experiment according to RCIQM. *Coloured ovals* denote events. *Dotted lines* denote the boundaries of the *backward light cones* of the measurement outcomes. *Arrows* denote causal connections. Backward causation from the R- and the L-measurement events convey information about the settings of the L- and the R-apparatus or the outcomes of the L- and R-measurement, depending on the kind of the RCIQM in question

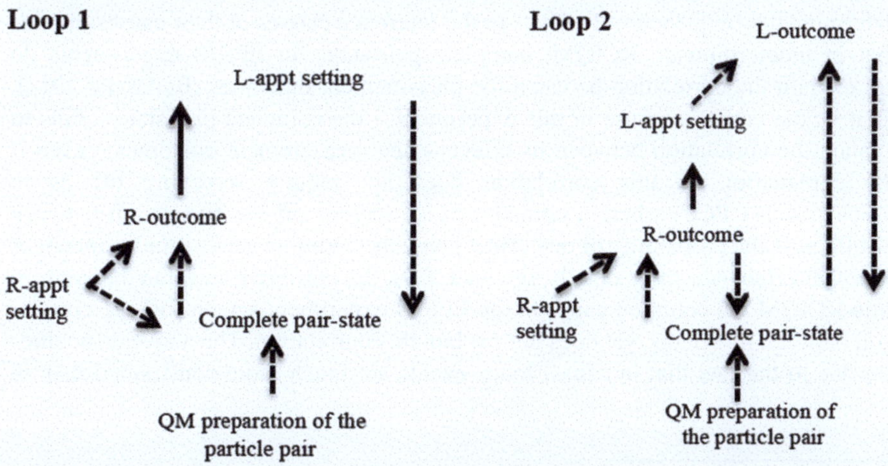

Fig. 8.7 The causal loops that indeterministic RCIQM predict in the contingent EPR/B experiment. Loop 1 depicts the causal loop that occurs in indeterministic RCIQM in which the settings of the measurement apparatuses influence the complete pair-state, and Loop 2 depicts the causal loop that occurs in indeterministic RCIQM in which the measurement outcomes influence the complete pair-state. *Solid arrows* denote deterministic causal connections and *dashed arrows* denote indeterministic causal connections. Some of the events in these loops have multiple partial causes. In such cases, the partial causes jointly determine the probability of the effect

1, the complete pair-state at the emission and the (fixed) setting of the R-apparatus jointly determine the probability of the R-outcome. The R-outcome determines the setting of the L-apparatus and this setting, the setting of the R-apparatus and the QM preparation of the particle pair jointly determine the probability of the complete pair-state at the emission[17]; wherein the 'QM preparation of the particle pair' is the preparation of the particle pair as prescribed by standard QM. In Loop 2, the complete pair-state at the emission and the (fixed) setting of the R-apparatus jointly determine the probability of the R-outcome. The R-outcome determines the setting of the L-apparatus, this setting and the complete pair-state at the emission jointly determine the probability of the L-outcome, and the L-outcome, the R-outcome and the QM preparation of the particle pair jointly determine the probability of the complete pair-state at the emission.

It is noteworthy that for a RCIQM to be local, QM wavefunctions should be interpreted as epistemic states, i.e. states of knowledge about physical systems or the conditions or fields that govern their behavior. Under such an interpretation, the singlet state is a state of information about the distribution of complete pair-states (for more details, see Sect. 8.8).

Although Loop 1 and Loop 2 are consistent, they pose a challenge for the predictive power of indeterministic RCIQM. In each of these loops, the theory assigns the probabilities that the loop's causes give their effects. Yet, these probabilities fail to determine the frequencies of effects in the reference class of their causes, independently of whether they are interpreted as objective single-case or long-run probabilities, or subjective probabilities.[18] Accordingly, indeterministic RCIQM fail to yield predictions for frequencies of outcomes in Loop 1 and Loop 2.[19] Further, the failure of the probabilities that causes give their effects to determine the frequencies of effects in the reference classes of their causes implies that in indeterministic RCIQM, complete pair-states in EPR/B experiments do not explain the correlation between the measurement outcomes (Berkovitz 2011). That is, the common cause in this experiment – the complete pair-state – fails to explain the correlation between its effects – the measurement outcomes – even if the explanation of this correlation does not require screening off as in Reichenbach's PCC; where a common cause screens off the correlation between its effects if the probability of one effect given the common cause is independent of the other effect (for more details, see Sect. 8.9). So, curiously, in some indeterministic RCIQM the common cause in the EPR/B experiment screens off the correlation between its effects, yet does not explain this correlation. The above difficulties are due to the fact that in causal loops events are much more constrained than in

[17]To simplify things, we ignore the question of backward influences on the QM preparation of the particle pair.

[18]For a review of interpretations of probabilities, see Hájek (2002/2012).

[19]In causal loops with only one indeterministic causal connection, the probabilities that causes give their effects determine the long-run frequencies of effects in the reference classes of their causes, but not the long-run frequencies of the causes or of the effects (Berkovitz 2008, 2011).

linear causation and that Loop 1 and Loop 2 have more than one indeterministic causal connection (Berkovitz 2008, 2011).

The question then arises: Do deterministic RCIQM fare better? In Sects. 8.8, 8.9 and 8.10, we shall consider the prospects of a deterministic local RCIQM, the causally symmetric Bohmian model. In Sects. 8.7 and 8.8 we introduce Bohmian mechanics and the causally symmetric Bohmian model, respectively.

8.7 Bohmian Mechanics

Unlike in standard QM, in Bohmian mechanics (BM)[20] wavefunctions of systems do not represent their states. Wavefunctions are states of a 'guiding field' that guides the trajectories of systems. Wavefunctions always evolve according to Schrödinger's equation and thus never 'collapse'. Particles always have definite positions, the so-called 'hidden variables'. The theory is deterministic. The positions of particles and their QM wavefunction at any given time jointly determine the particles' trajectories at all future times. Thus, the positions of systems and their QM wavefunction determine the outcomes of any measurements so long as these outcomes are recorded in the positions of some physical systems, as in any practical measurement. Wavefunctions govern the trajectories of systems according to the 'guidance equation', which expresses the velocity of a system at a time t, $\mathbf{v}(\mathbf{x}, t)$, in terms of its wavefunction at that time:

$$\text{(Guidance)} \quad \mathbf{v}(\mathbf{x}, t) = \frac{\hbar}{2im} \frac{\psi * \overleftrightarrow{\nabla} \psi}{\psi * \psi};$$

where m is the system's mass, $\mathbf{x}$ is the system's position configuration, $\hbar$ is Planck's constant, ψ is the system's QM wavefunction and $\psi *$ is its complex conjugate, $\overleftrightarrow{\nabla}$ stands for $\overrightarrow{\nabla} - \overleftarrow{\nabla}$, and the grad operators $\overrightarrow{\nabla}$ and $\overleftarrow{\nabla}$ act to the right and to the left, respectively.

Since BM is a deterministic theory, its predictions in individual cases are different from those of standard QM. Yet, the theory reproduces the statistical predictions of standard QM by postulating that the distribution of any possible position configuration $\mathbf{x}$ at time$t$, $\rho(\mathbf{x}, t)$, is given by the wavefunction at that time:

$$\text{(Distribution)} \quad \rho(\mathbf{x}, t) = \psi * \psi.$$

That is, the QM statistical predictions for measurement outcomes are obtained as weighted averages over all the possible position configurations (according to the weights prescribed by *Distribution*).

---

[20] See, for example, Bohm (1952), Bell (1987, Chaps. 4, 14, 15 and 17), Bohm and Hiley (1993) and Goldstein (2001/2013).

BM portrays the quantum realm as involving non-locality (Bell 1987, Chaps. 4, 14, 15 and 17, Goldstein 2001/2013, Berkovitz 2007/2016, Sects. 4, 5.3.1 and 7). For *Guidance* entails that in many-particle system in a non-separable wavefunction the velocity of a particle will typically depend upon the positions of the other, possibly distant, particles. Thus, in the EPR/B experiment the outcome of the first, say nearby, measurement influences the outcome of the distant measurement. This non-local influence accounts for the curious correlation between the distant measurement outcomes.

8.8 The Causally Symmetric Bohmian Model

The causally symmetric Bohmian model (CSBM) was developed by Rod Sutherland (2008). In this deterministic RCIQM, the velocity of a system depends on two wavefunctions, ψ_i and ψ_f, which are supposed to record or reflect the system's initial and final boundaries conditions, respectively. Sutherland does not specify the nature of these conditions and thus the exact ontological nature of the initial and final wavefunctions is not completely clear. The initial wavefunction of a system, ψ_i, is its QM wavefuction. ψ_i may be nonseparable, as the wavefunction of the particles in the EPR/B experiment. Yet, as we shall see below, in CSBM particles may also have separable initial wavefunctions in non-separable QM wavefunctions. The final wavefunction of a system, ψ_f, is independent of its initial wavefunction ψ_i and should not be confused with the time-evolved initial wavefunction to a later time. Both of these wavefunctions evolve according to Schrödinger's equation (i.e. they are solutions of the time-dependent Schrödinger equation). ψ_i evolves from past to future, reflecting a forward causal process, whereas ψ_f evolves from future to past, reflecting a backward causal process. The fact that both processes are governed by the same dynamical law implies, for example, that the interaction of a final wavefunction with a measurement apparatus and its evolution from the time after the measurement to the time before it, is similar to the interaction of the same initial wavefunction with the measurement apparatus and its evolution from the time before the measurement to the time after it. This means that before interacting with a measurement apparatus, a final wavefunction need not correspond to an outcome of the measurement carried out by this apparatus.

Sutherland proposes that *Guidance* and *Distribution* could be reformulated along the following lines, so as to yield CSBM:

$$\text{(Guidance}^{\text{S}}\text{)} \quad \mathbf{v}(\mathbf{x},t) = \frac{\mathrm{Re}\left[\left(\frac{\hbar}{2ima}\right)\psi_f^* \overleftrightarrow{\nabla} \psi_i\right]}{\mathrm{Re}\left[\left(\frac{1}{a}\right)\psi_f^* \psi_i\right]}$$

$$\text{(Distribution}^{\text{S}}) \quad \rho(\mathbf{x},t) = \text{Re}\left(\frac{1}{a}\psi_f^*\psi_i\right);$$

where Re(z) gives the real part of z, and a is a normalization factor:

$$\text{(Normalization)} \quad a \equiv \int_{-\infty}^{\infty} \psi_f^*(\mathbf{x},t)\psi_i(\mathbf{x},t)d^3\mathbf{x}.$$

That is, Sutherland proposes to substitute ψ_i for ψ and ψ_f^* for ψ^* in *Guidance* and *Distribution*. Given this substitution, the equations are no longer real and the total probability need not sum to one. Sutherland addresses this complication by taking the real part of the equations and introducing a normalization factor a.

While Bohmian mechanics (BM) reproduces the correlations between the distant measurement outcomes in the EPR/B experiment by postulating non-local influences between them, CSBM accounts for these correlations locally by a common-cause: the complete pair-state at the emission from the source – an event that occurs at the intersection of the backward light cones of the measurement outcomes. The complete pair-state is constituted by the positions of the particles and their initial and final wavefunctions, which, as we shall see below, are separable. The final wavefunctions of the particles between the spin measurements and the pair's emission from the source (i.e. after the interaction with the measurement apparatuses) are eigenvectors of spin quantities that correspond to the measurements (see *Predictions* below). The overlap of the initial wavefunction of the particle pair – the singlet state for spin – and the final wavefunction of the L-particle between the L-measurement and the emission from the source yields the separable initial wavefunction of the R-particle at the emission (see Fig. 8.8); and similarly, *mutatis mutandis*, for the production of the initial wavefunction of the L-particle. Formally, the separable initial wavefunctions of the particles between the pair's emission from the source and the measurements are obtained as follows:

$$\text{(SIW)} \quad \psi_i(x_1) = \frac{1}{N_1}\int_{-\infty}^{\infty} \psi_f^*(x_2)\psi_i(x_1,x_2)d^3x_2$$
$$\psi_i(x_2) = \frac{1}{N_2}\int_{-\infty}^{\infty} \psi_f^*(x_1)\psi_i(x_1,x_2)d^3x_1;$$

where $\psi_i(x_1,x_2)$ is the QM (non-separable) wavefunction of the particle pair, $\psi_i(x_1)$ ($\psi_i(x_2)$) and $\psi_f(x_1)$ ($\psi_f(x_2)$) are the initial and final wavefunctions of the L- (R-) particle, respectively, and N_j are normalization factors ensuring that the total probabilities are equal to 1.[21] Since the initial and final wavefunctions of both particles between the emission and the measurements are separable, the

[21]That is, the values of the N_j are such that $\int_{-\infty}^{+\infty} \psi_i^*(x_j)\psi_i(x_j)d^3x_j = 1$.

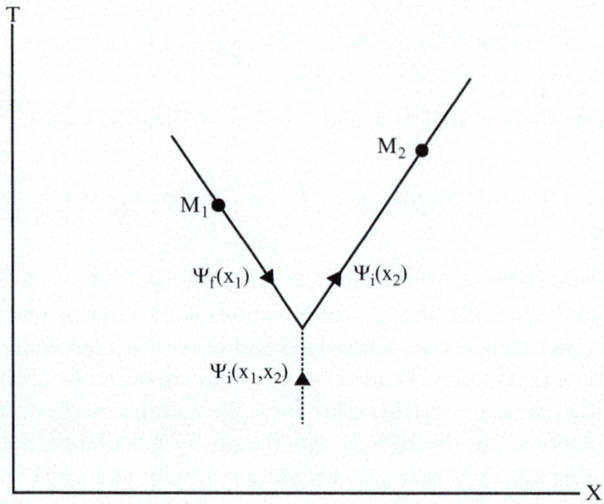

Fig. 8.8 Space-time causal diagram of the formation of the separable initial wavefunction of the R-particle in the CSBM model of the EPR/B experiment. The overlap of the initial non-separable wavefunction of the particle pair, $\psi_i(x_1, x_2)$, and the final wavefunction of the L-particle, $\psi_f(x_1)$, yields the separable initial wavefunction of the R-particle, $\psi_i(x_2)$. The initial separable wavefunction of the L-particle is formed in a similar way. M_1 and M_2 are the L- and the R-measurement, respectively, and x_1 and x_2 are the positions of the L- and the R-particle, respectfully

measurement outcomes are determined in a perfectly local way. Thus, CSBM provides a local, deterministic common-cause model of EPR/B experiment. The common cause – the complete pair-state at the emission, which is constituted by the particles' separable initial and final wavefunctions and their positions – screens off the correlation between the outcomes: the probability of the L-outcome given the complete pair-state (and the apparatus settings) is independent of the R-outcome; and similarly, *mutatis mutandis*, for the probability of the R-outcome. Accordingly, the joint probability of the outcomes given the complete pair-state (and the apparatus settings) factorizes into the probability of the L-outcome given the complete pair-state (and the apparatus settings) and the probability of the R-outcome given this state (and the apparatus settings). In fact, the CSBM model of the EPR/B experiment satisfies a stronger locality condition. The joint probability of the measurement outcomes given the complete pair-state (and the apparatus settings) factorizes into the product of the probability of the L-outcome given the complete state of the L-particle (and the setting of the L-apparatus) and the probability of the R-outcome given the compete state of the R-particle (and the setting of the R-apparatus).

In order for CSBM to be genuinely local, it is necessary to interpret non-separable QM wavefunctions, such as the singlet state, as epistemic states, i.e. states of knowledge about physical systems or the fields that guide them; for any interpretation of the QM wavefunctions of systems as their ontological states or states of fields that guide them would entail non-locality. As we shall see below, in CSBM QM wavefunctions of systems could be interpreted as states of information about the distribution of final wavefunctions and the distribution of separable initial wavefunctions. In particular, the QM wavefunction of the particle pair in the EPR/B experiment can be interpreted as a state that provides information about the probability distribution of the separable initial and final wavefunctions of the particles.

Unlike orthodox QM, measurements play no role in the dynamical laws of CSBM. The backward causal influences, which are realized by or reflected in final wavefunctions, are not triggered by measurements. Rather, these influences are part of the basic ontology of CSBM. Unlike more 'conventional' hidden-variables models of the EPR/B experiment, in the CSBM model of this experiment the QM wavefunction of the particle pair, i.e. the incomplete pair-state, does not determine the probability distribution of complete pair-states.[22] In this model, the distribution of complete pair-states at the emission from the source depends on the distribution of the final wavefunctions of the particles. Thus, in order to reproduce the statistical predictions of standard QM, CSBM has to be supplemented with a postulate that determines the distribution of the final wavefunctions in any given QM wavefunction and experimental set up. Sutherland proposes the following postulate:

> *Predictions*. Let ψ_i be the QM wavefunction of a (possibly multi-particle) system at some time t and ψ_f be the final wavefunction of the system at the same time. If ψ_f corresponds to one of the possible outcomes of a subsequent measurement on the system (like in the case of the particles' final wavefunctions between the emission from the source and the measurements in the EPR/B experiment), the conditional probability distribution of ψ_f given ψ_i, $\rho(\psi_f, \psi_i)$, is the following

$$\text{(Predictions)} \quad \rho(\psi_f, \psi_i) = |a|^2;$$

where the amplitude a is as defined in *Normalization*.

In the EPR/B experiment, the system is the particle pair and the pair's initial QM wavefunction, ψ_i, is the singlet state. Granted *Predictions*, the QM wavefunction determines the distribution of the separable final wavefunctions of the particles between the measurements and the pair's emission from the source, $\psi_f(x_1)$ and $\psi_f(x_2)$ (see Fig. 8.8); and granted *SIW*, the distribution of the final wavefunctions determines the distribution of the separable initial wavefunctions of the particles at the emission from the source, $\psi_i(x_1)$ and $\psi_i(x_2)$. Thus, granted *Predictions* and *SIW*, we obtain the distribution of complete pair-states in a given QM wavefunction of the particle pair. Now, although *Predictions* appeals to measurements, the dynamical laws of CSBM do not depend on measurements. *Predictions* could be thought of as an epistemic postulate: it provides information about the distribution of final

[22]The same is true for other local retro-causal interpretations of QM.

wavefunctions of systems, and accordingly information about the distribution of complete pair-states, in a given QM wavefunction.

While in BM there is only one probabilistic postulate – *Distribution* – in CSBM there are two independent probabilistic postulates – *Distributions* and *Predictions*. It is noteworthy that *Predictions* is sufficient for reproducing the statistical predictions of standard QM. Further, the information about the position configuration of the particle pair is unnecessary for screening off the correlation between the measurement outcomes in the EPR/B experiment: the initial and final wavefunctions of the particles constitute a local common cause that screens off the correlation between these outcomes. Yet, $Distribution^S$ is important for the ontology of CSBM and its solution to the measurement problem.

Like other RCIQM, CSBM predicts the existence of a causal loop in the contingent EPR/B experiment (see Figs. 8.5 and 8.6). This loop, henceforth Loop 3 (see Fig. 8.9), is different from Loop 1 and Loop 2 in two important respects. First, in Loop 3 all the causal connections are deterministic. Second, unlike Loop 2, in Loop 3 the measurement outcomes do not influence the complete pair-state at the source (see Fig. 8.10). Yet, due to *Predictions*, the measurement apparatuses send signals, carried by the final wavefunctions, that *correspond* to the measurement outcomes. That is, the final wavefunctions of the particles that emerge from the interactions with the measurement apparatuses and evolve back to the source are eigenfunctions of spins that correspond to the measurement outcomes. Thus, unlike RCIQM in which the measurement outcomes influence the complete pair-state, CSBM does not predict causal loops in 'conventional' EPR/B experiments. In Loop 3, the complete pair-state at the source and the setting of the (fixed) R-measurement apparatus jointly determine the R-outcome, the R-outcome determines the setting of the L-measurement apparatus, and the final wavefunctions from the L- and the R-apparatus and the QM preparation of the particle pair (and the position configuration of the particles) jointly determine the complete pair-state at the source.[23]

Unlike Loop 1 and Loop 2, in Loop 3 it is easy to figure out the frequencies of outcomes in the reference class of a given complete pair-state. And, given *Predictions and SIW*, it is also possible to predict the frequencies of the measurement outcomes in a given QM wavefunction. Thus, CSBM overcomes the predictive challenges that the indeterministic RCIQM we discussed in Sect. 8.6 encounter.[24] In Sect. 8.10, we shall discuss the question whether CSBM also overcomes the explanatory challenges that indeterministic RCIQM encounter. To prepare the ground for this discussion we need first to consider Reichenbach's principle of the common cause in the context of CSBM.

---

[23]The QM preparation of the particle pair and the particles' final wavefunctions jointly determine the measurement outcomes, independently of the specific position configuration of the particles. Thus, the position configuration does not play any explicit role in the analysis of Loop 3.

[24]The question whether indeterministic RCIQM in which the complete pair-state is determined by events/states that correspond to the measurement outcomes (but are not the outcomes themselves) could overcome these challenges, is beyond the scope this chapter.

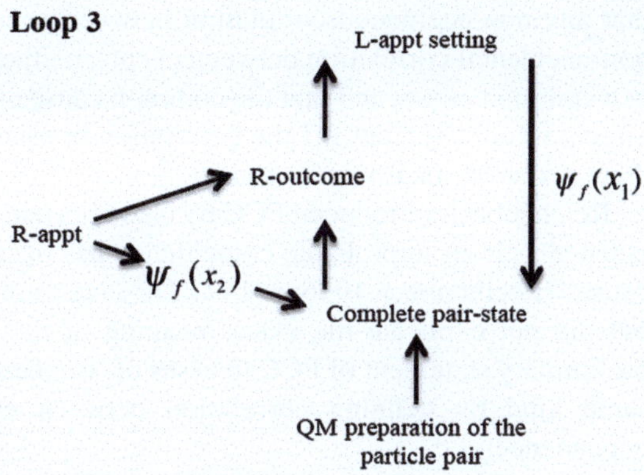

Fig. 8.9 The causal loop that CSBM predicts in the contingent EPR/B experiment (see Figs. 8.5 and 8.6). *Solid arrows* denote deterministic causal connections

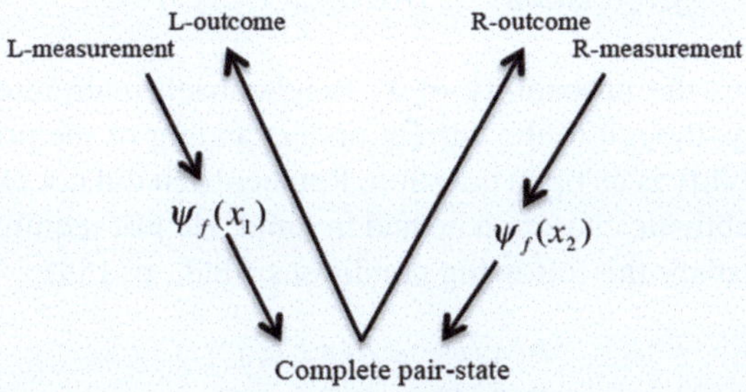

Fig. 8.10 The common-cause scenario in the EPR/B experiment according to CSBM. The common cause – the complete pair-state at the emission – is influenced by the particles' final wavefunctions that obtain from the interaction between the particles and the measurement apparatuses. Notably, the measurement outcomes do not influence the complete pair-state

8.9 On Reichenbach's Principle of the Common Cause

Recall that Reichenbach proposed that the direction of time could be reduced to the fork asymmetry in macroscopic phenomena – namely, the fact that in such phenomena all the open v-shaped causal forks have the same direction (see Fig. 8.1a). The kind of open causal forks that Reichenbach had in mind is a macroscopic situation in which: (i) there is a non-accidental correlation between two macroscopic events, *A* and *B*, neither of which causes the other; (ii) *A* and *B* are effects of a common cause, a macroscopic event *C*; (iii) *C* is probabilistic cause of *A* and *B*; and (iv) *C* screens off the correlation between *A* and *B*, and accordingly the probability of *A* and *B* given *C* factorizes: it is equal to the product of the probability of *A* given *C* and the probability of *B* given *C*. (i)-(iv) characterize Reichenbach's principle of the common cause (PCC).[25] The underlying ideas of this principle are

[25]Berkovitz (2000, 2002b) argues that PCC is not really a principle but rater a principle schema: for different specifications of the terms that appear in this schema – 'events', 'correlation', 'causation' and 'common cause' – we obtain different principles.

that any non-accidental correlation between events has a causal explanation, that non-accidental correlation between events, neither of which causes the other, is due to a common cause, and that the common cause explains the correlation between its effects when it screens off the correlation between them and, accordingly, renders them probabilistically independent.[26]

Reichenbach introduced PCC by characterizing correlation as 'improbable coincidence': "If an improbable coincidence has occurred, there must exist a common cause" (Recihenbach 1956, p. 157). He gave examples of 'improbable coincidence' but did not explicate the exact meaning of this term. He generalized the above preliminary statement of PCC to cases of 'frequent improbable coincidences' of the same kind by defining correlation between events as (positive) probabilistic dependence:

$$\text{(Correlation)} \quad P(A\&B) > P(A) \cdot P(B),$$

where $P(X)$ denotes the probability of X. Reichenbach interpreted probability as long-run frequency, though in the current understanding of the principle the interpretation of probability is an open question. Reichenbach did not explicate the exact nature of the probabilistic causation he had in mind. He just required that the causal fork of Fig. 8.1a satisfy the following conditions (ibid., p. 159):

$$\begin{array}{ll} \text{(ProbDep)} & P(A/C) > P(A/\sim C) \\ & P(B/C) > P(B/\sim C) \\ \text{(Factorizability)} & P(A\&B/C) = P(A/C) \cdot P(B/C) \\ & P(A\&B/\sim C) = P(A/\sim C) \cdot P(B/\sim C), \end{array}$$

where $P(X/Y)$ denotes the probability of X given Y, and *Factorizability* follows from *Screening Off*.

$$\begin{array}{ll} \text{(Screening Off)} & P(A/B\&C) = P(A/C) \\ & P(A/B\&\sim C) = P(A/\sim C) \end{array}$$

Here, probabilistic causation does not mean indeterministic causation: it is neutral between determinism and indeterminism. *Correlation*, *ProbDep* and *Factorizbility* define a 'conjunctive fork', "that is, a fork which makes the conjunction of two events A and B more frequent than it would be for independent events" (ibid.).

While Reichenbach intended PCC to apply to macroscopic phenomena, the principle has also been applied to microscopic phenomena. Indeed, Bell's (1987) arguments for quantum non-locality are based on the application of PCC to

[26]In fact, Reichenbach (1956, p. 158) talked about simultaneous events, but on the common understanding of PCC the principle is not limited to such events. For discussions of PCC, see for example Hitchcock (1997/2012, Sect. 2.3), Arntzenius (1999/2010), Sober (2001) and Berkovitz (2000, 2002b).

v-shaped forks open to the future that include both macroscopic and microscopic events or states – namely, forks with a microscopic common cause and macroscopic effects. Bell also generalized the principle to circumstances in which *Factorizability* is applied to a set of mutually exclusive common causes rather than a common cause and its absence.[27] Further, there is nothing in PCC *per se* to exclude its application to causal structures that involve backward causation.

Indeed, there have been various objections to PCC. Here are some notable examples. Fine (1981, 1986, 1989) denies that non-accidental correlations must have causal explanation. Sober (1988, 2001) argues that some non-accidental correlations between events, neither of which causes the other, have no common cause. Van Fraassen (1980), Cartwright (1989) and Chang and Cartwright (1993) argue that in indeterministic v-shaped common-cause forks, the common cause might fail to screen off the non-accidental correlation between its effects. And Sober and Eells (1986) argue that 'indirect' probabilistic common causes might fail to screen off the correlations between their effects.[28]

The objections above aim to challenge the idea that all non-accidental correlations between events, neither of which causes the other, share a common cause, or the idea that the common cause of such events screens off the correlation between them. Thus, these objections are irrelevant for the question of whether PCC obtains in CSBM; for in this interpretation of QM, the measurement outcomes in the EPR/B experiment share a common cause – the complete pair-state at the emission from the source – and this cause does screen off the correlation between the outcomes.

Assuming that PCC applies to both macroscopic and microscopic phenomena, and referring to Bell's generalized principle, the question arises: does this principle obtain in CSBM?

8.10 On Causal Explanation of Correlations in the Causally Symmetric Bohmian Model

The causally symmetric Bohmian model (CSBM) conforms to the letter of PCC but runs counter to its spirit. While the common cause of the measurement outcomes in the EPR/B experiment – the complete pair-state at the emission – screens off the correlation between them, it fails to explain this correlation! The correlation between the measurement outcomes is due to the correlation between the final wavefunctions of the particles. But the correlation between the final wavefunctions, which is postulated by *Predictions*, has no causal explanation: the separable final

[27]That is, in the local 'hidden-variables' models of the EPR/B experiment that Bell considered there are generally many different possible complete pair-states (which are compatible with the QM wavefunction of the particle pair), and Bell's generalized principle applies to all of them.

[28]Sober and Eells give as an example a case where an event C is a probabilistic cause of an event D which is a common cause of events A and B.

wavefunctions of the particles are neither linked by (a chain of) 'direct' causal connections, nor share a common cause. Indeed, in CSBM this correlation is a curious brute fact. Thus, while CSBM salvages locality and accordingly has better prospects of reconciling QM with relativity theory, it sacrifices an important motivation for RCIQM and more generally for local hidden-variables interpretations of QM: namely, to provide a local *causal explanation* of the curious quantum correlations between the properties of distant systems.

It may be argued that the common-cause scenario depicted by CSBM is different from the v-shaped common-cause fork open to the future that Reichenbach considered (see Fig. 8.1a). That is, it may be argued that, unlike the fork Reichenbach considered, in CSBM the joint effects influence each other through the common cause: the final wavefunction of the L-particle (between the L-measurement and the pair's emission from the source), which corresponds to the L-measurement outcome, influences the R-measurement outcome; and the final wavefunction of the R-particle (between the R-measurement and the pair's emission from the source), which corresponds to the R-measurement outcome, influences the L-measurement outcome (see Figs. 8.8 and 8.10). It is noteworthy, however, that while these final wavefunctions correspond to the measurement outcomes, they are neither the measurement outcomes, nor caused by them. Thus, in the CSBM model of the EPR/B experiment the effects of the common cause do not influence each other. In response, it may be argued that, while the effects of the common cause – the measurement outcomes – do not influence each other, the setting of the L-measurement influences the R-outcome through the final wavefunction of the L-particle, and the setting of the R-measurement influences the L-outcome through the final wavefunction of the R-particle. It is noteworthy, though, that the influence of the final wavefunction of each of the particles on the distant measurement outcome is irrelevant for the question of why the common cause fails to explain the correlation between its effects. First, these influences do not change the fact that the complete pair-state at the emission and the measurement outcomes constitute a v-shaped causal fork open toward the future (see Fig. 8.10); for the influence of the final wavefunction of each of the particles on the distant measurement outcome is mediated by the complete pair-state at the emission from the source (see Fig. 8.11). Second, Reichenbach (1956, Sect. 22) maintains that an intermediate cause between a cause and its effect should screen off the correlation between the cause and the effect. Accordingly, any influence that the final wavefunction of the L-particle (R-particle) has on the R-measurement outcome (L-measurement outcome) should be screened off by the complete pair-state at the emission from the source. And, indeed, the influence of each of the final wavefunctions on the distant outcome is screened off by this state. Thus, if *Screening Off*, and accordingly *Factorizability,* were a sufficient condition for a common cause explaining the correlation between its effects in a v-shaped causal fork, the complete pair-state at the emission would explain the correlation between the distant measurement outcomes in the CSBM model of the EPR/B experiment. In this model the failure of the complete pair-state to explain the correlation between the measurement outcomes is not due to the causal influences of the final wavefunctions on the

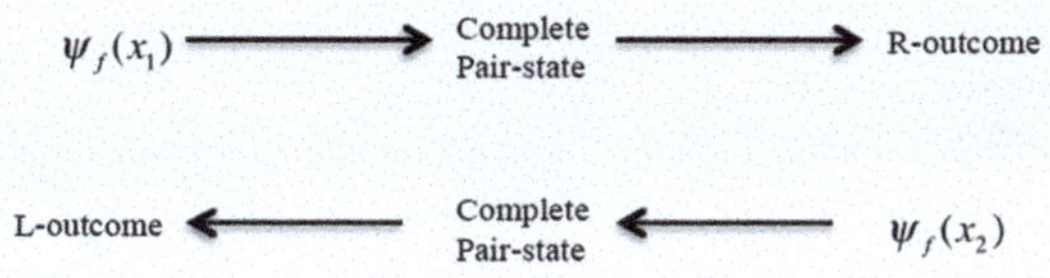

Fig. 8.11 The causal connections between the final wavefunctions and the measurement outcomes in the causally symmetric Bohmian model of the EPR/B experiment (see Figs. 8.8 and 8.10)

measurement outcomes, but rather due to the fact that the complete pair-state is not the source of the correlation between its effects. The source of the correlation between the measurement outcomes is the correlation between the final wavefunctions of the particles, and this latter correlation has no causal explanation.

Reichenbach did not take *Factorizability per se* to be a sufficient condition for a causal explanation of correlation. He analyzed two kinds of v-shaped causal forks. The first kind, which we discussed above, is open to the future and constituted by a common cause and its joint effects (see Fig. 8.1a). The second kind is open to the past and constituted by two causes and their joint effect (see Fig. 8.1b). As Reichenbach notes, a common effect *E* in the second kind of v-shaped fork may satisfy the probabilistic relations *ProbDep* and *Factorizability* with '*E*' substituted for '*C*'. But, obviously, even if *E* screened off a correlation between its causes *A* and *B*, it would not explain them. Reichenbach thought that the second law of thermodynamics excludes any non-accidental correlation that is not due to a common cause. Any non-accidental correlation between the causes *A* and *B* of a common effect *E* will be due to a common cause *C*. That is, *C*, *A*, *B* and *E* will form a double v-shaped causal fork, constituted by a common cause and a common effect (see Fig. 8.12). It is noteworthy, however, that at the microscopic realm the second law excludes neither the second kind of v-shaped fork, nor a third kind of v-shaped fork that is open to the future but is constituted by two causally independent causes and their past common effect (see Fig. 8.13). Further, the exclusion of v-shaped forks of the second and the third kinds does not imply that *Factorizability* shouldn't be a sufficient condition for a common cause explaining the correlation between its effects in forks of the first kind (see Fig. 8.1a). In fact, a common view has it that *Factorizability* in such forks is a sufficient condition for a common cause explaining the correlation between its joint effects. Accordingly, CSBM poses a new challenge for the idea that *Factorizability* is the key for common-cause explanations of correlations: *Factorizability* cannot even be a sufficient condition for a common cause explaining the correlation between its joint effects.

Finally, it is noteworthy that the causally unexplained correlations that CSBM postulate are not like the apparent inexplicable correlations that prevent the so-called causal paradoxes (see Sect. 8.4); for the latter correlations do have causal explanations (Dowe 2003). Consider, for example, the following version of the 'grandfather paradox', where Tim embarks on a time travel in order to kill his grandfather at a time before his mother was conceived. Tim waits in front of his

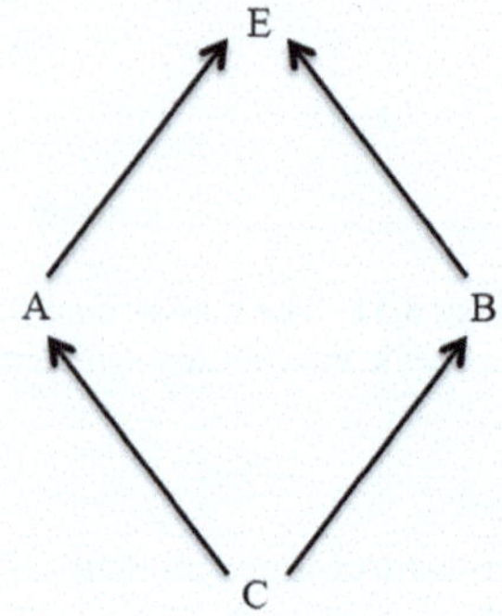

Fig. 8.12 Double fork constituted by a common cause and a common effect

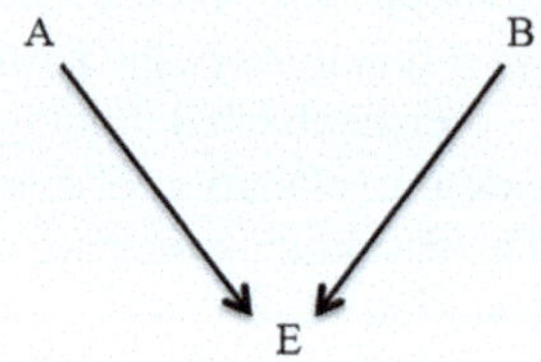

Fig. 8.13 A v-shaped fork open toward the future, constituted by two causally independent causes and their common past effect

grandfather's house. When grandpa leaves the house, Tim shoots. However, slipping on a banana skins, he misses and grandpa survives. Let *A* be Tim's attempt to kill his grandfather by shooting him, *B* be the fortuitous existence of a banana skin at the crime scene, *F* be Tim's shooting misses grandfather, *S* be grandfather's survival, and *T* be Tim's embarking on a time travel with the intent to kill his grandfather. *A* and *B* jointly cause *F*, *F* causes *S*, *S* causes *T* and *T* causes *A* (see Fig. 8.14); where by a 'cause' here we mean an event or state that in the actual circumstances causes its effect. For a detailed analysis of this causal loop, see Dowe (2003).

Multiple time travels to the past with the intent to kill grandfather seem bound to fail due to 'mundane' reasons like the existence of banana skins at the crime scene. Horwich (1987) argues that the coincidences between events like Tim's attempt to kill his grandfather and the existence of conditions that undermine his attempt are inexplicable. But, as it is not difficult to see from Fig. 8.14, Tim's attempt to kill his grandfather is an indirect effect of the existence of a banana skin at the crime scene, and accordingly the correlation between these 'events' has a causal explanation.

8.11 On Time, Causal Symmetry and Explanation in the Causally Symmetric Bohmian Model

In the causally symmetric Bohmian model (CSBM), the causal asymmetry at the macroscopic realm is compatible with the causal symmetry at the microscopic realm. Thus, this retro-causal interpretation of QM does not exclude theories of time in which the direction of time is determined by the direction of macroscopic

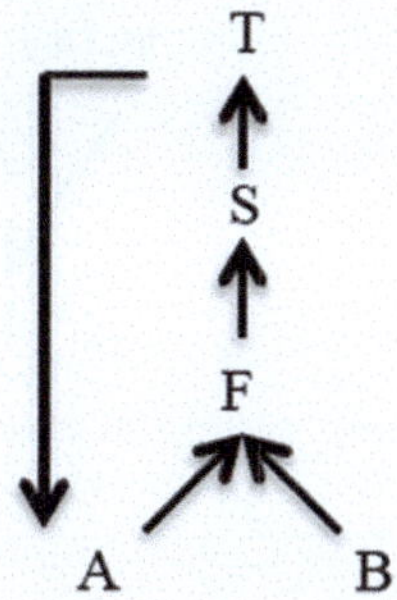

Fig. 8.14 The causal loop in the above version of the grandfather paradox. *Arrows* denote causal connections. The causal connection from *T* to *A* is backward in time

causal processes. In particular, it does not exclude Reichenbach's (1956) theory of the direction of time. Recall that, in this theory, the direction of time is determined by the direction of the open v-shaped macroscopic causal forks. Reichenbach argued that all such forks have the same direction and constitution – they are open to the future and are constituted by a common cause and its joint effects (see Fig. 8.1a) – and that this asymmetry supervenes on the asymmetry of thermodynamic processes.

Those who subscribe to Reichenbach's ideas about causal explanation in macroscopic phenomena expect that any correlation between macroscopic events, neither of which causes the other, will originate from a common cause. Further, they expect that the common cause will be a macroscopic event in the past of the correlated events, so as to create a v-shaped causal fork open to the future (see Fig. 8.1a), and that it will screen off the correlation between its effects. The expectation that in such causal scenarios the common cause be the origin of the correlation, will also be natural for those who maintain that common causes might fail to screen off the correlations between their effects, i.e. even for those who reject Reichenbach's view about how correlations are explained by common causes. Yet, as we have seen in Sects. 8.8 and 8.10, CSBM violates this expectation. Indeed, in the CSBM model of the EPR/B experiment the measurement outcomes have a common cause – the complete state of the particle pair at the emission from the source – that screens off the correlation between them. But, the common cause of these macroscopic events is not a macroscopic event. And while the common cause screens off the correlation between its effects, it does not explain it. The correlation between the measurement outcomes originates from the causally unexplained correlation between the particles' final wavefunctions, which propagate backward in time. Thus, CSBM poses a new challenge for the idea that screening off is the key for common-cause explanations of correlations: screening off cannot even be a sufficient condition for a common cause explaining the correlation between its joint effects.

It may be argued that the CSBM model the EPR/B experiment does not really violate PCC since this principle was intended to apply to macroscopic events. However, reflecting on local hidden-variables models of the EPR/B experiment, it seems more reasonable to expect *Screening Off*, and accordingly *Factorizability*, in the case of v-shaped forks open to the future with a microscopic common cause. It

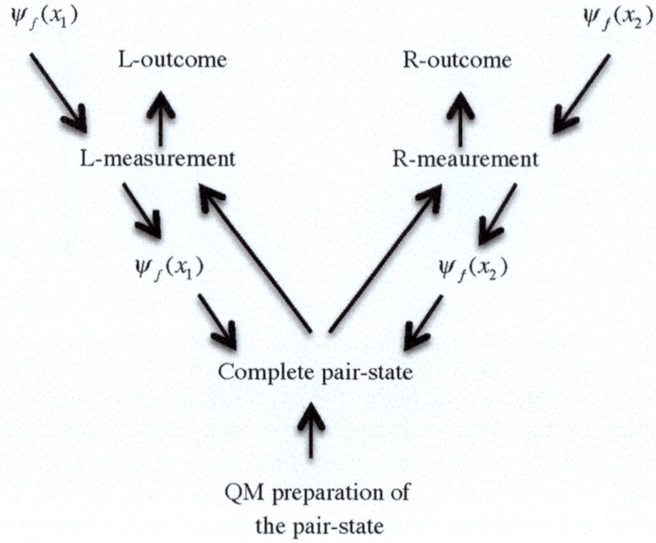

Fig. 8.15a The common-cause scenario in the CSBM model of the EPR/B experiment

may also be argued that backward causation does not exist at the macroscopic realm, and accordingly if PCC were limited to this realm, *Factorizability* could be a sufficient condition for a common cause explaining the correlation between its joint effects. Indeed, backward causation does not seem to exist in the macroscopic phenomena we are familiar with, but it is premature to conclude that it does not exist at all at the macroscopic realm. In any case, while the above violation of PCC is involved with backward causation, a similar violation could also occur in v-shaped causal forks open to the future that do not involve backward causation. The v-shaped causal fork that the CSBM model of the EPR/B experiment predicts is constituted by two macroscopic effects (the measurement outcomes) and a microscopic common-cause (the complete pair-state at the emission from the source). The microscopic common cause is determined by three partial causes (see Fig. 8.15a): a macroscopic event (the QM preparation of the singlet state), which is a forward cause, and two microscopic events/states (the particles' final wavefunctions between the measurements and the emission from the source), which are backward causes. The violation of PCC is due to the fact that the correlation between these miscroscopic partial causes has no causal explanation. But a similar violation could occur if these causes were forward rather than backward (see Fig. 8.15b). Yet, the question is whether the causal scenario of Fig. 8.15b could be theoretically motivated.

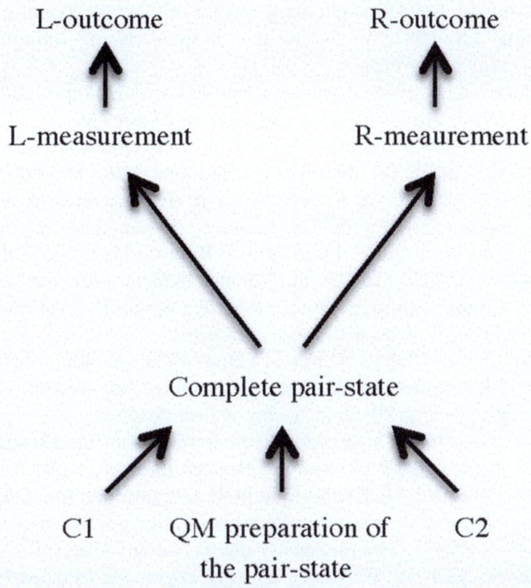

Fig. 8.15b A common-cause model of the EPR/B experiment in which C1 and C2 are correlated but causally independent of each other

Acknowledgment For comments on earlier versions of this chapter, I am very grateful to the audience at the Bordeaux conference "The Time of Nature, the Nature of Time", the editors Christophe Bouton and Philippe Huneman, and Aaron Kenna and Noah Stemeroff. The research for this paper was supported by a SSHRC Insight Grant.

List of Acronyms

BM:	Bohmian mechanics
CSBM:	causally symmetric Bohmian model
EPR/B:	Einstein-Podolsky-Rosen/Bohm
PCC:	Reichenbach's principle of the common cause
RCIQM:	retro-causal interpretations of quantum mechanics
QM:	quantum mechanics

References

Aharonov, Y. and E. Gruss 2005. *Two-time interpretation of quantum mechanics*. http://arxiv.org/pdf/quant-ph/0507269.

Aharonov, Y., and J. Tollaksen. 2007. New insights on time-symmetry in quantum mechanics. In *Visions of discovery: New light in physics, cosmology and consciousness*, ed. R.Y. Chiao, 105–151. Cambridge: Cambridge University Press. http://arxiv.org/pdf/0706.1232.pdf.

Albert, D.Z. 2000. *Time and chance*. Cambridge, MA: Harvard University Press.

Arntzenius, F. 1999/2010. Reichenbach's Common Cause Principle. In *The Stanford Encyclopedia of Philosophy* (Fall 2010 Edition), ed. Edward N. Zalta, URL = http://plato.stanford.edu/archives/fall2010/entries/physics-Rpcc/.

Aronson, J.L. 1971. On the grammar of 'Cause'. *Synthese* 22 (3–4): 414–430.
Bell, J.S. 1987. *Speakable and unspeakable in quantum mechanics*. Cambridge: Cambridge University Press.
Berkovitz, J. 1998. Aspects of Non-Locality II: Superluminal Causation and Relativity. *Studies in History and Philosophy of Modern Physics* 29 (2): 183–222.
———. 2000. The many principles of the common cause. *Reports on Philosophy* 20: 51–83.
———. 2001. On chance in causal loops. *Mind* 110 (437): 1–23.
———. 2002a. On Causal Loops in the Quantum Realm. In *Non-locality and Modality, Proceedings of the NATO Advanced Research Workshop on Modality, Probability and Bell's Theorems*, ed. T. Placek and J. Butterfield, 235–257. Dordrecht: Kluwer.
———. 2002b. On Causal Inference in Determinism and Indeterminism. In *Between Chance and Choice: Interdisciplinary Perspectives on Determinism*, ed. H. Atmanspacher and R. Bishop, 237–278. Exeter: Imprint Academic.
———. 2007/2016. Action at a Distance in Quantum Mechanics. In *The Stanford Encyclopedia of Philosophy* (Spring 2016 Edition), ed. Edward N. Zalta, URL = http://plato.stanford.edu/archives/spr2016/entries/qm-action-distance/.
———. 2008. On predictions in retro-causal interpretations of quantum mechanics. *Studies in History and Philosophy of Modern Physics* 39 (4): 709–735.
———. 2011. On Explanation in Retro-causal Interpretations of Quantum Mechanics. In *Probabilities, causes, and propensities in physics*, ed. M. Suárez, 115–155. Berlin/New York: Springer.
———. 2015. The propensity interpretation: A reevaluation. *Erkenntnis* 80 (Suppl. 3): 629–711.
Black, M. 1956. Why cannot an effect precede its cause? *Analysis* 16 (3): 49–58.
Bohm, D. 1951. *Quantum theory*. Englewood Cliffs: Prentice-Hall.
———. 1952. A suggested interpretation of the quantum theory in terms of 'hidden' variables, I and II. *Physical Review* 85 (2): 166–193.
Bohm, D., and B.J. Hiley. 1993. *The undivided universe: An ontological interpretation of quantum theory*. London: Routledge.
Butterfield, J.N. 1992. Bell's Theorem: What it takes. *British Journal for the Philosophy of Science* 43 (1): 41–83.
Cartwright, N. 1989. *Nature's capacities and their measurements*. Oxford: Oxford University Press.
Chang, H., and N. Cartwright. 1993. Causality and realism in the EPR experiment. *Erkenntnis* 38 (2): 169–190.
Costa de Beauregard, O. 1953. Une réponse à l'argument dirigé par Einstein, Podolsky et Rosen contre l'interprétation Bohrienne des phénomènes quantiques. *Comptes Rendus de l'Académie des Sciences* 236: 1632–1634.
———. 1977. Time symmetry and the Einstein paradox. *Il Nuovo Cimento* 42B: 41–64.
———. 1979. Time symmetry and the Einstein paradox – II. *Il NuovoCimento* 51B: 267–279.
———. 1985. On some frequent but controversial statements concerning the Einstein-Podolsky-Rosen correlations. *Foundations of Physics* 15 (8): 871–887.
Cover, J.A. 1997. Non-basic time and reductive strategies: Leibniz' theory of time. *Studies in History and Philosophy of Science* 28 (2): 289–318.
Cramer, J. 1980. Generalised absorber theory and the Einstein-Podolsky-Rosen paradox. *Physical Review D* 22: 362–376.
———. 1983. The arrow of electromagnetic time and generalized absorber theory. *Foundations of Physics* 13 (9): 887–902.
———. 1986. The transactional interpretation of quantum mechanics. *Reviews of Modern Physics* 58 (3): 647–687.
———. 1988. An overview of the transactional interpretation of quantum mechanics. *International Journal of Theoretical Physics* 27 (2): 227–236.
Dowe, P. 1992. Wesley Salmon's process theory of causality and the conserved quantity theory. *Philosophy of Science* 59 (2): 195–216.
———. 1995. Causality and conserved quantities: A reply to Salmon. *Philosophy of Science* 62 (2): 321–333.
______ .2003. The coincidences of time travel. *Philosophy of Science* 70 (3): 574–589.

Dummett, M. 1964. Bringing about the Past. *Philosophical Review* 73 (3): 338–359.

Elga, A. 2000. Statistical mechanics and the asymmetry of counterfactual dependence. *Philosophy of Science* 68 (3): S313–S324.

Fair, D. 1979. Causation and the flow of energy. *Erkenntnis* 14 (3): 219–250.

Faye, J. 2001/2015. Backward Causation. In *The Stanford Encyclopedia of Philosophy* (Winter 2015 Edition), ed. E. N. Zalta, URL = http://plato.stanford.edu/archives/win2015/entries/causation-backwards/.

Fine, A. 1981. Correlations and physical locality. In *PSA 1980*, ed. P. Asquith and R. Giere, vol. 2, 535–562. East Lansing: Philosophy of Science Association.

———. 1986. *The shaky game*. Chicago: The University of Chicago Press.

———. 1989. Do correlations need to be explained? In *Philosophical consequences of quantum theories: Reflections on Bell's theorem*, ed. J. Cushing and E. McMullin, 175–194. Notre Dame: University of Notre Dame.

Flew, A. 1954. Can an effect precede its cause? *Proceedings of the Aristotelian Society Supplement* 28: 45–52.

Frisch, M. 2013. Time and causation. In *A companion to the philosophy of time*, ed. H. Dykes and A. Bardon, 282–300. Oxford: Wiley.

Giere, R. 1973a. Review of Mellor's the matter of chance. *Ratio* 15 (1): 149–155.

———. 1973b. Objective single-case probabilities and the foundations of statistics. In *Logic, methodology and philosophy of science*, ed. P. Suppes et al., vol. IV, 467–483. Amsterdam/London: North-Holland Publishing Company.

Goldstein, S. 2001/2013. Bohmian Mechanics. In *The Stanford Encyclopedia of Philosophy* (Spring 2013 Edition), ed. Edward N. Zalta, URL = http://plato.stanford.edu/archives/spr2013/entries/qm-bohm/.

Gruss, E. 2000. A suggestion for a teleological interpretation of quantum mechanics. http://arxiv.org/pdf/quant-ph/0006070v2.pdf.

Hausman, D. 1998. *Causal asymmetries*. Cambridge: Cambridge University Press.

Hájek, A. 2002/2012. Interpretations of probability. In *The Stanford Encyclopedia of Philosophy* (Winter 2012 Edition), ed. Edward N. Zalta, URL=http://plato.stanford.edu/archives/win2012/entries/probabilityinterpret/.

Hitchcock, C. 1997/2012. Probabilistic Causation. In The Stanford Encyclopedia of Philosophy (Winter 2012 Edition), ed. Edward N. Zalta, URL = http://plato.stanford.edu/archives/win2012/entries/causation-probabilistic/.

Horwich, P. 1987. *Asymmetries in time*. Cambridge, MA: MIT Press.

———. 1995. Closed causal chains. In *Time's arrow today*, ed. S. Savitt, 259–267. Cambridge: Cambridge University Press.

Hume, D. 1777. *Enquiries concerning human understanding and concerning the principle of morals*, ed. L. A. Selby-Bigge, 3rd ed. rev. P. H. Nidditch. Oxford: Clarendon Press, 1975.

Humphreys, P. 1989. *The chance of explanation: Causal explanation in the social, medical, and physical sciences*. Princeton: Princeton University Press.

Kastner, R.E. 2012. *The transactional interpretation of quantum mechanics: The reality of possibility*. Cambridge: Cambridge University Press.

Kutach, D. 2001. *Entropy and counterfactual asymmetry*. PhD dissertation, Rutgers University.

———. 2002. The entropy theory of counterfactuals. *Philosophy of Science* 69 (1): 82–104.

———. 2007. The physical foundations of causation. In *Causation, physics and the constitution of reality: Russell's republic revisited*, ed. H. Price and R. Corry, 327–350. Oxford: Clarendon Press.

Lewis, D. 1973a. *Counterfactuals*. Oxford: Blackwell.

———. 1973b. Causation. *Journal of Philosophy* 70 (17): 556–567. Reprinted in Lewis (1986, Chap. 21).

———. 1979. Counterfactual dependence and time's arrow. *Noûs* 13 (4): 455–476.

———. 1986. *Philosophical papers*, Vol. 2. Oxford: Oxford University Press.

Loewer, B. 2007. Counterfactuals and the second law. In *Causation, physics and the constitution of reality: Russell's republic revisited*, ed. H. Price and R. Corry, 293–326. Oxford: Clarendon Press.

Maudlin, T. 1994/2011. *Quantum non-locality and relativity*. 3rd ed. Oxford: Blackwell.

Mellor, D.H. 1971. *The matter of chance*. Cambridge: Cambridge University Press.

———. 1981. *Real time*. Cambridge: Cambridge University Press.
———. 1995. *The facts of causation*. London: Routledge.
______. 1998. *Real time II*. London: Routledge.
Menzies, P. 2001/2014. Counterfactual Theories of Causation. In *The Stanford Encyclopedia of Philosophy* (Spring 2014 Edition), ed. Edward N. Zalta, URL = http://plato.stanford.edu/archives/spr2014/entries/causation-counterfactual/.
Menzies, P., and H. Price. 1993. Causation as a secondary property. *British Journal for the Philosophy of Science* 44 (2): 187–213.
Miller, D. 1996. Realism and time symmetry in quantum mechanics. *Physics Letters A* 222 (1): 31–36.
———. 2008. Quantum mechanics as a consistency condition on initial and final boundary conditions. *Studies in History and Philosophy of Modern Physics* 39 (4): 767–781.
Myrvold, W. 2016. Philosophical issues in quantum theory. In *The Stanford Encyclopedia of philosophy* (Spring 2017 Edition), ed. Edward N. Zalta. https://plato.stanford.edu/archives/spr2017/entries/qt-issues/.
Price, H. 1984. The philosophy and physics of affecting the past. *Synthese* 61 (3): 299–323.
———. 1994. A neglected route to realism about quantum mechanics. *Mind* 103 (411): 303–336.
———. 1996. *Time's arrow and Archimedes' point: New directions for the physics of time*. New York: Oxford University Press.
———. 2008. Toy models for retrocausality. *Studies in History and Philosophy of Modern Physics* 39 (4): 752–761.
Price, H., and B. Weslake. 2009. The time-asymmetry of causation. In *The Oxford handbook of causation*, ed. H. Beebee, P. Menzies, and C. Hitchcock, 414–443. Oxford: Oxford University Press.
Price, H. 2012. Does time-symmetry imply retrocausality? How the quantum world says “maybe”. *Studies in History and Philosophy of Modern Physics* 43 (2): 75–83.
Popper, K.R. 1990. *A world of propensities*. Bristol: Thoemmes.
Redhead, M. 1987. *Incompleteness, nonlocality and realism: A prolegomenon to the philosophy of quantum mechanics*. Oxford: Clarendon Press.
Reichenbach, H. 1956. *The direction of time*. Berkeley: University of California Press.
Reznik, B., and Y. Aharonov. 1995. Time-symmetric formulation of quantum mechanics. *Physical Review A* 52: 2538–2550. quant-ph/9501011.
Salmon, W. 1984. *Scientific explanation and the causal structure of the world*. Princeton: Princeton University Press.
Smith, N.J.J. 1997. Bananas enough for time travel? *British Journal for the Philosophy of Science* 48 (3): 363–389.
Sober, E. 1988. The principle of the common cause. In *Probability and causality*, ed. J.H. Fetzer, 211–228. Dordrecht: D. Reidel Publishing Company.
———. 2001. Venetian sea levels, British bread prices, and the principle of the common cause. *British Journal for the Philosophy of Science* 52 (2): 331–346.
Sober, E., and E. Eells. 1986. Common causes and decision theory. *Philosophy of Science* 53 (2): 223–245.
Suarez, M. 2013. Propensities and pragmatism. *Journal of Philosophy* 110 (2): 61–92.
Sutherland, R.I. 1983. Bell's theorem and backwards-in-time causality. *International Journal of Theoretical Physics* 22 (4): 377–384.
———. 1998. Density formalism for quantum theory. *Foundations of Physics* 28 (7): 1157–1190.
———. 2008. Causally symmetric Bohm model. *Studies in History and Philosophy of Modern Physics* 39 (4): 782–805.
Tooley, M. 1997. *Time, tense, and causation*. Oxford: Clarendon Press.
van Fraassen, B.C. 1980. *The scientific image*. Oxford: Clarendon Press.
Wheeler, J.A., and R.P. Feynman. 1945. Interaction with the absorber as the mechanism of radiation. *Reviews of Modern Physics* 17 (2–3): 157–181.
———. 1949. Classical electrodynamics in terms of direct interparticle action. *Reviews of Modern Physics* 21 (3): 425–433.

Chapter 9
The Representation of Time in Discrete Mechanics

Vincent Ardourel and Anouk Barberousse

Abstract The starting point of the chapter is a twofold observation:

(i) most current physical theories make use of a continuous parameter t that plays the role of time;
(ii) the current practice of modeling, namely, of computational modeling, makes use of a discrete parameter t_k (k = 1, 2, …, n) that plays the role of time, because computers cannot handle continuous quantities.

A "parameter" here is a symbol that does not play the same role as the other symbols in the physical theories or models: whereas the other symbols represent physical quantities, t (or t_k) just plays the role of that within which other quantities evolve. It is "time", but in a rather shallow sense.

The relationship between (i) and (ii) is not harmonious. In order to be solved by computers, continuous equations have to be replaced by discrete ones; however, more often than not, discrete equations are obtained by some discretization procedure (finite difference, Runge-Kutta, etc.) that generates various artefacts. None of these procedures is able to avoid problems of computational stability or robustness. The discrete t_k is thus both the very means by which the computational revolution can obtain and only a "second best" solution or even an obstacle to be bypassed. There also exists a more radical way to introduce discrete time within computational models: by using algorithms that are intrinsically discrete, as opposed to algorithms obtained after discretizing continuous equations. Many of the above-mentioned problems simply do not arise in computer simulations based on intrinsically discrete algorithms. The aim of the paper is to ask whether the intrinsically discrete algorithms make any difference as to the role of the time-symbol. The proposed answer is that no difference appears when t_k is still treated as a discrete

V. Ardourel (✉)
Université catholique de Louvain, Louvain-la-Neuve, Belgium
e-mail: vincent.ardourel@gmail.com

A. Barberousse
Université Paris-Sorbonne, Paris, France

C. Bouton, P. Huneman (eds.), *Time of Nature and the Nature of Time*, Boston Studies in the Philosophy and History of Science 326,
DOI 10.1007/978-3-319-53725-2_9

parameter, but that an important difference may appear within the formalisms in which t_k is considered on a par with other symbols, that is, as a "dynamical variable". In some versions of "discrete mechanics" (Lee 1983, 1987), the time-symbol thus plays a twofold role: it is both the usual (discrete) parameter within which other quantities evolve, and a physical quantity of its own, whose evolution is governed by the law of energy conservation.

The paper examines the epistemological consequences of this new role for the time symbol in physical theory.

9.1 Introduction

As an important innovation of the last decades, there exists nowadays an intrinsically discrete physical theory called Discrete Mechanics (DM). It is intrinsically discrete since it does not originate in the discretization of continuous equations but is rather built upon already discrete fundamental principles. DM is a new, variational theory of classical motion. Our leading assumption in examining the implications of DM for the representation of time is that DM is a "good" physical theory, at least as good as our best physical theories for all practical purposes, and even better, because it allows for easier and more precise predictions due to its straightforward implementation on computers. This assumption will be argued for below in the paper.

The purpose of the paper is to examine the consequences of adopting this discrete physical theory, and on the way we can represent time in physics. We focus on two main consequences. The first is the use of a discrete representation of time, the second is the representation of time as a dynamical variable rather than a parameter. In his seminal paper on DM, the Nobel Prize winner Tsung Dao Lee raises the following questions:

In place of treating time as a continuous parameter, we may ask:

1. Can time be a discrete parameter (discrete time formulation)?
2. Can time be discrete and treated as a bona fide dynamical variable (discrete mechanics)? (Lee 1983, p. 217)

We take on Lee's questions and analyze the implications of positive answers to both.

In the first section of the paper, we give a brief presentation of DM, emphasizing that it can be considered a genuine physical theory, not merely a numerical method. In Sect. 9.3, we argue that DM points out that the continuous representation of time is dispensable and discuss the implications of this claim. In Sect. 9.4, we explain why the time symbol is a dynamical variable in DM and present some implications of this original feature of DM.

9.2 Discrete Mechanics

In order to answer the questions we are interested in about the implications of using a discrete representation for time, we first present in some details the way scientists introduce discrete representations of time in their models. In Sect. 9.2.1, we review various discretization practices and in Sect. 9.2.2, we focus on the recent technique of variational integrators, which leads us to present Discrete Mechanics, the theory we are going to analyze further in the remaining part of the paper. Finally, in Sect. 9.2.3, we make clear that DM is not merely a numerical technique but has to be viewed as a discrete physical theory or, at least, as a discrete formulation of classical mechanics.

9.2.1 Discretizing Time in Computer Models

All attempts at building up a discrete theory of mechanics originate in the computer era, that is, in a context in which discrete models proliferate. In order to better capture the specificity of DM, it is useful to first present how time and other magnitudes are discretized in order to benefit from the machines' computational power.

For sure, computers are a tremendous help for solving the equations given by the physical theories. However, computational power has a cost, as using computers requires transforming the continuous equations into discrete ones because they can only deal with discrete values. As a result, contemporary modeling practice is throughout defined by the need for designing algorithms to transform continuous equations into discrete ones that can be solved by computer algorithms. In this section, we first present some common types of numerical integrators and we compare their respective merits, focusing on the more efficient ones, the variational integrators.

9.2.1.1 Types of Integrators

The transformation of continuous equations into discrete ones is obtained by numerical integrators. Among the different numerical integrators, we focus on the simplest ones: the forward/explicit Euler integrator and the backward/implicit Euler integrator. The first numerical integrator transforms the time derivative $dx/dt=y$ into the discrete equation $(x_{k+1}-x_k)/h= y_k$. With the second integrator, the time derivative $dx/dt=y$ is transformed into the discrete equation $(x_{k+1}-x_k)/h= y_{k+1}$.

Consider, as an example, the motion equations of a simple pendulum solved with these integrators. Two differential equations have to be transformed. The first one represents the definition of the angular speed $dq/dt=v$, where q is the pendulum's angle with the vertical. The second equation represents the evolution of the angular speed with respect to the size of the pendulum l and the acceleration of gravity g: $dv/dt = -g/l\ sin(q)$. For each integrator, both differential equations are transformed into discrete equations according to the previous rules. With the forward Euler integrator, the differential equations are transformed into the discrete equations:

$$(q_{k+1} - q_k)/h = v_k \quad and \quad (v_{k+1} - v_k)/h = -g/l \sin(q_k) \tag{9.1}$$

With the backward Euler integrator, they are transformed into:

$$(q_{k+1} - q_k)/h = v_{k+1} \quad and \quad (v_{k+1} - v_k)/h = -g/l \sin(q_{k+1}) \tag{9.2}$$

Let us introduce another numerical integrator to solve the simple pendulum: the variational/symplectic Euler integrator. It is a mix of the first two Euler integrators since the first differential equation is solved with the backward integrator and the second equation, with the forward Euler integrator. In this case, the differential equations are changed into:

$$(q_{k+1} - q_k)/h = v_{k+1} \quad and \quad (v_{k+1} - v_k)/h = -g/l \sin(q_k) \tag{9.3}$$

9.2.1.2 Comparison

Numerical tests of have been performed by Stern and Desbrun (2008) in order to compare the three Euler integrators. The result of the comparison is presented in Fig. 9.1, in which the phase space trajectories of the pendulum are computed for six different initial conditions for each integrator. Continuous Newtonian mechanics requires that the trajectories of the pendulum be closed curves since it is a conservative system. However, as we can see on the left, the trajectories obtained with the forward Euler integrator are increasing spirals due numerical instability. With the backward Euler integrator (in the middle), on the contrary, the trajectories are decreasing spirals due to numerical dissipation. Only the symplectic Euler integrator (on the right) exhibits the right behavior: the trajectories are closed curves, which means that the energy of the pendulum is conserved.

Hairer et al. (2006) performed similar numerical tests for the computed trajectories of the planets of a reduced solar system. Numerical computations of the motion of the five outer planets relative to the Sun, over a time period of 200 000 days, are carried out with the forward, backward and symplectic Euler integrators (Fig. 9.2). They lead to similar behaviours. With the forward integrator, the trajectories are unstable: the planets leave their stable orbits. With the backward integrator, the trajectories are not even periodic: the planets collapse on the sun.

As shown in the figures, variational integrators generally give better results than traditional ones. The main reason for this difference is that most traditional algorithms are non-symplectic: they do not preserve the geometric properties of the system's dynamics. This is especially inconvenient when the represented systems

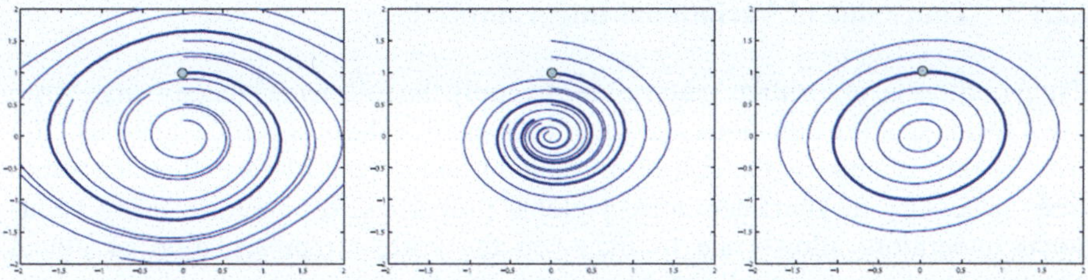

Fig. 9.1 The simple pendulum in the phase space (q, v) solved with three Euler integrators: forward Euler (*left*), backward Euler (*middle*), symplectic Euler (*right*). For each integrator, there are six different trajectories that correspond to different initial conditions. The bold lines in each of the three figures correspond to the same initial condition. (Stern and Desbrun 2008, p. 77)

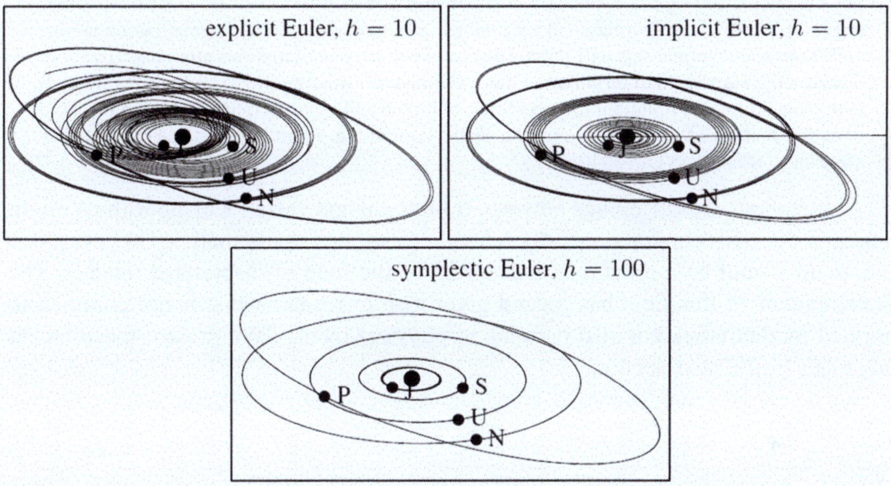

Fig. 9.2 The trajectories of the planets of a reduced solar system computed with three Euler integrators: forward Euler (at the *top left*), backward Euler (at the *top right*), symplectic Euler (*below*). A 10 day time step is used with the forward and backward Euler integrators while a 100 day step is used with the symplectic Euler. For each integrator, the initial conditions correspond to the positions of planets on September 5, 1994 at 0h00 (Hairer et al. 2006, p. 14)

are conservative. In this case, the problem of the traditional algorithms is that they cannot be conservative unless they include artificial mechanisms guaranteeing a not-too-incorrect resulting behavior. However, introducing a mechanism of this type is bound to distort the results. By contrast, variational integrators are symplectic: they naturally preserve the global geometry of the system's dynamics and do not need any artificial, parasitic mechanism to remain stable.

9.2.1.3 The Value of Variational Integrators

Using traditional algorithms requires one to apply the following strategy: first, write down local, specific equations and second, discretize these specific equations. The price of this strategy is the risk of introducing numerical artefacts that are sometimes necessary to guarantee correct global properties. By contrast, using variational integrators allows one to focus on the correct representation of global structures and invariants. Kang Feng, who has been a major contributor to the field of variational integration, realized in the 1970's that the choice between the two strategies has major consequences. As his brother emphasizes in the foreword of a Kang Feng's posthumous book:

> [Kang Feng] fully realized that different mathematical expressions for the same physical law, which are physically equivalent, can perform different functions in scientific computing [...]. In this way, for classical mechanics, Newton's equations, Lagrangian equations and Hamiltonian equations will show a different pattern of calculations after discretization. Because the Hamiltonian formulation has a symplectic structure, he was keenly aware that, if the algorithm can maintain the geometric symmetry of symplecticity, it will be possible to avoid the flaw of artificial dissipation of this type of algorithm and design a high-fidelity algorithm. (Feng and Qin 2010, p. x)

At first view, the difference between traditional and variational algorithms might appear a purely technical one, only relevant to applied mathematics. However, this judgment would be a reductive conception of the field of discretized models. The development of this field has indeed given rise to research that is not confined to applied mathematics but also pertains to physical theory. We present these recent advances in the next section.

9.2.2 Variational Integrators and Discrete Mechanics

Discrete Mechanics, which is our main focus in this paper, originates in two research trends. The first one is based on the need to design discrete integration algorithms, and the second one is what we shall call the "geometrical turn" of physical theories. Since the second half of the twentieth century, modern geometrical formulations of Lagrangian and Hamiltonian mechanics have been developed and, nowadays, differential geometry is the "mother tongue" of physical theories (Butterfield 2006a, b). As a result, it is common to consider the geometric structure of the investigated systems as very important. This is the reason why discrete models preserving this geometric structure are now favored over other models. As Stern and Desbrun have it,

> the very essence of a mechanical system is characterized by its symmetries and invariants. Thus preserving these symmetries and invariants (e.g., certain momenta) into the discrete computational setting is of paramount importance if one wants discrete time integration to properly capture the underlying continuous motion. (Stern and Desbrun 2008, p. 75)

Discrete mechanics results from the will to respect the global geometrical of mechanical behaviors and from the enhanced power of the induced integrators[1].

Under its Lagrangian form, classical mechanics can be treated from a geometrical point of view as it is grounded on a variational principle, the principle of least action. In Lagrangian mechanics, the trajectories of bodies are viewed as geodesic in a state space. Discrete mechanics also adopts a geometrical point of view to describe mechanical behaviors as it is but a discrete version of Lagrangian mechanics. Consequently, the link between variational integrators and discrete mechanics is strong, as emphasized by Stern and Desbrun in the following quote:

> The driving idea behind discrete geometric mechanics is to leverage the variational nature of mechanics and to preserve this variational structure in the discrete setting. [...] [I]f one designs a discrete equivalent of the Lagrangian, then discrete equations of motion can be easily derived from it by paralleling the derivations followed in continuous case. In essence, good numerical methods will come from discrete analogs to the Euler-Lagrange equations –equations that truly derive from a variational principle. (Stern and Desbrun 2008, p. 79)

Among the discrete versions of Lagrangian mechanics, some are more or less ad hoc, but DM is anything but ad hoc as its starting point is the discretization of the variational principle itself. Having such a sound foundation is the main virtue of DM since it enables its users to preserve the geometrical properties of variational mechanics. Consequently, the discrete motion equations derived from the discrete principle – the discrete Euler-Lagrange equations – also preserve the geometrical properties of mechanical behaviors. Most importantly, symplecticity and other physical properties like the conservation of momenta are preserved in DM.

In order to make clear how DM is linked with variational integrators, we introduce a few formulae in order to present how the simple pendulum is treated within DM. In the continuous Lagrangian mechanics, the equations of motion are derived from the minimization of the action S_c. It is defined as the integral over time of the Lagrangian L_c of the system: $S_c = \int L_c dt$. In the case of the simple pendulum, the Lagrangian is:

$$L_c = \frac{1}{2} m l^2 v^2 + m g l \cos(q) \tag{9.4}$$

with q the pendulum's angle with the vertical, $v = dq/dt$, the angular speed, m the mass of the body, l the length of the pendulum, and g the acceleration of gravity. The minimization of S_c leads to the equations of motion introduced in Sect. 9.2.1.1:

[1]We would like to emphasize that, at the beginning of the 1980s, Tsung-Dao Lee (1983, 1987) developed a Discrete Mechanics for another reason still. It was to solve the well-known divergence problems of Quantum Field Theory. He wanted to build up a Discrete Mechanics conceived as a first step toward fully discrete fundamental theories in which divergences could not occur. The idea was first to try to develop one discrete theory in order to generalize it to others. To our knowledge, it did not go very far toward QFT.

$$\frac{dv}{dt} = -g/l \sin(q) \tag{9.5}$$

Similarly, in DM, the discrete action S_d is defined as $S_d = \sum L_d h$ where L_d is the discrete Lagrangian of the system and S_d its sum over discrete time[2]. In the case of the simple pendulum, the discrete Lagrangian can be defined as:

$$L_d = \frac{1}{2} m l^2 v_{k+1}^2 + mgl\cos q_k \tag{9.6}$$

Where the pendulum's angle is q_k and the angular speed $v_{k+1} = (q_{k+1} - q_k)/h$. The minimization of S_d leads to the discrete equation of motion:

$$(v_{k+1} - v_k)/h = -\frac{g}{l} \sin q_k \tag{9.7}$$

We now see how DM is linked with variational integrators. Equation (9.7) and the definition of the discrete angular speed are precisely the discrete equations that correspond to the symplectic Euler integrator introduced in the Sect. 9.2.1.1.

To sum up, DM is a discrete version of classical mechanics with a discrete least action principle as first principle, from which the discrete motion equations are derived. These discrete equations are very important in a numerical context. They are variational integrators that, as we have seen in Sect. 9.2.1.2, lead to very stable computations.

A last point has to be made clear. Two slightly different versions of DM have to be distinguished: a "standard version" and an "extended version". Until now, we have focused on the "standard version". This version requires constant time step h. Time is represented as a series of equally spaced instants, $t_1, t_2, t_3 \ldots t_{k-1}, t_k \ldots$ *where* $t_2 - t_1 = t_3 - t_2 = t_k - t_{k-1} = \ldots = h$. By contrast, the "extended version" of DM requires a variable time step h_k. Within this version, the instants t_1, t_2, $t_3, \ldots t_k \ldots$ are not equally spaced.

The standard version is a particular case of the extended version for which the discrete time step h_k is constrained to be constant. This standard version is developed in order to build variational integrators, like the previous symplectic Euler integrator. It allows indeed for efficient numerical computation. The extended version is generally less easy to implement on computers. However, it leads to a more satisfying theory in the sense that it enables mechanical systems to have more conserved quantities. Like the standard version, the extended version preserves symplecticity and the momenta. But in addition, it preserves energy *exactly*.

[2]We would like to emphasize that there is no single way to define a discrete Lagrangian for a mechanical system. The choice of another discrete Lagrangian for the simple pendulum leads to a different equation of motion that corresponds to another variational integrator. In other words, it can be admitted that there are as many DMs – or versions of DM – as ways to define discrete Lagrangians. We go back to this point in the conclusion of the paper.

J.E. Marsden, one of the major creators of DM, makes clear how to go from the standard version to the general theory:

> Unlike the standard discrete variational mechanics [. . .] however, we extend the framework to include time variations in addition to the usual configuration variable variations, as in Lee (1983, 1987). [. . .] [W]e obtain an extra equation which exactly ensures preservation of a quantity we can identify as the discrete energy. In this way, both the definition of the discrete energy and the fact that it is preserved arise naturally from the variational principle. (Marsden and West 2001, p. 463)

As we have seen in this section, energy is reasonably well conserved when the symplectic Euler integrator based on the standard version of DM is used. This is one of the main virtues of such integrators with respect to the traditional ones. However, energy is *not exactly* conserved within the standard version. Energy oscillates around a mean value. By contrast, within the extended version, energy is *exactly* conserved. In the remaining sections of the paper, we focus on the extended version as the fundamental version of DM, the standard version being a simplified version for practical needs.

9.2.3 DM as a Physical Theory

As should be clear from Sect. 9.2.2, the principles governing the development of discrete mechanics are very different from the recipes allowing physicists to implement their models on a computer. The strategy is not: write out the equations first, and then find out a way to solve them on the computer, but rather: let us begin with discrete fundamental equations and find out the solutions of the specific equations within this general framework. Therefore, we argue that discrete mechanics is a genuine discrete physical theory rather than a mere set of numerical models or numerical techniques. This claim agrees with the terminology used by one of the first creators of DM, the 1957 recipient of the Nobel Prize of Physics T.D Lee, who calls DM a "new theory" (Lee 1987, p. 844). In this section, we review some reasons why DM can be considered a genuine physical theory.

First, DM possesses an empirical content. Its equations can be interpreted as providing information about physical phenomena. In particular, the discrete Euler-Lagrange equations can be interpreted as describing the dynamics of mechanical systems. We present below the cases of two elementary physical systems, a falling body subjected to the gravitational potential and a harmonic oscillator (Sect. 9.3.3.2). We will see in these examples how DM allows one to represent the temporal evolution of physical systems.

Current developments of DM strongly suggest that DM is empirically adequate as continuous classical mechanics is. In other words, DM and continuous mechanics are empirically equivalent. The first indication of the equivalence is that the equations of DM tend to the equations of continuous mechanics when discrete time steps go to zero (D'Innocenzo et al. 1987, p. 246). Therefore, one can assume that if the discrete time steps are taken sufficiently small, the predictions of DM cannot be distinguished from those of continuous mechanics. We do not have space enough to discuss this point. However, to make this point clear, let us take the example of the harmonic oscillator system. It can be proved that the frequency of the oscillator within DM differs from the frequency within continuous mechanics by a factor h^2 (D'Innocenzo et al. 1987, p. 250). Consequently, if the time step h is taken sufficiently small, the frequencies are as close to each other as we want and, thus, cannot be distinguished by any physical measure[3].

Secondly, let us focus on the way its theoretical statements are linked to each other. DM has its own first principle: a principle of least discrete action. The laws of DM, like the discrete equations of motion and the discrete law of conservation of energy, are deduced from this first principle. Other properties like the symplecticity of the equations of motion are also deduced form this first principle. As a result, DM is a physical theory from the point of view of its deductive architecture: it is a set of hierarchically ordered theoretical statements.

Thirdly, allowing for exact results for some of its equations, DM provides another argument in favor of its being a genuine physical theory. For instance, it is possible to exactly solve, without relying on numerical computation, the equations of motion of a harmonic oscillator. As we show in Sect. 9.2, exact results are important in physical theories because they enable scientists to describe and explain the behavior of elementary physical systems.

Even though DM possesses some important features of physical theories, it is unclear whether it is an independent theory from continuous classical mechanics or rather a formulation thereof. If the latter, DM's first principle and fundamental laws would be discrete versions of the principles and the laws of continuous classical mechanics rather than autonomous statements. The only difference between DM and continuous classical mechanics would be notational and would by no means affect empirical content.

The claim that DM has the same empirical content as continuous classical mechanics but is differently formulated is consistent with the way DM is used in physics. The main use of DM is indeed the numerical study of the behavior of mechanical systems. Consequently, contributors to the development of DM define it explicitly as "a *formulation*[4] of mechanics in discrete-time that is based on a discrete analogue of Hamilton's principle, which states that the system takes a trajectory for which the action integral is stationary" (Lee et al. 2009, p. 2001). But DM being a formulation of classical mechanics rather than an independent theory of classical motion does not mean that it does not call for analysis. On the contrary, the existence of different, however equivalent, formulations of a same theory calls for a specific analysis.

---

[3]We emphasize that there is not a unique DM but a family of DM depending on the (initial) value of the discrete time step. Thus, when we claim that DM and continuous are empirically equivalent, we mean more precisely that there exists at least one DM that is empirically equivalent to continuous mechanics.

[4]Our emphasis.

Even though they have the same empirical content, different formulations of the same theory do not say exactly the same about their common domain of phenomena. They describe them with the help of different equations, requiring different *concepts*. For example, the Newtonian formulation of mechanics requires the concept of force whereas the Hamiltonian formulation requires the concept of energy. Even though this distinction is not empirically relevant, in the sense that no physical measure could make any difference, it may be relevant from a theoretical or metaphysical point of view. Therefore, in order to discuss the metaphysical consequences of a physical theory, attention has to be paid to its formulations.

For instance, in the case of continuous classical mechanics, J. North argues that, even though they are empirically equivalent, Lagrangian and Hamiltonian formulations have "irreconcilable differences":

> Lagrangian and Hamiltonian mechanics may be equivalent for the purpose of doing classical mechanics. Nonetheless, there are important differences between them. There are differences in *structure*. (North 2009, p. 72)

The two formulations are based on different state spaces. The state of a mechanical system is described by the position q and its derivative dq/dt within the Lagrangian formulation while it is described by q and its conjugate momentum p within the Hamiltonian formulation. The state spaces do not have the same mathematical structure: the former is a Riemannian manifold and the latter a symplectic manifold. Even though they are connected by an isomorphism (North 2009, p. 84), the difference is by no means negligible:

> Modern geometric formulations of physics suggest that there is more to a theory's empirical content than its set of dynamically possible histories. There is also the statespace in which those histories are traced out. And there is the structure of that space. The equivalence of theories is not just a matter of physically possible histories, but of physically possible histories *through a particular statespace structure*. Hamiltonian and Lagrangian mechanics are not equivalent in terms of that structure. This means that they are not equivalent, period. (North 2009, p. 79)

The equivalence is thus only empirical, but does not extend to other aspects.

Similarly, the state space of DM, and more precisely the state space of discrete Lagrangian mechanics, is different from the state space of continuous Lagrangian mechanics. As there is no velocity variable dq/dt within DM, the state spaces have different structures. In DM, there are only discrete positions at successive instants q_k and q_{k+1}. The notion of speed is not fundamental; rather, motion is represented by series of discrete positions. Therefore, there are also "irreconcilable differences" between DM and continuous Lagrangian mechanics. The differences might be even more irreconcilable than in the case discussed by North since, as Marsden and West emphasize, the state space of DM "is not locally isomorphic to the [. . .](traditional) Lagrangian state space" (2001, p. 473).

To sum up, even if their predictions cannot be distinguished by physical measures, DM and continuous Lagrangian mechanics are not equivalent throughout. In the remainder of the paper, we discuss the implications of the use of DM for the representation of time.

9.3 What Does It Imply to Use a Discrete Representation of Time?

As made explicit in Sect. 9.1, the use and development of DM relies on the adoption of a discrete representation of time. In this section, we answer the question: What does it imply to use a discrete representation of time when doing physics, that is, when trying to describe, predict and explain physical phenomena? We study three possible answers to this question in turn. The first tentative answer is that it implies using equations that are not as fundamental as continuous equations. We dismiss this answer. The second answer is to dispense with the mathematical continuum. We also dismiss this answer and explain why. The last answer we discuss is that it implies that it is possible to dispense with the continuous representation of time in physics. As the meaning of this answer is difficult to capture, we first endeavor to give a precise meaning to the dispensability of the continuous representation of time in physics. Then we give some arguments showing that this answer is highly plausible.

9.3.1 Discrete Equations and Fundamental Laws

It is generally assumed that differential equations, which are continuous, are more fundamental than difference equations, which are discrete. This opinion relies on the belief that difference equations necessarily derive from differential equations. Even though difference equations are pervasive in current physics as well as in other scientific domains, they are commonly considered as inessential substitutes of the continuous equations.

For sure, most of the discretized equations that are used on a day-to-day basis are derived from general, continuous equations associated with a lot of restricting, idealizing, and approximating hypotheses. They are thus very specific to the model at hand. However, as it should be clear from Sect. 9.2.2, not all difference equations are model-specific. DM's difference equations are perfectly general equations, as discrete mechanics is not devoted to finding out the solutions to specific-purpose models. Here is the way Lee presents the contrast between differential and difference equations within DM:

> For more than three centuries we have been influenced by the precept that fundamental laws of physics should be expressed in terms of differential equations. Difference equations are always regarded as approximations. Here, I try to explore the opposite: Difference equations are more fundamental, and differential equations are regarded as approximations. (Lee 1987, p. 859)

It is clear from this quote that using a discrete representation of time does not imply using less fundamental equations within DM. In Lee's theory, difference equations are assumed to be the first equations of the theory. Accordingly, differential equations derive from difference equations. They result from a limit procedure on difference equations, when the discrete time steps tend to zero. Thus, while in continuous mechanics differential equations are fundamental and difference equations result from a discretization procedure, in DM difference equations are fundamental and differential equations result from a limit procedure.

9.3.2 Discrete Equations and Mathematical Continuum

Now that we have dismissed the claim according to which using a discrete representation of time necessarily implies using non-fundamental equations, we turn to the question whether using a discrete representation of time implies dispensing with the mathematical continuum altogether. The answer to this question is simple and factual, as DM, although using a discrete representation of time, does not dispense with real numbers. On the contrary, the restriction to rational numbers would amount to a dramatic loss in predictive and explanatory power. In order to argue in favor of these two points, we first show how DM makes use of real numbers. Then, we briefly describe an older version of DM whose purpose was precisely to dispense with the mathematical continuum, in order to make the limits of this previous version clear.

In the version of DM we focus on, the discrete time symbol is defined as $t_k = kh_k$ where k is a natural number and $k<N$, with N fixed, and the time steps h_k are real numbers (Marsden and West 2001, p. 370); as a result, the "instants" t_k are also real numbers. It is important to recall that when a variable takes its value within a discrete set, it is not necessarily rational for all that. The discrete set in question may be composed of real numbers. Thus, the difference between a discrete representation of time and a representation of time by rationals only is best expressed in terms of the cardinality of the sets of instants. In the continuous representation, a time interval is represented by an uncountable infinite set of instants. On contrary, in the discrete representation, no representation of intervals is available, but only representations of sets of time points, represented by discrete sets of real numbers.

Let us complete this factual point about the discrete representation of time not being restricted to rational numbers by pointing out that real numbers play a truly important role within DM for another reason yet: the proof of certain general results in DM actually requires the use of real numbers and mathematical operators defined on real numbers. For instance, a discrete Noether theorem can be proved within DM by relying on real numbers. It is an important theorem since it states that physical quantities are conserved when a mechanical system satisfies some symmetry. More precisely, the Noether theorem refers to differentiable symmetries of the system's Lagrangian. For instance, if a mechanical system behaves in the same way regardless of where it is located, the system is symmetric under translations in space. By the Noether theorem, the linear momentum of the system is conserved. Further, if a system is rotationally symmetric, the Noether theorem states that its angular momentum is conserved. As far we know, the Noether theorem and its proof, in

DM as well as in conventional mechanics, require using differential calculus, and, as such, they rely on real numbers[5].

In order to better appreciate the importance of real numbers in DM, we now present the limitations that are inherent to an older version of DM whose project was to precisely dispense with real numbers. It was indeed Donald Greenspan's purpose to develop a discrete physical theory that would appeal to real numbers in none of its descriptions or computations. During the 1970s and 1980s, Greenspan developed what he called a "discrete mechanics" (Greenspan 1973, p. 10) that is different from the theory we focus on in this paper. His discrete mechanics consists in reformulating traditional Newtonian mechanics with a discrete representation of time. In Greenspan's discrete mechanics, time is represented by a discrete parameter t_k where $t_k = kh$, k being a natural number and h a rational number. The differential equation expressing Newton's second law is replaced by a difference equation with a discrete acceleration and a discrete force $m\boldsymbol{a}_k = \boldsymbol{F}_k$. The discrete acceleration $\boldsymbol{a}_k$ is defined as the ratio $(\boldsymbol{v}_{k+1} - \boldsymbol{v}_k)/h$, with $\boldsymbol{v}_k$ the discrete velocity vector, and the discrete force $\boldsymbol{F}_k$, as a function of a discrete potential energy. More precisely, the discrete force is written as:

$$\boldsymbol{F}_k = \frac{V(\boldsymbol{r}_{k+1}) - V(\boldsymbol{r}_k)}{\boldsymbol{r}_{k+1} - \boldsymbol{r}_k} \frac{\boldsymbol{r}_{k+1} + \boldsymbol{r}_k}{r_{k+1} + r_k} \tag{9.8}$$

with $V(\boldsymbol{r}_k)$ the discrete potential energy and $\boldsymbol{r}_k$ the discrete position vector at the instant t_k.

Greenspan and LaBudde show that, under few assumptions, all the main conservation laws of the traditional Newtonian classical mechanics can be established within their version of discrete mechanics, for discrete linear momentum, energy and angular momentum are exactly conserved (LaBudde and Greenspan 1974).

For Greenspan and LaBudde, the best argument in favor of their version of a discrete mechanics is that exact conservation of discrete quantities, like energy, allows for more stable numerical computations. Since energy and momenta are constrained to be equal to their initial values, they cannot diverge and thus cannot lead to unstable computations.

Greenspan argues that the development of his discrete mechanics shows that scientists can study physical phenomena without relying on the mathematical continuum. At the beginning of his book on discrete mechanics, he emphasizes:

> The concept of infinity and the consequential concepts of limit, derivative, and integral are reasonable for the pure mathematical study of real numbers and real functions, but are *not* reasonable for the modeling of physical phenomena. [...]
>
> Dynamical behavior [of mechanical systems] will be studied entirely in terms of arithmetic, or more precisely, in terms of high-speed arithmetic, for it is the availability of the modern digital computer which will make our approach both reasonable and practical. The dynamical equations of our models will be difference equations which, whether linear or nonlinear, well easily be solvable. Thereby, it is hoped that if an applied

[5]For details on the Noether theorem, see (Butterfield 2006b).

> scientist is willing to learn the simple language of a computer, then he need be equipped only with the rudimentary mathematical knowledge of arithmetic and algebra in order to study highly complex physical phenomena. (Greenspan 1973, p. 4)

According to Greenspan, differential equations and real numbers can be eliminated from the study of mechanical systems. As the dynamics of mechanical systems is described by DM, its equations of motion are solved on computers, and as computers can only deal with natural and rational numbers, real numbers are not required to describe physical phenomena. Real numbers can be approximated on computers but this requires the use of rational or decimal numbers. Instead, one does not need this transformation in solving on computers the equations of DM formulated with rationals.

We disagree with Greenspan's evaluation of the capacities of his version of DM. By choosing to only use rational numbers, Greenspan severely restricts the expressive power of difference equations, for even though difference equations are by definition discrete, they may in general accept any real number as a solution (or any complex number in some cases). The restriction to rationals enables one to approximately solve Greenspan's DM equations on computers, but it is impossible or, at least, very difficult to solve them exactly or analytically without using the real numbers. Now, as we shall emphasize in Sect. 9.3.3.2, exact results are extremely important in order to assess the performance of a physical theory.

Second, the only use of rational numbers may be not sufficient to describe physical phenomena even on computers. Greenspan himself acknowledges that, within his DM "classical continuous mathematics will be used [...] in the study of stability, where properties of sets of rational numbers can be derived most easily by considering these numbers as subsets of the real number systems" (Greenspan 1973, p. 4). As a result, Greenspan's DM turns out to be unserviceable unless relying on real numbers. Thus, the development of Greenspan's DM does not imply that physics can dispense with the mathematical continuum for describing and explaining the behavior of mechanical systems.

9.3.3 The Continuous Representation of Time and Dispensable Representations

Let us now present the answer we favor to the question of what the discrete representation of time in DM implies. According to this answer, the discrete representation of time in DM implies that physics can dispense with the continuous representation of time. In order to argue in favor of this claim, we first give some details about the conditions at which a physical theory can dispense with the continuous representation of time. The main condition is that the theory dispensing with the continuous representation of time should be *usable* by physicists, that is, it should enable physicists to describe, predict and explain physical phenomena. Second, we show that DM satisfies this condition and thus argue that the use of a discrete representation of time in DM implies that the continuous representation of time is dispensable in physics.

9.3.3.1 Usable Theories

We begin our inquiry into the question whether the continuous representation of time is dispensable in physics by recalling that the notion of "dispensable" is stronger than the notion of "eliminable". In the original debate on the dispensability of various mathematical entities, a mathematical entity is said to be dispensable in a theory if it can be eliminated from the theory without any loss in its scope, the resulting theory being a *good* one (cf. Colyvan 2001).

First, we suggest to extend the question of dispensability to mathematical representations, or mathematical tools of representation. The original debate focused on mathematical entities. However, in this paper, we are not interested in the question whether mathematical entities like numbers, sets, functions do exist but only in analyzing the notion of dispensability. There is no reason, in our view, to restrict this notion to entities.

Secondly, while "dispensability" is not "eliminability", it is unclear in the original debate, as Colyvan himself emphasizes, what a "good" theory is supposed to be. To remedy this deficiency, we suggest, that an entity or a mathematical representation can be said to be dispensable in a theory if it can be eliminated from the theory without any loss in its scope, the resulting theory being usable: it should be able to allow for description, prediction, and explanation of phenomena. Equipped with the notion of a usable theory, we can safely discard formal reconstructions of theories that (i) dispense with certain entities and (ii) are logically equivalent to the original ones, but (iii) cannot be used to investigate phenomena. These theories do not fall within the scope of the interesting theories for our purpose.

However, we reject Colyvan's criterion according to which the resulting theory should be *preferable* to the original theory, or at least more attractive. To our mind, it is not necessary to compare the resulting theory to the original one in order to decide whether an entity or a mathematical representation is dispensable in the former. The comparison constitutes a different, independent step in the analysis. For example, it seems that the notion of "force" can be eliminated from classical mechanics. Hamiltonian mechanics only seems to require the notion of "energy" as a fundamental concept (Wilson 2007, p. 175). If this is true, we would certainly be inclined to say that the notion of force is dispensable even if it is hard to say – and maybe not decidable – whether Newtonian or Hamiltonian mechanics is preferable. Maybe they are equally attractive or the one is preferable in some contexts and the other is preferable in other contexts. However, in our view, the notion of "force" is said dispensable in classical mechanics because it is eliminable and the resulting theory, Hamiltonian mechanics, is a usable theory.

9.3.3.2 DM Is a Usable Theory

Let us now examine whether the version of DM we focus on in this paper can be said to be usable. Our main argument in favor of this claim is that the equations of motion for (at least some) mechanical systems can be given explicit solutions (D'Innocenzo et al. 1987). Explicit solutions are those expressed as functions of time. For instance, the equations of motion for a body subjected to a constant force and the harmonic oscillator system can be given explicit solutions. Let us briefly present these solutions (see Appendix 1 for more details).

Continuous Newtonian mechanics describes the phenomenon of falling bodies on Earth with the Newton's second law of motion, which is a vectorial differential equation:

$$md^2\mathbf{r}/dt^2 = m\mathbf{g} \tag{9.9}$$

The product of the mass by the acceleration on the left, where $\boldsymbol{r}$ is the position vector, is equal to the gravity force $m\boldsymbol{g}$ on the right. The x and z components of this vectorial equation have each an explicit solution, which are:

$$x(t) = v_{x,i}t + x_i \quad and \quad z(t) = -\frac{1}{2}gt^2 + v_{z,i}t + z_i \tag{9.10}$$

where x_i, z_i, $v_{x,i}$, $v_{z,i}$ the initial conditions for position and speed, and g the gravity constant. Similarly, in DM, we have, as explicit solutions for the two components:

$$x_k = v_{x,i}t_k + x_i \quad and \quad z_k = -\frac{1}{2}gt_k^2 + (v_{z,i} + gh)t_k + z_i \tag{9.11}$$

where there is in addition the discrete time step h. The harmonic oscillator can also be solved explicitly within DM. In that case, the solution is:

$$x_k = A_d \cos(\omega_d t_k) + B_d \sin(\omega_d t_k) \tag{9.12}$$

where t_k is the discrete time, A_d, B_d two constants depending on the discrete time step h, and ω_d, the discrete pulsation also function of h. These two examples illustrate that some equations of motion can be solved explicitly within discrete mechanics.

The fact that some equations have explicit solutions within DM is remarkable because most equations of motion, both in continuous mechanics and in DM, do not have any explicit solutions. In continuous mechanics, the explicit solutions to most differential equations are indeed unknown or impossible to derive. However, even if only very few explicit solutions are known, scientists value their existence because an explicit solution provides them with a good understanding of the mechanical phenomenon. For instance, knowing the explicit solution of equation of motion of a falling body better enables one to grasp the phenomenon than only knowing a numerical solution on computer. That is because explicit solutions are both *general* and *exact*, which are two features that make them explanatory, as emphasized by Paul Humphreys:

> The switch from analytic mathematics to numerical mathematics [. . .] has one immediate consequence: we often lose both the great generality and the potentially exact solutions that have traditionally been desired of scientific theories. (Humphreys 2004, p. 64)

Let us review these two properties in turn.

Generality Explicit solutions are general in the sense that within them the position, as well as other variables, is a function of the relevant variables and parameters, thus requiring no specification of their numerical values. The explicit solutions express a relation between possibly all particular values of the parameters and variables of the function. For instance, in the case of the falling body problem, the solution x_k represents the trajectory whatever the numerical values of the gravity constant g, or the initial conditions x_i and v_i. It is a single formula describing a large range of different physical phenomena, for instance a falling body on Earth, or on the Moon, and with any value for its initial speed.

Generality, as a property of mathematical functions, is a key component in the explanatory role of explicit solutions. Explicit results provide explanations of physical phenomena in the sense that they show how the variations of some physical variables imply the variations of the other variables.

Exactness Explicit solutions being exact also accounts for their explanatory power. Exact solutions are such that, when put in place of the equations, they lead to an identity. For instance, $\sqrt{2}$ is an exact solution to the equation $x^2-2=0$ since, when put in place of x, the identity $0=0$ can be derived. In the same way, the explicit solutions of the falling body system with DM, Eq. (9.11) are exact solutions of the motion equations of DM (Eq. (9.18, 9.19) and (9.21, 9.22)) since, put in place of them, they lead to the identity $1=1$.

Because there are exact, explicit solutions can be used to deduce other results about the system. Consequently, other features or properties of the system can be known and better understood. For instance, consider again the free falling body. As said above, the equations of motion for this system can be solved explicitly within DM: the positions x_k and z_k are functions of t_k. With these explicit solutions (11), it is first possible to derive the trajectory of the falling body. After some algebraic manipulations, the time t_k is eliminated and the altitude z_k is expressed as a function of the distance x_k (see Appendix 2):

$$z_k(x_k) = -\frac{g}{2v_{x,i}^2}(x_k - x_i)^2 + \frac{v_{z,i} + gh}{v_{x,i}}(x_k - x_i) + z_i \tag{9.13}$$

In this case, the deduced trajectory is a parabola. This provides a geometrical (vs. dynamical) representation of the phenomenon. It exhibits global information about the position of the falling body, enabling us to grasp the general behavior of falling bodies.

Second, specific, quantitative features of the trajectory can be deduced from the trajectory equation, increasing our knowledge of the system (for details, see Appendix 2). For instance, the highest position can be deduced via a few mathematical operations. It is the position where the partial derivative z_k with respect to x_k vanishes. It is reached at

$$x_k^* = v_{x,i}(v_{z,i} + gh)/g + x_i \quad \textit{and} \quad z_k^* = (v_{z,i} + gh)^2/(2g) + z_i \tag{9.14}$$

It is also possible to deduce the distance x_k at which the falling body reaches the ground. Such deductions increase the explanatory power of the explicit solutions of the motion equations in enabling us to broaden our knowledge of the system.

MD's ability to provide explicit solutions to some mechanical problems is a strong argument to consider it as a usable theory. In particular, the existence of explicit solutions gives its principles the explanatory power that is expected from usable physical theories. Needless to say, MD being a discrete theory ensures that it can easily yield predictions, because it speaks the same language as computers.

Having argued that MD is a bona fide usable physical theory, we can claim that its existence, together with its predictive and explanatory powers, show that the continuous representation of time is dispensable in physical theories. In the next section, we discuss the implications of this claim.

9.3.4 Dispensability and the Structure of Time

Can we draw any conclusion about the structure of time from the dispensability argument we have just presented? Relying on existing discussions on this topic in the literature (Carnap 1966, Newton-Smith 1980, Maddy 1997), it seems that it should be so. In this section, we briefly review how this theme has been analyzed so far and present our own view.

The starting point of discussions about the relationship between the representation of time and the structure of time is the following observation. The use of the usual *t*parameter in classical mechanics and other continuous theories is commonly associated, at least implicitly, with the assumption that the structure of time is continuous, in the following sense:

> If time is continuous the set of all instants under the ordering given by being temporal before will be isomorphic to the reals or to some sub-sequence of the reals. (Newton-Smith 1980, p. 114)

This assumption seems to be generally accepted. For instance, in their textbook in classical mechanics, Kibble and Berkshire adopt it as an explicit *hypothesis* of the theory (2009, p. 2). In Suppes' axiomatic reconstruction of classical mechanics, the statement according to which time is continuous (Suppes 1957, p. 294) is presented as an axiom of the theory. In the chapter entitled "Indispensability and Scientific Practice" in her *Naturalism in Mathematics*, Maddy also asserts that the use of the real parameter t in physical theories presupposes that time is continuous (1997, p. 155).

If one assumes that adopting the continuous representation of time is tantamount to hypothetizing that it has a continuous structure, as do Kibble and Berkshire, Maddy, Newton-Smith, and Suppes, it is clear that one should admit that the dispensability of the continuous representation has an impact on what is possible to claim about the structure of time. For instance, Newton-Smith claims, as far as we understand him, that the dispensability of the continuous representation of time implies that the structure of time is empirically undecidable (1980, p.126). More precisely, for the sake of his argument, he introduces two physical theories whose only difference resides in the way they represent time. The first one represents time by a parameter defined on real numbers and the second by a parameter defined on rationals. He shows that these theories are empirically equivalent. From this result, he concludes that the structure of time is empirically undecidable, thus relying on a property of the *representation* of time to argue about the *structure* of time[6].

Can Newton-Smith's argument be transposed to DM so as to conclude that the dispensability of the continuous representation of time in DM leads to a similar undecidability result about the structure of time? Let us first emphasize that Newton-Smith does not distinguish between using the continuous representation of time and using real numbers in the representation of time. His inquiry focuses on the difference between reals and rationals as to their usefulness, or indispensability, in physics[7]. On the contrary, as we have made explicit in Sect. 9.2, DM being a discrete theory does not forbid it to use real numbers. This is a first important difference between the case imagined by Newton-Smith and the DM case, implying that Newton-Smith' argument cannot be transposed without important transformations to the DM case.

Admitting that we were willing to work out the relevant transformations, would it be appropriate to transpose Newton-Smith's argument? If it were the case, one could conclude from the study of DM that "time is continuous" is an undecidable statement. However, in our view, the very question whether "time is continuous" is undecidable is not legitimate. Let us make this point clear. Newton-Smith's

[6]Newton-Smith then asks: "How ought we to respond to this undecidability result?" (1980, p. 126). According to him, the statement "time is continuous" can be interpreted as an empirical statement whose truth value is impossible to assess because there is no evidence to be had about it – this is "the Ignorance response". However, it can also be interpreted as a non-empirical statement describing a theoretical framework within which to describe empirical facts – this is "the Arrogance response". We do not discuss these interpretations since, as we make clear below in this Section, we deny that any conclusion about the undecidabilty of the structure of time can be drawn from physical theories, like DM, which represent time as discrete.

[7]His line of inquiry thus continues Carnap's, who examined the following question: "Because irrational numbers are always the results of calculations, never the result of direct measurement, might it not be possible in physics to abandon irrational numbers altogether and work only with the rational numbers? That is certainly possible, but it would be a revolutionary change." (Carnap 1966, p. 88–89) As clear in the main text, this question is irrelevant to our own inquiry about the implications of the discrete representation of time in DM.

argument in favour of the undecidability of the structure of time seems to presuppose that the way temporal parameters are used in physical theories puts constraints on the way we should conceive of the structure of time. Here is the answer Newton-Smith proposes to the question "why do we tend to regard time as continuous?":

> [it] is quite simply that the best physical theories we have in fact constructed of the physical world require in their mathematical formulation a time parameter that ranges over the elements of the real number system. [...] We take individual real numbers as denoting instants and project back onto the interval a non-denumerably infinite number of extensionless parts or instants, one corresponding to each real number in the interval. That is, our belief in the continuity of time [...] arises from our projecting onto the world the richness that is present in the mathematical system which we have found to date to be essential to the construction of viable physical theories. (Newton-Smith 1980, p. 118)

Consequently, according to Newton-Smith, "if our best theories should in the end turn out to involve representing time by a discrete timevariable, we would have at least as much reason to regard time as discrete as we now have for regarding it as continuous" (ibid.). Here, we see that Newton-Smith is willing to read the structure of time off the physical theories. However, such an assumption is shaky; accordingly, we reject it. Indeed, we do not feel committed to regard time as "as a non-denumerably infinite number of extensionless parts or instant" even if physical theories represent time by the continuous parameter *t*. The reason why physical theories do so might be that they are not interested in investigating the structure of time and just use time as an element of the general framework in which the descriptions of phenomena hold.

Maddy also denies that the use of a continuous temporal parameter in physical theories commits us to claim that time is continuous. Her argument is slightly different. As we have seen above in the main text, she claims that representing time as continuous in physical theories amounts to hypothesizing that time is continuous. However, she argues that this hypothesis, and more generally "the physical structural assumptions underlying the application of the mathematics are not held to the same epistemic standard as ordinary physical assumptions" (Maddy 1997, p. 156). Even if the hypothesis of the continuity of time is actually made in physical theories, it is not endorsed by scientists. We agree with Maddy's conclusion. However, we reach it in a different way, by denying from the start that representing time by a continuous parameter amounts to hypothesizing that time is continuous. Accordingly, we claim, *pace* Newton-Smith, that the dispensability of the continuous representation of time is not sufficient to argue that the structure of time is undecidable.

At first glance, before any philosophical enquiry, the possible answers to the question whether time is continuous seem to be "yes" or "no". Thanks to Newton-Smith's work, it has been discovered that the answer may also be "undecidabe". In our view, the logical space is even more extended for there is a preliminary question to be asked: is the question legitimate? After having opened up the logical space, we close it by claiming that the question is illegitimate, thus making the former answers irrelevant. Our argument is that the use of a particular representation of time in physical theories (continuous, discrete or even another one) is not relevant to the investigation of the structure of time. Neither Carnap, Newton-Smith, nor Maddy seems to consider the possibility we have mentioned. However, it allows one to understand why DM is viewed by its authors as having no implication whatsoever for an inquiry on the structure of time.

9.3.5 *Conclusion of Section 9.3*

Our aim in this section has been to answer the question: What does it imply to use a discrete representation of time in physics? We have examined two answers that we found implausible. First, we have shown that adopting a discrete representation of time does not imply having to use equations that are not as fundamental as continuous equations. In DM, the use of a discrete representation of time does not imply that discrete equations are model-specific and inevitably derived from fundamental, differential equations. In DM, the fundamental equations of motion are themselves discrete. Second, we have shown that the use of a discrete representation of time does not entail dispensing with the real numbers. For us, it is an important consequence. As far as we know, the distinction between representing time as continuous and using real numbers to represent time has not been introduced yet in the literature. We have made clear that in DM, time is represented as discrete even if instants are defined on real numbers.

We have then argued that an actual implication of the use of a discrete representation of time in physics is that physics can accordingly dispense with the continuous representation of time. In order to make explicit the meaning of this claim, we have first introduced the notion of a *usable* physical theory. It is a theory which enables one to actually describe, predict, and explain physical phenomena. Second, we have distinguished our dispensabilist implications from more ancient, similar claims, namely Carnap's and Newton-Smith's. In our view, the question of the (continuous or discrete) structure of time cannot be legitimately asked when investigating how time is represented in (continuous and discrete) physical theories. We have suggested that the statement "time is continuous", besides being judged true, false, or undecidable, can also be judged irrelevant.

9.4 Time Symbols and Their Interpretation

The aim of this section is to present another important feature of DM with respect to the representation of time, besides its being discrete. Within DM, the time symbol can be either a parameter, as it is the case in most physical theories, or a variable. This state of affairs is distinctive of DM. We investigate its consequences. In order to do so, we first present how it can be that the time symbol is a dynamical variable in DM. In order to study what we can draw from this fact, we analyze the distinction between a variable and a parameter and how it applies to the time symbol. Finally, we discuss what difference it makes to treat the time symbol as a dynamical variable in a physical theory.

9.4.1 The Time Symbol as a Dynamical Variable in DM

In this section, we present the reason why the time symbol has to be considered a dynamical variable vs. a parameter in discrete mechanics. The parameter-variable distinction is not as obvious as it seems to be. A preliminary version of this distinction is that dynamical variables vary with respect to parameters, whereas parameters refer to the quantities with respect to which variables vary. In the following sections, we shall make this distinction more precise, but we begin with the presentation of the situation in DM.

To put it briefly, in DM, it is necessary to represent time as a dynamical variable unless energy conservation is lost. This is due to the Ge-Mardsen theorem whose content, in a nutshell, is the following. It states that when using discrete variational principles, that is, variational principles involving discrete time and space, it is impossible to have two desirable properties together if the time step is kept constant. These desirable properties are (i) the conservation of geometric properties, like symplecticity, and (ii) energy conservation[8] (see Appendix 3 for more details).

So the Ge-Mardsen theorem seems to block up from the start the very possibility of building up satisfying discrete models – if the time step is kept constant. Or, seen from the other side, the theorem entails that time-step adaptation (i.e. variability of the time-step) is needed to have symplectic-energy-momentum conservation. Thus, when one wants to preserve energy conservation in variational discrete models, the distribution of the time steps should be determined by minimizing the discrete action. More precisely, as emphasized by D'Innocenzo et al.:

> In the continuous case, energy conservation is a consequence of Newton's equations for conservative systems for which the Lagrangian does not depend explicitly on time. This is not automatic in an ad hoc discrete mechanics, on the contrary, only by treating both $\boldsymbol{r}$ and t as dynamical variables does one obtain both Newton's law and discrete energy conservation laws from the least-action principle. Therefore the asymmetry present in continuous

[8]"Given the importance of conserving integrals of motion and the important role played by the Hamiltonian structure in the reduction procedure for a system with symmetry, one might hope to find an algorithm that combines all of the desirable properties: conservation of energy, conservation of momenta (and other independent integrals), and conservation of the symplectic structure. However, one cannot do all three of these things at once unless one relaxes one or more of the conditions. [...] It is interesting to note that when adaptive time steps are used, the arguments above no longer apply and indeed in this case it is possible to find integrators that are, in an appropriate sense, symplectic, energy preserving and momentum preserving, as shown in (Kane et al. 1999)." (Marsden 2009, p. 178).

> mechanics (where time is not treated on the same basis as the spatial coordinates) is eliminated. (D'Innocenzo et al. 1987, p. 246)

The sense in which time is represented by a discrete dynamical variable vs. a discrete parameter is thus that within DM, the time symbol is a variable with respect to which the action must be minimized. This means that its values are not arbitrary in the following sense: they are not determined by any external rule, like the rule $t_{k+1} = t_k + h$ that is used in traditional discrete models such as models based on backward or forward Euler integrators (Sect. 9.2.1.1). The values of the time variable are instead determined by the evolution of the other variables.

Now that we have presented the reason why the time symbol is a discrete dynamical variable in DM, we turn back to the analysis of the variable-parameter distinction in the following section. This will allow us to draw some conclusion about the time symbol being a dynamical variable in DM.

9.4.2 *Variables and Parameters*

In order to better capture the distinction between variables and parameters, we begin by briefly presenting a type of model in which the distinction is quite straightforward, namely causal models. Then we turn to the distinction as it holds in the general case. From this background, we move to the way time is represented in various physical theories. A last step will be necessary before examining further the status of the time symbol(s) in DM: we shall describe how time is represented in traditional discrete models.

9.4.2.1 Variables and Parameters in Scientific Models

Because the distinction between variables and parameters is difficult to capture, we first present a case in which it is reasonably clear:

> The distinction between variables and parameters is well defined only in the context of a given model. It is largely a matter of the aims of modelling, a consequence of the somewhat arbitrary but unavoidable division between what is endogenous and what is left exogenous and what is considered as changeable and what fixed. Causal relations are change relating and what are perceived to be the relevant causal relations depends, to some extent, on what kinds of possible changes are thought to be relevant. If the modeller is interested in how some features of a system are determined by other features of a system, she endogenizes these features as dependent variables. If some features are taken for granted or possible changes in these features are thought to be unlikely or irrelevant, they are left as parameters. (Kuorikoski 2012, p. 368)

In causal models, a parameter thus describes the background against which variables vary, whereas variables are used to represent the causal relations one is interested in. It may be noted that according to the model's purpose, a quantity may be sometimes represented as a variable, and by a parameter when the modelers is interested in another question.

Let us now turn to physical models. Not all physical theories represent causal relationships; however, the distinction between variables and parameters still holds in non-causal theories, with the same general meaning. It reflects the different ways the represented quantities may depend on each other. A variable can depend on the variations of a parameter, whereas a parameter has its own source of variation, which is independent of the variation of the other represented quantities.

To sum up, a variable is a symbol of a magnitude whose variations are the topic of the formulae; a parameter helps formulae to say something about the variations of variables.

9.4.2.2 The Case of the Time Symbol

Equipped with this understanding of the distinction between variables and parameters, we now turn to the fate of the time symbol in physical theories, where it is a parameter more often than a variable. For instance, the time symbol is a parameter in classical mechanics, statistical mechanics, quantum mechanics, in hydrodynamics, electromagnetism. As these theories describe the evolution of physical phenomena over time, time is in the background, so to speak. It is not an object of investigation in itself. By contrast, in cosmology, the time symbol can be a variable: it can vary with the scale factor or with temperature. Thus, in contemporary physical theories, there is no uniform representation of time: it is usually a parameter, but it is a variable in some theories.

The general situation in physics is that the physical theories do not have anything to say about time proper. They just use it as an element of the general framework in which the descriptions of phenomena hold. The symbol t has a life of its own in the equations, so to speak. It is independent of the physical laws in the sense that it does not enter into the equations otherwise than in "going further", or "stepping further", that is, in going from one instant to the next.

Two remarks may complement this brief description. (i) Things were different at the beginning of physical theorizing. As is well-known, Newton devoted a few pages of the *Principia* to a discourse about space and time. In his view, a physical theory like his own had to say something about space and time: he conceived space and time as proper objects of physical theorizing and his discourse about them as a necessary component of his theory. (ii) Within modern theories, fragmented as they are, it is striking that the theories which represent time by a parameter are deliberately at odds with the theories that say something about time. For instance, in classical mechanics as well as in non-relativistic quantum theory, time is represented as entirely separated from space. It is as if these theories did not take seriously what relativity (special and general) has to say about the nature of time. For sure, the formal and empirical relationships among these theories are relatively clear and the separation of space and time in non-relativistic theories does not raise any problem of consistency, let alone empirical problems. However, it is notoriously difficult to translate these mathematical relationships into conceptual relationships. Even though the intertheoretical relationships are well-controlled, this does not imply that the conceptual relationships are as controlled. There is a sort of blind spot here: the concept of time in classical mechanics, say, is clearly different from the concept of time in relativity, and the difference cannot be reduced to any approximation relation.

To sum up, there is no uniform representation of time in physical theories, past or modern. The representation of time is even so diverse that it is difficult to articulate the roles played by the symbols representing time in the various theories.

9.4.2.3 The Time Symbol(s) in Discrete Models

In order to complete our review of the various ways time is represented in physics, we now inquire into the way time is represented in traditional discrete models, that is, discrete models not relying on DM.

In traditional discrete models, the discrete symbol t_k is always a parameter in the sense that it is not an object of investigation but rather a vehicle of the investigation. The discrete variables evolve from one t_k to t_k+h etc., where $t_k=kh$ and h a time step. This "*k*" corresponds to the second representation of time discussed by Lesne (this volume, Chap. 4, Sect. 4.5). In these models, "stepping further" means going from one instant to the next[9]. The main difference with continuous models lies in the fact that the set of possible values for tk is discrete rather than continuous[10]. Such continuous time-symbol t is the third representation of time in Lesne's classification (*ibid.*).

There is another difference, however. Contrary to the continuous time-symbol t, the time-symbol t_k is associated to two different temporal notions, namely t_k and k. The discrete time-symbol t_k plays the same role as the continuous time-symbol t: it represents the time over which phenomena occur. k, on the other hand, represents the "time in the model", so to speak. It represents the time over which the representations of the phenomena occur. The rule "$k \rightarrow k+1$", by which k is "stepping further", governs the evolution of symbol t_k in traditional discrete models.

[9]There are further aspects to the use of time symbols in discrete models. On the one hand, there are numerical methods with "stage" like the Runge-Kutta methods. Here, intermediate discrete instants $t_{k,i}$, where i is an integer, are introduced between two consecutive instants t_k and t_k+h. Two levels of time discretization are thus used. On the other hand, there are discrete models with adaptive discrete time symbols: the time step between two consecutive discrete time symbols t_k and t_{k+1} is variable. At each discrete instant, the time step is adapted according to criteria of computational accuracy.

[10]All these notions differ from "the time of the simulation", i.e., the time spent by the computer to solve the discrete model's equations. In the remaining part of the paper, we shall put this notion aside.

Let us take the example of a free particle. With a continuous model, its motion can be described by the equation $x=v_it+x_i$, where x represents the position of the particle, v_i and x_i the initial conditions, and t the time over which the particle is moving. Within a discrete model, the motion of the particle can be described by the equation $x_k=v_it_k+x_i$. Like t, t_k represents the time over which the particle is moving. It can be expressed in seconds, minutes, days etc. depending on the phenomena. t_k is defined as the product of h by k, where h is the discrete time step and k, an integer. k is a mathematical parameter that marks the discrete instants; it does not represent the time over which phenomena happen. However, the increment "+1" in the passage from k to $k+1$ determines the evolution of discrete time from the instant t_k to its successor t_{k+1}. Depending of the size of the time step h, the same increment "+1" corresponds to different evolutions of the time parameter t_k, e.g., one microsecond, one second, one day etc. Thus, because the rule "$k \rightarrow k+1$" governs the evolution of the discrete time t_k, we call it the "time of the model". It is the clock, so to speak, that makes the discrete time parameter t_k evolve[11]. In the following section, we shall investigate the fate of both temporal notions in the framework of DM.

9.4.2.4 The Time Symbol in DM

In order to analyze the time symbol t_k as it is used in DM, it seems to us useful to introduce some distinctions concerning the temporal notions that are at play in physical theories and models. They are meant to summarize and develop the abovementioned points about time symbols.

(i) We first have to mention the ordinary time, that is, the time in which the investigated phenomena take place, or the time over which the states of the investigated systems change. Let us name it "time*".
(ii) In physics, time* is represented by a time-symbol written either t, or, in discrete models and DMs, t_k. As emphasized about, t and t_k are often parameters.
(iii) It is also important to include the "time of the model" in the discussion, i.e., the time over which the *representations* of the phenomena change, from one

---

[11] We interpret k in a temporal way, as the "time of the model". However, it seems that one might also interpret it in a spatial way since the change "$k \rightarrow k+1$" governs also the evolution of the position x_k to x_{k+1}. Nevertheless, at least within DM, k seems to be more a temporal parameter than a spatial one. One reason would be that, to our knowledge, the parameter k shares with t_k an order property: when *k<k+1* one cannot have $t_k>t_{k+1}$. In other words, k and t_k have the same direction. On contrary, while *k<k+1*, one can have $x_k<x_{k+1}$ as well as $x_k> x_{k+1}$. Besides, when discrete equations are implemented on computers, there might be other kinds of "clocks". For example, the computation of the coordinates (x_{k+1}, t_{k+1}) might be implemented in a device *before* the computation of the coordinates (x_k, t_k). There would be a clock that commands to use $k+1$ before k. Since we do not discuss implementation issues and focus on the equations of DM, we leave aside this point.

state to another. In continuous models and theories, there is no difference between the representation of time* and the time of the model: the same symbol t holds for both. In discrete models and in DM, by contrast, it is noted as k, k being an integer. k cannot be but a parameter.

The main difference between DM and other discrete models is that in DM, the way the "time of the model" goes from t_k to t_{k+1} is determined by the equations of the investigated system rather than by a rule that is unconnected to the system's equations. Therefore, t_k is a variable relative to the parameter k. As a result, the representation of time has a dual nature in DM: whereas time* is represented by t_k, the "time of the model" is represented by k. This builds up a sharp contrast with both continuous models and traditional discrete models.

9.4.3 *Discussion*

Let us now analyze the implication of the dual representation of time in DM. We first examine whether it is indispensable in DM. Then we introduce a new distinction within the notion of representation, from which we argue that the indispensability of the dual representation of time may be interpreted in a deflationary fashion.

9.4.3.1 Is the Dual Representation of Time Indispensable in DM?

How are we to interpret the difference between k and t_k within DM? Two options are available. The first one is to claim that the difference between k and t_k is a necessary feature of DM and cannot be dispensed with. This would imply that discrete mechanics allows for new insights about time: according to this interpretation, the fact that t_k varies with the other physical variables says something about the nature of time*. The second option is to consider the difference between k and t_k as an artefact of DM.

The second option looks disappointing because it does not do justice to the fact that within DM, t_k actually depends on the system's dynamics. In order to consider this as an artefact of the representation being discrete, some explanation has to be given why and how an artefact could transform a symbol that is very commonly a parameter into a variable. Now, this explanation looks difficult to find out. Accordingly, in the following, we focus on the first option.

9.4.3.2 Representing by vs. Representing as

According to the first above-presented option, the difference between k and t_k is a necessary feature of DM, which seems to imply that DM has something new to say about time*. To what extent is this implication sound? In this section, we investigate whether adopting the first option necessarily involves jumping to the conclusion that DM says something new about the structure of time*.

In order to do so, let us introduce a distinction within the concept of representation. In what follows, we shall call "a representation" any concrete item (like ink imprints) that is *used* as a representation by an agent, and understood as such by another agent, the receiver. According to this definition, a representation relates four items: the user of the representation, its content, the receiver, and the medium of the representation (e.g., line-drawing, painting, newspaper article, etc.) Here, we focus on the relationship between the content of the representation and the medium. In some cases, this relationship is based on physical interactions, like in photographs, but in many cases, it is only based on conventions. For instance, one may decide that this pencil represents a railway. In this case, the railway is represented by the pencil. The pencil, associated with the appropriate convention, is the medium of the representation. However, the medium of the representation may also play another role. Let us take the classical example of the French politician Thiers represented as a pear (see Fig. 9.3).

In this caricature, Thiers is not represented by a pear but as a pear. This difference may be subtle, but it is important to understand it in order to rightly interpret the meaning of representations. If someone understood the caricature of Thiers as meaning that he is a gardener, it would be a gross error[12]. The difference between "represented by" and "represented as" is central in the way we use pictures and symbols to communicate and share meanings.

Prima facie, the difference between "represented by" and "represented as" seems relevant to the interpretation of symbols in physics. Let us now suggest how this distinction is usually applied to the symbols that are usually used in physics.

On the one hand, when the physical quantity X is represented *by* the symbol x, the symbol x holds for X in the picture provided by the theory in which x is used. "x" is not only a letter; in physics, it is always presupposed to be the name of a variable, taking its values within this or that set of numbers (the integers, the reals, or a discrete set of the reals, etc.). The set may be chosen because it is supposed to be faithful to the nature of X, or for commodity reasons. As an example of the second case, X can be represented by a symbol taking its value in a discrete (resp. continuous) set of numbers without any implication on the discrete (resp. continuous) nature of X.

On the other hand, representing a physical quantity X as an A or a B, for instance, as a variable or as a parameter, involves interpreting X as endowed with the properties of the As or Bs. For instance, representing time *as* a variable involves interpreting time as *dependent* on other variables, whereas interpreting it *as* a parameter involves interpreting it as *independent* of other variables. Representing

[12]In the nineteenth century, pear is known in France to be a metaphor for bourgeois monarchy. This caricature shows thus the affinity of Thiers with this political regime.

Fig. 9.3 "Fleurs, fruits et légumes du jour – La poire – M. Thiers [Flowers, fruits and vegetables of the day – Pear – M. Thiers]", Alfred Le Petit (1871). © Picture RMN-Grand Palais – M. Bellot

X as an A or a B thus involves a (more or less implicit) a claim about the nature of X.

9.4.3.3 Representing by, Representing as, Variables, and Parameters

The above examples seem to imply that physical quantities are represented *as*variables or parameters. This seems to be the common way to use the variable-parameter distinction. In this last section, we want to suggest that it is also possible to interpret "variable" and "parameter" as naming different *mediums of representation* rather than as saying anything about the nature of the represented quantities. Accordingly, we claim that some physical quantities can be represented *by* variables or parameters, in the very same way as they can be represented *by* real numbers.

Let us now present the implications of our claim. According to it, representing a physical quantity *by* a variable does *not* involve any claim about its nature. It does not involve that it is intrinsically linked to other variables, for instance. To put it in other words, in our view, variables and parameters are just representational tools, deprived of any further content or presupposition. In particular, t_k being a variable in DM, whereas k is a parameter whose role is similar to the role of the usual continuous parameter t in continuous theory, does not imply that DM reveals that time* is endowed with a double nature.

We do not generalize our conclusion from the variable t_k in DM to other variables and parameters in physical theories. As already discussed in the Sect. 9.3.4, time plays a special role in physical theories. They do not generally investigate time. That is why we claim that time should be viewed as *represented by* rather *represented as* variables and parameters. However, we do not extend this claim to other physical magnitudes.

To put it in a nutshell, our claim is that the indispensable, dual representation of time* in DM has not implication about the nature of time*. Indispensability is no guide toward ontological conclusions, but rather a guide toward a complete analysis of the means we use to represent quantities in physics.

9.4.4 Conclusion of Section 9.4

In this section, we have first presented the reasons why t_k can be said to be a dynamical variable within DM, whereas t is usually a parameter in continuous physical theories. In order to better appreciate the meaning of t_k being a dynamical variable, we have then discussed the distinction between variables and parameters as it is used in empirical sciences. At last, we have argued that the dual representation of time in DM, even though it is indispensable, has no ontological implications if one accepts that it is possible to represent a physical quantity *by* a variable (resp. a parameter) instead of representing it *as* a variable (resp. a parameter).

9.5 Concluding Remarks

Our aim in this paper has been to investigate the consequences of adopting a discrete physical theory on the representation of time. By doing so we have found out that the fundamental opposition between a continuous and a discrete representation has to be supplemented with three other oppositions.

The first one is the opposition between a discrete representation only using natural and rational numbers (as focused on by Carnap and Newton-Smith) and a discrete representation by means of discrete sets of real numbers, as currently used in DM. The availability of the second type of discreteness radically changes the terms of the old discussion about the dispensability of the continuous representation of time, as we have shown in Sect. 9.3. In particular, we have argued that one can dispense with the continuous representation of time in physical theories. However, we have emphasized that one should not draw any consequence from this result on the structure of time. "Is the structure of time continuous?" is sometimes an irrelevant question when one focuses on how time is represented in physical theories.

The second opposition is between variables and parameters. This opposition is widely used, but its interpretation varies from one domain to the other. We have tried to make clear its meaning in physical theories in Sect. 9.4.

The third opposition we have uncovered is between two sorts of representing, namely representing by and representing as. It is commonly assumed that when a physical quantity is represented *by* a discrete symbol, i.e., a symbol taking its value in a discrete set of numbers, there is no reason to interpret such a use as meaning that the quantity is conceived of as discrete. Representing *by* a discrete symbol does not indicate any wish to foster any metaphysical assumption. By contrast, it is commonly assumed that representing a physical quantity *as* a dynamical variable implies that this quantity plays a certain role among other relevant quantities. Representing *as* a dynamical variable fosters substantial assumptions about the quantity so represented. In Sect. 9.4, we have shown that these common assumptions are misleading. The current state of DM induces us to claim that there is no reason to associate substantial assumptions with the act of representing *as* a variable or a parameter. The variable/parameter opposition should be viewed as of the same type as the continuous/discrete opposition: it does not involve any interpretation advice. Accordingly, it is more appropriate to use the expression "representing *by* a variable (resp. a parameter)" instead of "representing *as* variable (resp. a parameter)".

The conclusion we have just summed up not only applies to DM, but virtually to every other physical theories as well, even continuous ones. It sheds light on the richness of the practices of representing time in physics and helps distinguish between aspects that only pertain to representational skills and aspects that are linked with scientific hypotheses about time.

Acknowledgements We wish to thank the participants of the conference "The Time of Nature, The Nature of Time" for comments and discussion. We are most grateful to Christophe Bouton and Philippe Huneman for helpful comments on previous drafts of this paper. We would like also to thank Paul Humphreys for precious suggestions and remarks on the paper. The authors are grateful to Andrew McFarland for a thorough language check of the manuscript.

Appendices

Appendix 1: Discrete Mechanics and Some Applications

Let us call A_d the action in Discrete Mechanics (for details see (D'Innocenzo et al. 1987; Marsden and West 2001, Chap. 5)):

$$A_d = \sum_{k=0}^{N-1} L_d \cdot (t_{k+1} - t_k) \tag{9.15}$$

with L_d the discrete Lagrangian. The principle of least action states that: $\delta A_d = 0$. It results in the following twofold discrete Euler-Lagrange equations (DEL). The first discrete Euler-Lagrange equation is:

$$(t_k - t_{k-1}) \frac{\partial L_d(t_{k-1}, q_{k-1}, t_k, q_k)}{\partial q_k} + (t_{k+1} - t_k) \partial \frac{L_d(t_k, q_k, t_{k+1}, q_{k+1})}{\partial q_k} = 0 \tag{9.16}$$

The second discrete Euler-Lagrange equation is:

$$\frac{\partial}{\partial t_k}[(t_k - t_{k-1}) L_d(t_{k-1}, q_{k-1}, t_k, q_k)] + \frac{\partial}{\partial t_k}\left[(t_{k+1} - t_k) L_d(t_k, q_k, t_{k+1}, q_{k+1})\right] = 0 \tag{9.17}$$

Let us follow d'Innocenzo et al. for the choice of the discrete Lagrangian:

$$L_d(q_k, q_{k+1}, h_{k+1}) = \frac{1}{2} m \left(\frac{q_{k+1} - q_k}{h_{k+1}}\right)^2 - V\left(\frac{q_{k+1} + q_k}{2}\right)$$

with $h_{k+1} = t_{k+1} - t_k$. Under these conditions, let us solve (i) the free particle system, (ii) the one dimension falling body problem, and (iii) the one dimension harmonic oscillator system.

(i) The *free particle* is the system with $V = V_0$. The DEL become:

$$m \frac{v_{x,k+1} - v_{x,k}}{h_{k+1}} = 0 \tag{9.18}$$

$$\frac{1}{2}mv_{x,k+1}^2 = \frac{1}{2}mv_{x,k}^2 \tag{9.19}$$

with $v_{x,k} = \frac{x_k - x_{k-1}}{t_k - t_{k-1}}$. The solution of the equations is:

$$x_k = v_{x,i}t_k + x_i \tag{9.20}$$

with the initial conditions $t_0=0$, $x_i=x_0$, $v_{x,i}=(x_1-x_0)/h_1$.

(ii) The one dimension falling body problem is the system where $(z_k, z_{k+1}) = mg\frac{z_{k+1}+z_k}{2}$. Thus, the DEL are:

$$v_{z,k+1} - v_{z,k} = -\frac{g}{2}(t_{k+1} - t_{k-1}) \tag{9.21}$$

$$\frac{1}{2}mv_{z,k+1}^2 + mg\frac{z_{k+1}+z_k}{2} = \frac{1}{2}mv_{z,k}^2 + mg\frac{z_k + z_{k-1}}{2} \tag{9.22}$$

with $v_{z,k} = \frac{z_k - z_{k-1}}{t_k - t_{k-1}}$. We follow the resolution of d'Innocenzo et al. (1987) with different notations. The solution of the equations is $t_k - t_{k-1} = h$ and:

$$z_k = -\frac{1}{2}gt_k^2 + (v_{z,i} + gh)t_k + z_i \tag{9.23}$$

with the initial conditions $t_0=0$, $z_i=z_0$, $v_{z,i}=(z_1-z_0)/h_1$.

(iii) The one dimension *harmonic oscillator* system is the system where $V(x_k, x_{k+1}) = \frac{1}{2}K\left(\frac{x_{k+1}+x_k}{2}\right)^2$. Thus, the DEL are:

$$m(v_{x,k+1} - v_{x,k}) = -\frac{K}{4}[(x_k + x_{k-1})h_k + (x_{k+1} + x_k)h_{k+1}] \tag{9.24}$$

$$\frac{1}{2}mv_{x,k+1}^2 + \frac{1}{2}K\left(\frac{x_{k+1}+x_k}{2}\right)^2 = \frac{1}{2}mv_{x,k}^2 + \frac{1}{2}K\left(\frac{x_k + x_{k-1}}{2}\right)^2 \tag{9.25}$$

Following d'Innocenzo et al. (1987) with different notations (in particular see (Cieslinski and Ratkiewicz 2006), we have:

$$x_k = x_0 \cos(\omega_d t_k) + \frac{x_1 - x_0 \cos(\omega_d h)}{\sin(\omega_d h)} \sin(\omega_d t_k)$$

with $\omega_d = 1/h \tan^{-1}\left(\frac{\omega h}{1-\omega^2 h^2/4}\right)$ *and* $\omega = \sqrt{K/m}$.

Appendix 2: Trajectory of a Falling Body in DM

Since $x_k = v_{x,\,i} t_k + x_i$ (see Eq. (9.20) in Appendix) then, $t_k = \frac{x_k - x_i}{v_{x,i}}$. Hence, put it in the Eq. (9.23) in Appendix, we derive the equation of the trajectory:

$$z_k = -\frac{1}{2} g t_k^2 + (v_{z,i} + gh) t_k + z_i \; = -\frac{g}{2 v_{x,i}^2} (x_k - x_i)^2 + \frac{v_{z,i} + gh}{v_{x,i}} (x_k - x_i) + z_i$$

Now, we can deduce the highest position of the body. It is the position where the partial derivative z_k with respect to x_k vanishes: $\frac{\partial z_k}{\partial x_k}\left(x_k^*\right) = 0$. Hence,

$$x_k^* = v_{x,i}(v_{z,i} + gh)/g + x_i \quad \textit{and} \quad z_k^* = z_k\left(x_k^*\right) = (v_{z,i} + gh)^2/(2g) + z_i$$

Appendix 3: Ge-Marsden Theorem

We report here the Ge-Marsden theorem and its proof as they are formulated in the original paper:

> We recall that there are algorithms which exactly preserve energy, some of which also preserve other quantities [...]. However, *these algorithms cannot be symplectic*, according to the following result of Ge:
>
> Let H be a Hamiltonian which has no conserved quantities (in a given class $\mathcal{H}$, for example analytic functions) other than functions of H. That is, if $\{F,H\}=0$, then $F(z)=F_0(H(z))$ for a function F_0. Let $\Phi_{\Delta t}$ be an algorithm which is defined for small Δt and is smooth. *If this algorithm is symplectic, and conserved H exactly, then it is the time advance map for the exact Hamiltonian system up to a reparametrization of time*. In other words, approximate symplectic algorithms cannot preserve energy for nonintegrable systems.
>
> This result is in fact easy to prove. The algorithm being symplectic, is generated by a dependent function $F(z,t)$, which we assume belong to $\mathcal{H}$. Since $\Phi_{\Delta t}$ preserves H, and H is assume to be time independent, F commutes with H, and so $F(z)=F_0(H(z))$. It follows that the hamiltonian vector fields of F and H are parallel, so their integral curves are related by a time reparametrization. (Ge and Marsden 1988, p. 135).

As far we understand this result, while the Hamiltonian system is assumed to be nonintegrable – i.e that an exact solution cannot be constructed – if the symplectic algorithm exactly preserves energy, the solution of the algorithm would be the exact solution of the system modulo a reparametrization. This contradicts the assumption according to which the system is nonintegrable (see also (Ge 1991, p. 380; Marsden 2009, p. 178)). However, as Marsden (2009, p. 179) emphasizes, if the time step is variable, the previous result does not hold and symplectic algorithms can exactly preserve energy.

References

Butterfield, J. 2006a. On symplectic reduction in classical mechanics. In *The handbook of philosophy of physics*, ed. J. Earman and J. Butterfield, 1–131. North Holland: Elsevier.

———. 2006b. On symmetries and conserved quantities in classical mechanics. In *Physical theory and its interpretation*, ed. W. Demopoulos and I. Pitowsky, 43–99. Dordrecht: Springer.

Carnap, R. 1966. *Philosophical foundations of physics*. New York/London: Basic Books.

Cieslinski, J.L., and B. Ratkiewicz. 2006. On simulations of the classical harmonic oscillator equation by difference equations. *Adv. Difference Equ.*: 40171.

Colyvan, M. 2001. *The indispensability of mathematics*. Oxford: Oxford University Press.

D'Innocenzo, A., L. Renna, and P. Rotelli. 1987. Some studies in discrete mechanics. *European Journal of Physics 8*: 245–252.

Feng, K., and M. Qin. 2010. *Symplectic geometric algorithms for hamiltonian systems*. Heidelberg/Dordrecht/London/New York: Springer.

Ge, Z. 1991. Equivariant symplectic difference schemes and generating functions. *Physica D 49*: 376–386.

Ge, Z., and J.E. Marsden. 1988. Lie-Poisson Hamilton-Jacobi theory and Lie-Poisson integrators. *Physics Letters A 133* (3): 134–139.

Greenspan, D. 1973. *Discrete models*. London: Addison-Wesley Publishing Company.

Hairer, E., C. Lubich, and G. Wanner. 2006. *Geometric numerical integration: Structure-preserving algorithms for ordinary differential equations*. Berlin/Heidelberg: Springer.

Humphreys, P. 2004. *Extending ourselves: computational science, empiricism, and scientific method*. New-York: Oxford University Press.

Kane, C., J.E. Marsden, and M. Ortiz. 1999. Symplectic-energy-momentum preserving variational integrators. *Journal of Mathematical Physics 40*: 3353–3371.

Kibble, T.W.B., and F.H. Berkshire. 2009. *Classical mechanics*. London: Imperial College Press.

Kuorikoski, J. 2012. Mechanisms, modularity and constitutive explanation. *Erkenntnis 77* (3): 361–380.

LaBudde, R.A., and D. Greenspan. 1974. Discrete mechanics-A general treatment. *Journal of Computational Physics 15*: 134–167.

Lee, T.D. 1983. Can time be a discrete dynamical variable? *Physics Letters 122B* (3–4): 217–220.

———. 1987. Difference equations and conservation laws. *Journal of Statistical Physics 46* (5–6): 843–860.

Lee, T., M. Leok, and H. McClamroch. 2009. Discrete control systems. In *Encyclopedia of complexity and systems science*, ed. Robert A. Meyer, 2002–2019. New York: Springer.

Maddy, P. 1997. *Naturalism in mathematics*. Oxford: Clarendon Press.

Marsden, J.E. 2009. *Lectures on mechanics*. Cambridge: Cambridge University Press.

Marsden, J.E., and M. West. 2001. Discrete mechanics and variational integrators. *Acta Numer 10*: 357–514.

Newton-Smith, W.H. 1980. *The structure of time*. London: Routledge & Kegan Paul.

North, J. 2009. The structure of physics: A case study. *Journal of Philosophy 106* (2): 57–88.

Stern, A. and M. Desbrun. 2008. Discrete geometric mechanics for variational time integrators. In: *Discrete differential geometry: An applied introduction*, Siggraph 2006 Course Notes, Chap. 15.

Suppes, P. 1957. *Introduction to logic*. New York: Van Nostrand Reinhold.

Wilson, J. 2007. Newtonian forces. *The British Journal for Philosophy of Science 58*: 173–205.

Part III
Dealing with Deep Time (Paleontology)

Chapter 10
Paleontology: Outrunning Time

John Huss

Abstract In this paper, I discuss several temporal aspects of paleontology from a philosophical perspective. I begin by presenting the general problem of "taming" deep time to make it comprehensible at a human scale, starting with the traditional geologic time scale: an event-based, relative time scale consisting of a hierarchy of chronological units. Not only does the relative timescale provide a basis for reconstructing many of the general features of the history of life, but it is also consonant with the cognitive processes humans use to think about time. Absolute dating of rocks, fossils, and evolutionary events (such as branching events on the tree of life) can be accomplished through the use of radiometric dating, chronological signals extractable from fossil growth patterns, and the "molecular clock." Sometimes these different methods of absolute dating, which start from largely independent assumptions and evidentiary bases, converge in their temporal estimates, resulting in a consilience of inductions. At other times they fail to agree, either because fossils and molecules are giving temporal information about different aspects of nature and should not be expected to agree, or because of flawed assumptions that give rise to an inaccurate estimate. I argue that in general, despite the fact that it can be difficult to integrate disparate kinds of evidence, the principle of total evidence should be applied to the dating of evolutionary events. As a historical science, paleontology studies past events we cannot observe directly. This raises questions of epistemic access, meaning that due to the fragmentary nature of the fossil record we may find ourselves without access to the relevant traces to adjudicate between rival hypotheses about the past. The problems and prospects of epistemic access are explored through a case study of the reconstruction of the colors of dinosaurs. The paper closes with a reflection on the Darwin-Lyell metaphor of the fossil record as a highly fragmentary history book, and a call for a reconsideration of the book metaphor in favor of a systems view of the geologic and fossil records.

J. Huss (✉)
Department of Philosophy, The University of Akron, Akron 44325-1903, OH, USA
e-mail: jh76@uakron.edu

C. Bouton, P. Huneman (eds.), *Time of Nature and the Nature of Time*, Boston Studies in the Philosophy and History of Science 326,
DOI 10.1007/978-3-319-53725-2_10

10.1 Introduction

Within evolutionary biology, time has long been considered the province of the paleontologist. Strictly speaking, this is not true. Time itself is just as elusive for the paleontologist as it is for anybody else. Paleontologists do not traffic in time itself, but rather in events arrayed in time, in processes occurring over time, and in techniques for the measurement and representation of time. That said, paleontology does have something distinctive to contribute to our understanding of the deep past, namely its ability to draw justifiable inferences about the temporal sequence of events in the history of life. Although absolute dating based on radioactive decay of atomic nuclei and other physical methods have made it possible to get better and better estimates of the ages of rocks and fossils, paleontology throughout its history has reconstructed the history of life without explicit consideration of duration, but rather by using the relative positions and presumed temporal sequence of fossil-bearing strata, and the biological affinities among the fossil remains those strata contain. Despite refinements in absolute dating, and the introduction of molecular methods, the fossil- and event-based relative timescale has remained the backbone of the paleontological conception of time.

Time in paleontology presents a menu of philosophical issues: the phenomenology of time, the nature of relative and absolute time, and the epistemic issues arising out of attempts to reconstruct the past from its present-day traces. We will begin by contemplating the phenomenology of time in the work of the practicing paleontologist. From research in the psychology of time perception, we know that our sense of time is event-based (Resnick et al. 2012). Thus our general sense of time arises out of the first person perspective we have on our lived experience, augmented by reflection on the events of history. Yet these senses of time are inadequate to paleontology. Consider that the oldest known fossils on Earth, of anaerobic bacteria, were found in Australian rocks dated at 3.4 billion years before present (Wacey et al. 2011). The timespan over which the events studied by paleontologists has occurred is so vast as to merit its own designation, Deep Time (McPhee 1981; see also Ricqlès's Chap. 11 in this volume). Deep Time dwarfs the timescales humans are cognitively equipped to deal with. It is immeasurably long compared to the entire span of human history. So the question is whether anyone can truly grasp Deep Time. Efforts to do so nearly always make use of appropriately scaled spatial metaphors. It is a standard part of the undergraduate training of geologists to produce some spatially extended model (e.g., a ruler, a roll of toilet paper, a piano keyboard) such that the vastness of Deep Time can be compared to the mere hair's breadth (or less) of human history that blemishes the end of it. Spatializing time serves two roles. Existentially, the exercise puts us in our place: as a species, we are a relatively recent addition to Earth's biosphere. Cognitively and heuristically, space often stands in for time itself in the scientific work of the paleontologist. In diagrams, time is represented by an axis. In the field, time is represented by the succession of strata.

If you think about it for a moment, there doesn't seem to be anything special about the vastness of time. Space is vast too. Just like Deep Time, vast spatial distances used in astronomy such as the parsec (30,856,774,879,000,000 m) are also difficult to conceive of without some sort of heuristic. Strangely enough, the most intuitively graspable spatial measure used to render astronomical distances tractable—the light year—does so through an implicit time-for-space substitution. While most of us may not be able to fathom such an immense distance, we certainly have a better handle on just how long (or short) a year is. Space and time are both vast. We tame them by scaling them down to human dimensions.

Yet the spatializing of time is not simply a taming technique. The objects that paleontologists study—fossils—and the matrix in which they are embedded—rocks—are spatially extended records of processes occurring over time (see Chap. 11 in this volume). Because the geologic strata in which fossils are found are formed by a process of sedimentary deposition, younger layers sit atop older layers (Steno's Law [Miall 2004]), and as depositional environments of the past migrated due to shifts in water level and the sinking, rising, and filling of sedimentary basins, there are vertical successions of environments that were once next to one another (Walther's Law [Ebbighausen and Korn 2013]). Of course, certain of these shifts result in old deposits being eroded away too, so the geologic record is also a record of missing time. The paleontologist clambering up and down a rock outcrop to collect fossils or to note their positions is also, in a metaphorical sense, traveling through time. Indeed, the field paleontologist is always oriented to her position in time, and even in cases where rock sequences have been overturned by Earth's tectonic processes, is always keenly aware of whether she is moving backwards or forwards through geologic time. Aware of intervals of time unrepresented by rock, she also learns to "mind the gap" as they say.

Fossils too embody a spatialization of temporally extended biological processes. Consider the fossilized tusks of extinct mammals such as mastodons and gomphotheres (Fox and Fisher 2004). Like the trunks of trees, tusks contain growth rings. The chemical and isotopic signatures of these annual increments allow paleontologists to reconstruct life history, diet, and environment. Snails and corals also grow in such a way that their fossilizable hard parts embody a record of the processes that gave rise to them (Gould 1970). As we shall see, these fossils can also be used to test dates determined via radioactive decay.

Thus, the basic parameters of the human condition, of being a being whose kind recently arrived on the scene and is passing through time, combined with the nature of the objects of paleontological study, make space-for-time substitution a fundamental part of the science of paleontology. In what follows I will discuss: (1) the relative time-scale from the perspective of cognitive science; (2) the relationship between molecular and fossil-based estimates of the timing of evolutionary events; (3) the problem of epistemic access to past events and processes; and (4) Lyell's and Darwin's book metaphor for the fossil record.

10.2 Fossils, Time, and the Human Mind

A complete history of the geologic time scale is beyond my scope here, but no one would doubt that the geologic time scale has its present form for historical reasons (see Tassy's Chap. 12 in this volume). William Smith (1799) worked out the stratigraphy of England and Wales based on the lithologies of the rock layers as well as their contained fossils. Smith recognized that strata could be ordered according to the fossils they contained—the Principle of Faunal Succession (Bjonerud 2005). He also discovered that in cases where the strata were jumbled or missing, fossils were the key to correlating far-flung local sequences into one overall geologic column (Winchester 2002). Georges Cuvier and Alexandre Brongniart (1808) established the stratigraphic sequence within the Paris Basin based on similar principles (see Chap. 11 in this volume).

Over time, stratigraphers worked out the succession of strata locally in various places around the world, and out of the ongoing attempt to correlate local stratigraphy globally came the stratigraphic record (see Chap. 12 in this volume). Paleontology has always played an important role in temporally correlating one local section with the next, as fossil species are identified that are reliable and geographically widespread markers of particular intervals in earth history (these are known as "index fossils"). That the geologic time scale is a product of history is also clear from the current and former names of units of rock and their corresponding units of geologic time. "Cretaceous" evokes the chalk deposits of the Paris Basin studied by Cuvier and Brongniart. The last stage in the Cretaceous—the Maastrichtian—is named for the stratotype found at Maastricht, Netherlands. The Devonian recalls Devonshire, England, where it was first defined (Sedgwick and Murchison 1839; Rudwick 1985). Until they were recently respectively renamed Paleogene and Neogene, the Tertiary and Quaternary periods were vestiges of John Phillips's (1829) periodization: Primary-Secondary-Tertiary-Quaternary. It is important to note that these historical boundaries coincide with evolutionary and stratigraphic discontinuities in the fossil and rock records, often but not always global in scale, and not always strictly contemporaneous at all locations (Newell 1959, 1966). In other words, the geologic time scale reflects not only the human history of its construction, but also the events comprising the history of the earth (e.g., flooding of continents) and of life (e.g., mass extinctions). The relative time scale is fundamentally historical in these two very different senses.

The question is why, given improvements in radiometric dating and better ability to correlate stratigraphic sequences worldwide using additional geophysical data such as the record of reversals in the earth's magnetic polarity recorded in rocks (magnetostratigraphy), geologists and paleontologists continue to cling to the geologic time scale. Why not move to a system of absolute dating of rocks, fossils, and events? In other words, instead of speaking of a Devonian trilobite, why not give the most accurate and precise estimate possible—in years—of when the particular specimen (or species) lived? Can't paleontology just "go digital"? The answer is yes, it can. But even so, there are still good reasons for retaining the relative time scale. Despite all of the history of science frozen into the names of the subdivisions of the geologic time scale, it would be a mistake to explain the persistence of the geologic timescale as nothing more than a curiosity. Not only does its structure reflect major events in the physical and biological history of the earth, but its staying power can also be explained by cognitive science, as we shall see in the next section.

10.3 Paleontology and Geologic Time: The Relative Time Scale

Humans are cognitively ill-adapted to reasoning about interval scales of linear time, but are well adapted to thinking of time in terms of events (Resnick et al. 2012). Thus it should not be surprising that even though multiple means of absolute dating of rocks and fossils have been developed, paleontologists and geologists still use the traditional, event-based relative time scale (i.e., the geologic time scale) to situate events in time, and are constantly trying to date event boundaries more precisely. In other words, the relative time scale tends to remain more or less fixed. The dates of the defining event boundaries—evolutionary events such as mass extinctions of cosmopolitan fauna or geological events such as episodes of widespread continental flooding—are what get revised. For example, the Cretaceous-Paleogene (K-Pg) boundary is defined in the marine stratigraphic record by the extinction of certain species of foraminifera (a microscopic shelled marine organism), and debates over how to define the boundary stem from the problem that foraminifera are not found in deposits from terrestrial environments, in which case it may be argued that instantaneous traces of some alternative event—such as shifts in ocean geochemistry or magnetic shifts recorded in rocks—be used to define the boundary (see Keller 2011 for discussion). Yet what one does not see are proposals to do away with the event-based timescale. Rather, absolute dates are used to improve the temporal resolution of what is essentially an event-based timescale.

Two features of the geologic time scale are worth pointing out in this regard: it is hierarchical and it is event-based. Eons are composed of eras, which are composed of periods, which are composed of epochs, which are composed of stages, and the boundaries between these subdivisions are defined by events in earth history. Its event-based and hierarchical nature helps explain why the geologic time scale is still in use: it happens to correspond to one cognitive model—the category adjustment model—of how humans retrieve information when making temporal estimates. In the category adjustment model, a person retrieves information at the level in a hierarchy of categories that seems appropriate given the question, in light of the location of adjacent boundaries (Huttenlocher et al. 1988). They then use whatever metrical information they may have to adjust their estimate relative to a known category boundary (Resnick et al. 2012). For example, confronted with the task of estimating when in geologic time a given event occurred, a person will default to the nearest boundary of which they are aware, which may be at a coarser scale in the hierarchy than required by the question (Resnick et al. 2012). If we restrict our expectations about "grasping the magnitude of geologic time" to being able to provide accurate estimates of the dates of events in the fossil or rock records, then clearly the hierarchical system of the geologic time scale fits well with a tried and true cognitive strategy. In working with students to give them a better sense of events in the history of life, pedagogical research has shown that students give more accurate estimates when they are taught to treat intervals of geologic time as event-bounded categories within which can be embedded finer scale event-bounded categories (Resnick et al. 2012). If the cognitive scientists are right, then the geologic time scale is too useful to be discarded anytime soon. Nonetheless, for purposes of calculating rates of evolution and extinction, and for testing models of extinction dynamics, it is crucial to be able to improve the accuracy and precision of the ages of fossils, measured in years (Erwin, 2014). In fact, efforts in this regard which involve collaboration among geochemists, geophysicists, geochronometers, paleontologists, and molecular biologists (such as the EARTHTIME project: earthtime.org) provide a fine example of scientific unification that does not involve reducing one scientific theory to another, but rather is a form of methodological integration (Grantham 2004).

10.4 Absolute Time

10.4.1 Radiometric Dating

Radiometric dating relies on the decay of radioactive isotopes whose decay constants (which can be expressed in terms of the time it takes for fifty percent of the parent isotope to decay—the isotope's "half-life") are known. It works by identifying systems, assumed to be chemically closed, containing radioactive isotopes and their daughter products, and uses the ratio between parent isotope and daughter isotope to compute the age of the dated object (often mineral grains from volcanic ash beds). Carbon-14, Uranium-Thorium, and Potassium-Argon are well known systems, whose mode of occurrence and half-lives suit them each to different age ranges. Dates established using radiometric methods are where the buck stops in dating events in the geologic past. But are such dates accurate? How do we know? Is it possible that physical changes in the universe cause alterations in the rate of radioactive decay, possibly by altering the underlying physical constants? Is it possible that changes in pressure and temperature, such as occur in many geological processes, alter decay constants? Other than uniformitarian assumptions about the rate of radioactive decay for various isotopes, how do we know that radioactive decay constants are, well, constant?

10.4.2 Using Fossils to Check Radiometric Dates

Ever since John Wells (1963) first proposed the idea, it has been an ongoing project to develop methods that take advantage of growth features of fossils to check the accuracy of radiometric dates. This is because many marine invertebrates incorporate annual, seasonal, (possibly) monthly, and daily growth bands into their skeletons/shells. This holds out the prospect that information about astronomical time can be extracted from fossils. Working with exquisitely preserved corals found in rocks of Devonian age, Wells (1963) proposed a way to use daily growth bands as a (cheaper) alternative to radiometric dating for estimating the absolute age of fossils and the strata in which they are found. Wells started from the assumption that tidal forces continually put the brakes on the earth's rotation around its polar axis, lengthening the day, and thus that the further into the past we look, the faster the rotation of the earth and the greater the number of days in a year. Astrogeophysicists can calculate the rate at which tidal friction has slowed down the earth's axial rotation (about 2 s per 100,000 years), which yields the number of days per year at points in the past. Wells counted the number of daily growth bands for particularly well-preserved coral specimens from the Middle Devonian of New York and Ontario. The counts ranged between 385 and 410—not very precise—but consistent with a year of 393–396 days as calculated based on astrogeophysical models of changes in the rate of the earth's rotation. This provided proof in principle that growth bands on fossils could be used to date the strata in which the fossils were found, though subsequent research has emphasized some of the practical and methodological difficulties of using fossils in this way, including the possibility that the growth bands may not be a strictly daily phenomenon (Scrutton and Hipkin 1973; Hughes 1985). Nonetheless, paleontologist Stephen Jay Gould (1970, p. 104) considered it a noteworthy reversal of the disciplinary pecking order: "Somehow, the thought that eminent physicists (Runcorn 1966a, b) are studying humble corals does wonders for our self-respect."

Let us assume that some of the difficulties with inferring astronomical time from fossils can be (biologically and mathematically) worked out (Sisterna and Vucetich 1994). Perhaps the growth lines reflect lunar time, or some other calculable periodicity (Pompea et al. 1979). Earth-moon dynamics from the deep past can be independently estimated from sedimentary successions of thinly bedded sandstones, siltstones and mudstones known as rhythmites, which are thought to record the effects of lunar tide periodicities on sedimentation (Mazumder and Arima 2005). We then would have independent lines of evidence for the age of certain strata and their contained fossils, a consilience of inductions. As Forber and Griffith (2011) have argued—following in the footsteps of William Whewell (1840, 1858)—consilience has evidential import in sciences such as paleontology and geology. The estimation of the age of a stratum from growth bands in one of its contained fossils relies on a set of auxiliary assumptions completely independent from those of radiometric dating (paleoastronomy based on invertebrate fossils and that based on rhythmites do share at least one auxiliary assumption in common—

the constancy of Newton's universal gravitational constant G [Mazumder and Arima 2005]). In any case, to the extent that past dates can be estimated by multiple means whose auxiliary assumptions differ substantially, or are entirely independent, the resulting date garners the evidential support conferred by a consilience of inductions (Whewell 1858; Wimsatt 1981; Forber and Griffith 2011).

10.5 Rocks and Clocks: Competition vs. Consilience

Consilience is inferentially powerful, and at times sets hearts aflutter, but independently derived quantities do not always agree (not necessarily a bad thing, so long as we can localize the error, as Wimsatt [1981] has pointed out in his discussions of failures of robustness). This has been the case in the so-called "rocks and clocks" debate over the timing of evolutionary events. Paleontology and molecular biology can both be used to estimate the timing of branching events in the history of life (see Huneman's Chap. 14 in this volume). Paleontology relies on hypothesized evolutionary relationships and dates of fossil occurrences (based ultimately on radiometric dates of the geologic record), whereas molecular biology uses molecular sequence data and estimated substitution rates (calculated using dated fossil occurrences!). The resulting estimates sometimes differ greatly. For example, one molecular clock estimate (Kumar and Hedges 1998) placed the divergence time of modern eutherian mammals 64 million years earlier than the corresponding estimate based on the fossil record (Foote, et al. 1999).

The choice of the term "molecular clock," as proposed by Emile Zuckerkandl and Linus Pauling (1965), was meant to suggest that amino acid substitutions in lineages of molecules (whether DNA or other informational molecules, or "semantides" as they put it) accumulate at a steady but stochastic rate, ticking not like a metronome, but rather more like a Geiger counter (Wilson et al. 1987). Zuckerkandl and Pauling (1962, 1965) observed that the degree of difference between hemoglobin molecules in different species (several species of primates, horse, sheep, pig, shark, bony fish, lungfish, worm) is proportional to their evolutionary distance. On that basis they proposed the molecular evolutionary clock, partly to be provocative, but primarily because they saw the possibility of bringing together molecular biology, morphological evolution, and paleontology (Morgan 1998). As a rough calibration they placed the most recent common ancestor of human and horse somewhere in the "Cretaceous or possibly Jurassic period," resulting in a substitution rate of one amino acid substitution per 11–18 million years (Zuckerkandl and Pauling 1962, p. 201). They knew that there would be differences in the rate of evolution for different molecules, and suspected that the rate of molecular evolution might vary, but Pauling was fond of presenting the clock concept using simple equations that assumed greater constancy than they knew to be the case empirically (Morgan 1998). In addition, they recognized that "back substitutions"—a mutation back to the original base or nucleotide—would go undetected, leading to an underestimate of the number of substitution events that had taken place (Morgan 1998). While it has been said that Zuckerkandl and Pauling did not discover the molecular clock but rather assumed it (Easteal et al. 1995), Pauling pointed out some years later that they did check their molecularly derived age estimates against those that paleontologists had derived based on macroscopic characters (Morgan 1998). At the level of imprecision at which they were working (recall how rough was their placement of the horse-human ancestor), the overall agreement between paleontological dates and molecular dates constituted consilience. Moreover, at a time when the reconstruction of the timing of evolutionary events relied exclusively upon a fragmentary fossil record, Zuckerkandl and Pauling's research constituted proof in principle that molecules could serve a useful role in reconstructing the evolutionary past.[1]

As molecular biology, phylogenetics, and paleontology have become increasingly concerned with quantitative rigor in the handling of data to place evolutionary events in absolute time, there has often been conflict between temporal estimates based on the fossil record, and those derived from molecular data. Part of the problem may be sociological, as paleontologists and molecular biologists (respectively) may be more familiar with the strengths of their own data and methods, and hence distrustful or even dismissive of those of their disciplinary counterpart, but there is also the general problem—the weight of evidence problem—of determining how to assign an appropriate weight to fundamentally different kinds of (sometimes conflicting) evidence (Douglas 2010). The problem becomes especially acute when data fail to agree. If consistency is sought, it is always possible to assign outliers a weight of zero, and to work only with those data that are consistent, but this is epistemically suspect, as consistency is no guarantee of accuracy, and throwing away some data merely because they disagree with other data smacks of *ad hoc*kery (Brochu et al. 2004). Moreover, if we accept or reject a given molecular date solely because it agrees or fails to agree with current estimates, this is tantamount to saying we gain no new information from molecular data, a dubious proposition (Bromham 2006).

There are many reasons why molecular dates and those based on the fossil record may fail to agree. One reason is that the molecular clock does not generally tick at a constant rate, or even at a stochastically constant rate (i.e., it is not analogous to radioactive decay after all [Wilson et al. 1987]). There is variation in substitution rate within and between branches, with some genetic sequences more functionally constrained than others. Another reason is that, to the extent that the molecular clock can be calibrated (i.e., that the average substitution rate can be estimated) based on dates established independently (geologically or paleontologically), there has been disagreement both about which dates to use and how to use them (Conroy

[1]Perhaps surprisingly, although the molecular clock is often associated with the neutral theory of molecular evolution (Dietrich and Skipper 2007), Zuckerkandl and Pauling were operating within a selectionist framework (Morgan 1998). Essentially, they assumed that functional constraints on substitution were sufficiently steady as to allow substitution rate to serve as an approximate indicator of elapsed time (Morgan 1998). For more on the neutral model and its influence on paleontology, see Huss (2009).

and van Tuinen 2003). Due to the incompleteness of the fossil record, the evolutionary origin of a taxonomic group (clade) nearly always pre-dates its first fossil. A given group is at least as old as its oldest correctly identified and assigned fossil, which represents a hard minimum constraint on the age of the group (Benton et al. 2009). That said, establishing a reliable and accurate phylogeny from fossil evidence can be quite difficult. This is both because diagnostic traits are often absent in the earliest representatives of an evolutionary group (because they were not yet evolved), and due to the fragmentary nature of fossil remains more generally. Furthermore there is disagreement over which methods of phylogenetic reconstruction are most likely to yield the "true" evolutionary tree.

Just as students of molecular evolution have attempted either to develop more realistic models of evolution or to use methods that are as insensitive as possible to deviations from the model of evolution assumed (Huelsenbeck 1995), paleontology has increasingly emphasized mathematically modeling the incompleteness of the fossil record, such that first and last fossil appearances are interpreted appropriately in estimating the timing of evolutionary events (Benton et al. 2009). When there is a large temporal gap between a molecular estimate of the age of a group and its oldest known fossil, assumptions about preservation, diversification, and extinction rates can be used to calculate how plausible it is to have such a large temporal gap in preservation (Foote et al. 1999). Ultimately, however, there has been, until recently, too strong a partisan attempt on the part of (some) proponents of molecular clock methods to downplay the usefulness of fossil evidence, and for (some) paleontologists to point to problems with molecular clock methods and the model of evolution on which they are based (Donoghue and Benton 2007). The situation finds a parallel in debates over using stratigraphic information to constrain the topology of phylogenies based on character state data (Smith 2000; Fisher et al. 2002; Grantham 2004; see also Tassy's Chap. 12 in this volume). In the end, paleontology should have taught us a valuable lesson by now: evidence is too hard to come by to be thrown away. If our methods are causing us to discard evidence, then new methods should be devised that adhere to the Principle (trumpeted by Bayesians but more broadly applicable) of Total Evidence (Sober 2009). Increasingly this is occurring, as new methods play to the strengths of morphological, stratigraphic, and molecular data in dating the time-tree of life. The fossil record provides hard minimum and soft maximum ages to help calibrate molecular phylogenies (Benton et al. 2009). In turn, relaxed clock methods, which do not assume constancy of the molecular clock, instead allow multiple models with freely "evolving" substitution rates to compete, with the nodes of the branching tree constrained by dates from the fossil record (Hasegawa et al. 2003). Since it is automatically the case that allowing substitution rates to vary will improve goodness of fit to the data, model selection criteria, such as the Akaike Information Criterion (Forster and Sober 1994), must be used to find those models that will achieve the best fit with the fewest independently estimated substitution rates (Hasegawa et al. 2003). Overfitting the model to the data is to be avoided (Forster and Sober 1994). In addition, assessment of molecular dates should not be viewed purely as model selection in the statistical sense. The assumptions of models of molecular evolution also need to be empirically tested and consistent with what is known of the underlying biology, and sensitivity of date estimates to violations of those assumptions needs to be assessed (Bromham 2006).

It is worth pausing for a moment to consider the significance of the apparent failure of consilience between fossil and molecular dates. In this case the problem appears to have been a failure to recognize that fossil data and molecular data should not be expected to estimate the same quantity. While it had long been assumed that the failure of concordance of fossil-dated calibration points and branching points determined using molecular methods was undesirable, the standard for consistency in these cases is simply that the fossil date be younger than the molecular estimate, as fossils do not record branching points (Benton et al. 2009). Instead the fossil calibration point corresponds to the oldest phylogenetically secure member of the clade, dated with a precision corresponding to that of the fossil-bearing stratum in which it is found. Even so, there are conditions under which well-dated fossil remains could provide evidence against a molecular date. Suppose we have a taxon that is readily fossilized, and that the rocks of the appropriate age and environment exist to be studied, but there are no fossils found in the gap between the molecular estimate of the date of origin of the clade and the oldest known fossil from that clade. We would have good reason to suspect that the molecular estimate is overestimating the age of the clade (Foote et al. 1999; Grantham 2004).

10.6 The Importance of Absolute Dating for Paleontology

As mentioned previously, paleontology has reconstructed much of the history of life on earth using only a relative time scale. As long as the aim was to record the succession of taxa (extinctions and originations); to correlate their occurrences from place to place; to study their morphology, characteristic modes of preservation, ecological or sedimentological context, and taxonomic affinities; an absolute time scale was not needed (nor was one available). With the advent of the diverse, independent to quasi-independent methods of absolute dating mentioned earlier, and international collaboration among geochronometers, such as the EARTHTIME initiative, it is now becoming possible to begin to study events in the history of life with much greater precision and accuracy, and with this ability comes a novel capacity to ask different sorts of questions (Erwin 2014).

Paleontologist Douglas Erwin has said that to really understand the dynamics of evolution, diversification, and extinction, it is necessary to "go digital" and start using absolute dates to compute rates of evolution and extinction and to "deconvolve" specific events around mass extinction boundaries (Erwin 2014). Part of the problem comes from binning the fossil record into the units of the geologic time scale, making time resolution coarser than it needs to be. The limitation of saying a fossil is of, say, late Permian age is that all fossils from the late Permian will be binned together as if they co-existed when in fact the late

Permian spanned millions of years and any given late Permian fossil ultimately was collected from a (perhaps) more precisely datable stratigraphic bed within it. Take mass extinctions for example. Ever since Raup and Sepkoski (1984) found a 26.2 million year periodicity in the fossil record of mass extinctions, an extraterrestrial origin (such as an asteroid impact) for mass extinctions has been hypothesized. Let us set aside the possibility that the durations of stratigraphic stages themselves contain a periodic signal and thus that the periodicity of mass extinctions may be an artifact of placing last appearances in temporal bins (Stigler and Wagner 1987, 1988). Pursuit of the mass extinction periodicity hypothesis has led paleontologists to assume a uniform, recurring cause for mass extinctions, and therefore to direct their efforts toward finding certain kinds of selectivity in the traits and habitats of organisms that survived and perished at mass extinction boundaries. But increasingly it looks as if not all of the mass extinctions had extraterrestrial causes, and attention has turned to each individual extinction and its ecological dynamics (Shen et al. 2011). Here is where improved dating of extinction events becomes critically important. In order to test hypotheses that, for example, there was a disruption in the web of ecological relations, it is absolutely vital to know in some detail which taxonomic groups became extinct in what order around event boundaries (Erwin 2014). Only in this way does it become possible to differentiate between common cause explanations in which all taxa are driven directly extinct and ecological cascades in which directly caused species deletions percolate through trophic webs to cause further extinctions (Erwin 2014). Improved resolution of the timing of evolutionary events can help distinguish between rival hypotheses and improve our epistemic access to the past. The problem of epistemic access to the past is an important challenge for paleontology, one that goes beyond ascertaining the dates of evolutionary events, but to the extraction of biological features from fossil remains, a topic to which I now turn.

10.7 Epistemic Access to the Past: The Colors of the Dinosaurs

Does the study of the past, especially the deep past, present epistemic problems distinct from those of sciences that study the here and now? If so, what is the best way to characterize them? For philosophers who have turned their gaze to paleontology, a consensus has emerged that the key to understanding its epistemic challenges involves grappling with the underdetermination of theory by data (Cleland 2002; Turner 2007; Jeffares 2010; Forber 2009; Forber and Griffith 2011). Carol Cleland (2002) set the terms of the debate by invoking a pervasive time asymmetry of nature that historical sciences such as paleontology exploit in drawing causal inferences. Cleland points out that in causal sequences in general, earlier states of affairs are vastly overdetermined by later states of affairs. For example, to infer that a baseball had shattered a window would require not every single trace of the event but perhaps a few appropriately positioned shards of glass and some fibers of horsehide (Cleland 2002). Cleland's view supports epistemic optimism about inference in historical science. With a superabundance of physical traces of effects, all that need be done is to find a disjoint combination of just the right ones—a "smoking gun"—to eliminate all but one of a number of competing causal hypotheses (Cleland 2002).

Derek Turner (2007) takes issue with Cleland's analysis of historical science. He points out that many questions raised by historical science will forever go unanswered. This is because the passing of time generally is accompanied by the degradation of evidence, and at times by the total loss of evidence (but see Sober and Steel 2014). Information-destroying processes abound (Turner 2007). Thus, even if Cleland is right about the time asymmetry of causation, for many events in the past there will remain rival hypotheses underdetermined by the available evidence. Moreover, in many cases background theories give us good reason to believe that new evidence is not forthcoming—and will not be forthcoming—to break ties between rival hypotheses (Turner 2007). Turner labels this the problem of *local underdetermination*, and it underwrites a certain epistemic pessimism about inference in historical science.

As an example of a question that will forever remain unanswered, Turner chose, "What color were the dinosaurs?" As is now widely known, paleontologists have begun to reconstruct the colors of certain dinosaurs on the basis of fossilized melanosomes, microscopic pigmentation bodies previously interpreted as traces of bacteria. In choosing his example, Turner clearly did not expect this development, and, if we are to be honest, probably most paleontologists did not expect it either. We could easily chalk this up to an ill-chosen example of an unanswerable question—if he could replay life's tape I am sure Turner would have chosen a different example—but I think we should focus our attention on what we can learn from it.

Jeffares (2010) has suggested that the mistake lies in focusing exclusively on the fact that, as time passes, evidence of the past is constantly being lost to information-destroying processes. Rather, he argues, we should think of the possibilities of reconstructing the past against the background of the totality of scientific knowledge. New theoretical, empirical, and technological developments in far-flung fields can lead us to recognize that traces exist where we may not have thought to look. In fact, it is noteworthy that in his analysis, Turner, looking to taphonomy as one of the main background theories in paleontology, characterizes it as the study of processes that distort or degrade the fossil record, whereas it could just as easily be thought of as the science of unlocking information in the fossil record (Bergmann et al. 2012). This holistic position of Jeffares gives support to epistemic optimism about our ability to know the past.

Forber (2009) has analyzed local underdetermination (*sensu* Turner 2005a) and concludes that in a large class of cases, including many in paleontology, our epistemic situation is often such that we are confronted by a set of rival hypotheses, each of which is empirically equivalent relative to a specific data set (i.e., not to all presently available data, nor to all potential data) but that we are in no position at the present time to say what our epistemic state will be in the future. We simply do not know what evidence may or may not be forthcoming. Forber calls this type of underdetermination *contrast failure* and leaves open the possibility that the condition may be temporary.

Forber and Griffith (2011) focus on the issue of epistemic access to the past. They emphasize that one way to gain epistemic access to the past is to, in a sense, triangulate on an estimate of a particular quantity by utilizing evidentiary inferences starting from independent bodies of evidence and sharing as few auxiliary assumptions as possible. As an example, they cite various estimates of the size of the Chicxulub impact crater (caused by the asteroid thought to have been responsible for the end-Cretaceous mass extinction) that agree with each other despite utilizing distinct data sets and relying on independent auxiliary assumptions: a *consilience of inductions* (Whewell 1840, 1858).

Wimsatt (1981, 1987) has not written extensively on epistemic issues in the historical sciences, but his remarks on robustness analysis are a helpful adjunct to discussions of consilience. Forber and Griffith (2011) emphasize the epistemic support garnered by independent lines of evidence that do not overlap (or overlap as little as possible) in their auxiliary assumptions, as each line of evidence will be somewhat immune to criticisms leveled against another line. Yet Wimsatt's concept of robustness emphasizes the extent to which certain derivations will be relatively insensitive to the particular auxiliary assumptions made. Thus, multiple lines of evidence may converge because their different independent auxiliary assumptions do not bias the derivation so greatly that convergence fails to occur. In fact, if each independent line of evidence made some different, faulty assumption, and yet they somehow happened to converge on the same (accurate) estimate of some quantity, this would seem to require an enormously improbable coincidence. This lends plausibility to Wimsatt's view. Also, the existence of multiple independent lines of evidence increases the probability that at least one of the evidentiary lines does not contain a fatal error. A thorough discussion of robustness and its relationship to consilience is beyond the scope of this chapter. Wimsatt (1981) hints at it.

In light of these different characterizations, let us return to the colors of the dinosaurs. How should we understand this example? As a point of clarification it should be noted that birds are part of the dinosaur clade. In common parlance, when most of us speak of dinosaurs, we are excluding birds, instead referring to non-avian dinosaurs, and these are the kinds of organisms whose reconstructed feather colors are under discussion. The primary evidentiary basis for the inferred colors of dinosaur feathers consists of the microscopic fossilized bodies that have been interpreted as melanosomes. Their size, shape, orientation, and distribution in well-preserved fossils is similar to that of feathers in living birds (Fig. 10.1), and to those of fossilized bird feathers as well.

I will focus here on the reconstruction of plumage color patterns in *Anchiornis huxleyi* (Li et al. 2010; Fig. 10.2). A specimen of this non-avian dinosaur was collected from Late Jurassic strata in the Liaoning Province of China. Its feathers

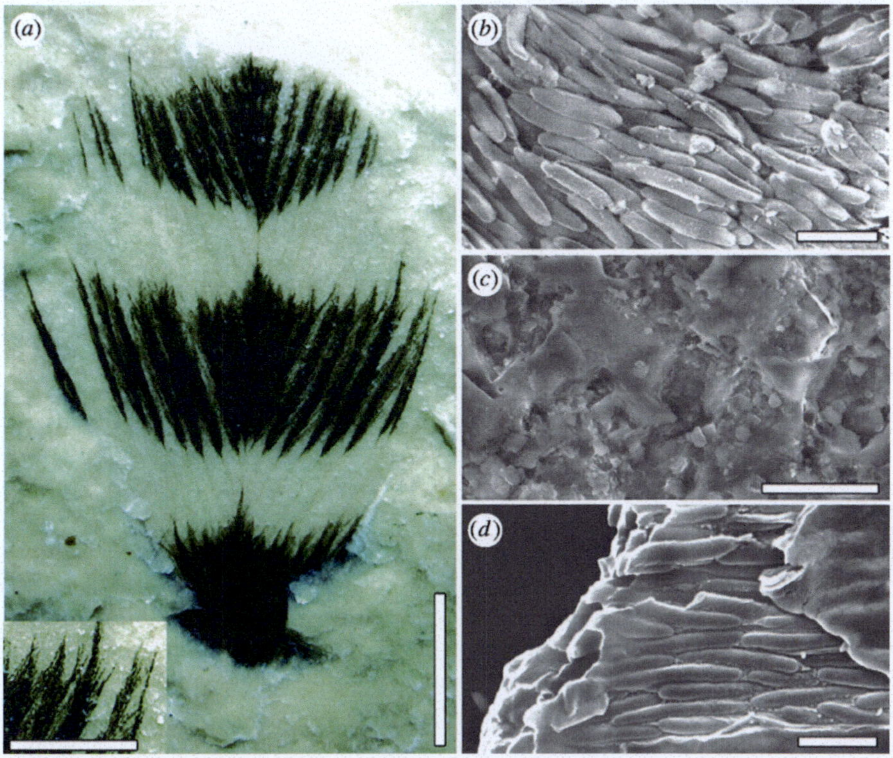

Fig. 10.1 (**a**) Fossil bird feather from Crato Formation, early Cretaceous, Brazil; (**b**) photomicrograph of dark band from fossil, showing what are interpreted as eumelanosomes; (**c**) photomicrograph of rock matrix; (**d**) photomicrograph of dark band of redwing Blackbird feather, showing eumelanosomes (Vinther et al. 2008; used with permission)

were sampled using a scalpel, mounted, and viewed using a scanning electron microscope (Li et al. 2010; Fig. 10.2).

Measurements were taken of the size, shape, and orientation of melanosomes. In order to reconstruct plumage color patterns, the researchers compiled a training set of melanosome measurements from a phylogenetically diverse set of living birds. These measurements were analyzed using discriminant analysis. Discriminant analysis works by identifying the set of measured variables that best separates samples into two or more classes, in this case, color classes: rufous (red), black and gray. The discriminant analysis was "trained" to correctly assign the feathers of extant birds to color classes with an accuracy of 88.9–91.7% (Li et al. 2010). Once the discriminant analysis had been trained on the feathers of extant birds (whose human-perceived color categories are known), the same analysis was used on the measurements of sampled feathers from *Anchiornis huxleyi* (Li et al. 2010). The discriminant analysis "predicted" (in the words of the authors) a combination of red, gray, and black feathers.

Fig. 10.2 An artist's reconstruction of *Anchiornis huxleyi* (from Li et al. 2010). Used with permission

Several sorts of auxiliary hypotheses, of varying degrees of reliability, play a role here, but immediately we can see the problem that Forber (2009) and Jeffares (2010) pointed out with local underdetermination, as articulated by Turner (2007): we were not in a position in 2007 to say what our epistemic state would be in 2010. Perhaps the most important auxiliary assumption here is that the microscopic bodies are melanosomes. Could they be fossilized bacteria instead? It has recently been argued that this possibility cannot be ruled out. In experiments, bacteria on decaying feathers align in a fashion indistinguishable from that of melanosomes (Moyer et al. 2014), but Vinther (2015, p. 6) has argued that these bacteria are far larger than any known melanosome, fossil or living. Ultimately the argument that the bodies are melanosomes rests on their distinctive morphology, their occurrence in feathers, and their similarity in size and shape to those found in feathers in extant birds. Further research on the fossilization of melanin and melanosomes will certainly shed further light on these issues (Colleary et al. 2015).

A second auxiliary assumption that stands out is the assumption that the size, shape, and distribution of melanosomes found in the fossil feathers reflects what these attributes were in life. In fact, the fossilized melanosomes are typically smaller than their corresponding voids (Vinther 2015, p. 5). Since melanosome shape (aspect ratio) is one of the most important variables in color prediction, isometric shrinkage would not affect the predicted color, but distortion of aspect ratio would (Vinther 2015, p. 5).

A third auxiliary assumption is the uniformitarian assumption that the size, shape, and distribution of melanosomes in *Anchiornis huxleyi* feathers yield the same human-perceived color as those in feathers of extant birds. It is important to differentiate three distinct bases for this assumption: analogy, homology and physiochemistry. Reconstructions based simply on simple analogy have a poor track record (Turner 2005b). If the scientists were using modern birds simply as an *analog* for *Anchiornis huxleyi*, this would be the weakest form of support for the color inference. A somewhat better supported inference would have a phylogenetic basis (homology). But here we have a combination of support from both phylogeny and physiochemistry. *Anchiornis huxleyi* feathers are homologous with those of modern birds, and the optics, physics, and chemistry that give rise to human-perceived color in modern bird feathers is well-understood. We have very little reason to believe that the interaction of light with melanosomes of a certain shape was any different optically in the Jurassic from what it is today. That said, there are additional factors in feather coloration—structural iridescence and other pigments besides melanins—and at this stage it is unclear whether these can be recovered from fossil feathers (Vinther 2015, p. 8; but see Li et al. 2012 for a discussion of fossil evidence of iridescence). Yet, the visual effects of co-expressed pigments might well be "masked" in parts of feathers where melanin is present due to its strong light absorption (D'Alba et al. 2012).

Forber and Griffith (2011) argued that a consilience of inductions provides epistemic support to the extent that inferences do not share auxiliaries. They point out (p. 3) that apparent disconfirmations of historical hypotheses by extant traces can often be explained away by shifting blame to suspect auxiliaries, thus the benefit of having multiple independent lines of evidence. Yet the flip side of this observation is also important. The epistemic support of a consilience of inductions is not undermined by overlap in their "secure" auxiliaries (e.g., fundamental laws of physics and chemistry). This means that special attention should be paid to identifying suspect auxiliaries and seeking alternate lines of evidence that do not share those assumptions. An example of this in the present context is the use of synchrotron X-ray fluorescence and X-ray absorption techniques, which are capable of detecting trace metals (organically chelated copper) associated with melanins (Bergmann et al. 2012) even after the melanosomes have been lost to dissolution. The problem is that organically chelated copper is not diagnostic of melanin, as other organic compounds common in nature (such as porphyrins which are common in feathers) are also capable of chelating copper (Vinther 2015).

The prospect of reconstructing the colors of dinosaurs has generated a great deal of excitement, but in our present epistemic state, some pieces of the puzzle are more secure than others. Color patterning (patterns in the presence and absence of melanosomes in particular feathers) is a fairly secure inference, the actual colors somewhat less so. Yet it is premature to say that we will never know the color of (certain) dinosaurs. One promising independent line of evidence for fossil feather coloration comes from fossil feathers encased in amber that apparently retain their original color (McKellar et al. 2011).

10.8 The Book Metaphor

Epistemic access to the past is made all the more difficult by the fact that not only do the events of interest lie at a temporal remove, but the geologic and fossil record are incomplete. Thus, it is not surprising that at least since Lyell, the dominant metaphor for the geologic record has been that of a multivolume history book—with much of the story missing. Darwin picks up on this in the *Origin*, where he writes:

> For my part, following out Lyell's metaphor, I look at the natural geological record, as a history of the world imperfectly kept, and written in a changing dialect; of this history we possess the last volume alone, relating only to two or three countries. Of this volume, only here and there a short chapter has been preserved; and of each page, only here and there a few lines. Each word of the slowly-changing language, in which the history is supposed to be written, being more or less different in the interrupted succession of chapters, may represent the apparently abruptly changed forms of life, entombed in our consecutive, but widely separated, formations. On this view, the difficulties above discussed are greatly diminished, or even disappear. (Darwin 1859, pp. 310–311)

The difficulties to which Darwin is referring are rehearsed throughout Chapter Nine of the *Origin* from which this passage is drawn: the rarity of transitional forms in the fossil record (a.k.a. missing links), the (apparent) abruptness with which new species appear, and the absence of fossil-bearing strata below what was in Darwin's day called the Silurian, which included what we now call the Cambrian (since then, fossils have been found in older rocks).

The implications of the book metaphor are clear. With incompleteness at every scale—from volume to chapter to page to line to word—it is no wonder that all changes in fossil form appear to be sudden and abrupt. The long chain of intermediate living forms expected under Lyellian gradualism, and presupposed by Darwin, are words in the missing lines, or on the missing pages, or in the missing chapters or volumes, of the history of life. While the large-scale patterns of the fossil record find an explanation in Darwin's evolutionary theory, the absence of the expected fossil intermediates and the abrupt appearance of fossils at the base of the Silurian period find their explanation in Lyell's geologic theory.

The incompleteness of the geologic record and with it, the incompleteness of the fossil record, are familiar commonplaces. Sedimentary rocks come in packages, bounded by diachronous surfaces of time unrepresented by rock. The patchy nature of the fossil record led to the oft-expressed lament that evolution always seemed to be happening "off-stage," somewhere other than where paleontologists actually find the fossils. In fact Niles Eldredge (1971) and later Stephen Jay Gould, in arriving at their theory of punctuated equilibrium, which borrowed greatly from Ernst Mayr's (1963) theory of allopatric speciation, "read" the fossil record in exactly this way (Eldredge and Gould 1972). According to punctuated equilibrium, the fossil record we see is one primarily of stasis within large populations. Speciation occurs when small subpopulations become reproductively isolated—by becoming geographically isolated according to Mayr—accumulating mutations and then eventually replacing the parent population (for more on punctuated equilibrium, see Chap. 14 in this volume). On this view, evolution—in the form of speciation—really is happening elsewhere. As the architects of punctuated equilibrium write:

> Since speciation occurs rapidly in small populations occupying small areas far from the center of ancestral abundance, we will rarely discover the actual [speciation] event in the fossil record (Eldredge and Gould 1972, p. 96)

Punctuated equilibrium was certainly received as a controversial and revolutionary theory (Sepkoski 2012, p. 137ff), perhaps largely because for a change paleontologists were seen as contributing to evolutionary theory rather than (merely) documenting evolutionary history (Maynard Smith 1984). Yet there has always been some disagreement over just how "radical" and "non-Darwinian" the theory of punctuated equilibria is: can it be easily accommodated within neo-Darwinism (Mayr 1982), or is it proposing some fundamentally different process involving "systemic mutations" and "hopeful monsters" reminiscent of Goldschmidt (1940), as Gould (1977) thought? But if we set evolutionary theory off to the side for a moment, and compare the way Darwin read the fossil record with the way Eldredge and Gould did, we see a stark contrast. Darwin looked at the fossil record and saw broad scale evidence for evolution, but explained away the absence of intermediate life forms by appealing to Lyell's metaphor of the missing words, lines, pages, chapters, and volumes of history. One of the significant features of Eldredge and Gould's theory of punctuated equilibria is that they interpret the face value fossil record in evolutionary terms. The absence of intermediates is not something to be explained away by the incompleteness of the fossil record, but rather what we might expect if most of the evolutionary action is taking place among small populations on the periphery of ancestral geographic ranges, and that species are in morphological stasis throughout much of their history. Small populations are statistically less likely to leave fossil remains, and gene flow within large populations keeps them in morphologic stasis. Regardless of whether one regards punctuated equilibrium as a Darwinian or non-Darwinian evolutionary process, Eldredge and Gould's reading of the fossil record is different from what Darwin's was.

10.9 Closing the Book Metaphor?

Taking the punctuated equilibrium of Eldredge and Gould (1972) as a point of departure, it may be time to consider the limitations of Darwin's book metaphor (Peters 2011). David Sepkoski (2012) has rightly shown that the metaphor of reading (and re-reading) the fossil record has had heuristic value both for paleontologists and for those who wish to understand their practice. Indeed, new readings of the fossil record are constantly being proposed, and at the very least require revisions to Darwin's book metaphor.

It is time to devote serious study to marine sedimentary rocks devoid of fossils as something other than missing pages from the book of life. Shanan Peters (2007) points out, channeling the voice of the late Jack Sepkoski, that for many fossil-free strata, rather than reflecting the loss or dissolution of once present fossils, it may be the case that in some of these units "nobody's home" (Peters 2007, p. 165). Intensive study of such barren intervals may be helpful in understanding Earth history in terms of systematic interactions among temperature, precipitation, salinity, circulation, sedimentation, and geochemistry in marine environments (Peters 2007). In other words, it may facilitate a systems view of Earth and life history. Moreover, while there is a strong correlation between the amount of sedimentary rock available for sampling and the sampled diversity of fossil taxa over geologic time, it is not necessarily the case that measured fossil diversity is simply a result of how much rock there is to sample (Peters and Heim 2011).

An alternative hypothesis is that perturbations to the earth system are a common cause of both changes in sedimentation and extinction rate (Peters and Heim 2011). Long periods of time unrepresented by rock, such as the Great Unconformity in the Grand Canyon, where the 525 million year old Tapeats Sandstone sits atop the 1740 million year old Vishnu Schist, are not just hiatuses, but rather records of what were surely massive environmental disruptions (Peters and Gaines 2012). Peters and Gaines hypothesize that the massive influx of eroded sediments that bequeathed us the Great Unconformity altered marine geochemistry to the point where biomineralization could occur, leading to a large scale diversification in shelled fossils in the Cambrian period, a phenomenon known as the Cambrian explosion (Peters and Gaines 2012). Intriguingly, molecular dates have placed the origins of marine animals prior to this period (Erwin 2015). If these early animals lacked mineralized skeletons, then the biomineralization hypothesis of Peters and Gaines (2012) may explain why there was a time lag between the early diversification of animals and their appearance in the fossil record.

In summary, barren zones in the fossil record may not be missing pages from the book of life, as Lyell and Darwin envisaged, but may actually be a record of the absence of life from those environments. Likewise, hiatuses in sedimentation, and intervals of time unrepresented by rock, while they may be described as missing pages from the book of life, are also records of perturbations of the earth system that may have played a major causal role in the diversification and extinction of the biota. As Peters (2011, p. 14) put it:

> Geologists would do well to move beyond Darwin's view of the incompleteness of the rock record and remember the words of the French composer Claude Debussy: "music is the space between the notes."

10.10 Conclusion

Paleontologists, as students of the history of life, grapple both with the immensity of time and with the fact that their objects of study lived in the past, often the quite remote past. One factor in the study of the past is the construction of a framework and methods for dating past events as accurately and precisely as possible (Erwin 2014). Fossils have played, and continue to play, a central role in this, not only by serving as the basis for the relative timescale and allowing global stratigraphic correlation, but also by virtue of being, in a sense, "time recorders," literally embodying periodic signals (Wells 1963). Studying the history of life on earth requires drawing reliable inferences about past organisms, events, and processes from present day traces, including molecules in living organisms (Easteal et al. 1995). Progress in understanding the past is limited by the available traces, but the history of science has shown that at any given time, it simply cannot be known what traces will prove to be evidentially relevant (Jeffares 2010). Inferences about the past must be indemnified against certain sorts of risks. Auxiliary assumptions involved in drawing inferences are particularly suspect, meaning that multiple inferential chains that do not share suspect auxiliary assumptions improve the reliability of inferences (Forber and Griffith 2011). So do inferential chains whose auxiliary assumptions are relatively secure. While it is true that in a sense, paleontology is outrunning time, trying to develop new and better ways to access evidence about the deep past despite the fact that traces of the past are lost to information-destroying processes (Turner 2007), it should also be borne in mind that those information-destroying processes are part of the dynamic, living earth system, and feed back into history of life itself (Peters and Gaines 2012). The earth may be like a book—an autobiography (Bjonerud 2005)—but if it is, then the missing words, pages, and chapters are a part of the story itself.

Acknowledgments This paper began as a talk given at the University of Bordeaux, and benefited from feedback received from my fellow symposiasts. My thanks go out to Christophe Bouton, Sharon Cebula, Patrick Forber, Todd Grantham, Philippe Huneman, Shanan Peters, David Sepkoski, Matt Shawkey and Joanna Trzeciak for their comments on the manuscript, and to John Alroy and Douglas Erwin for discussion. I also thank Carl Mendelson at Beloit College for introducing me to the idea of Deep Time.

References

Benton, M.J., P.C.J. Donoghue, and R.J. Asher. 2009. Calibrating and constraining molecular clocks. In *The Timetree of Life*, ed. S.B. Hedges and S. Kumar, 35–86. Oxford: Oxford University Press.

Bergmann, U., P.L. Manning, and R.A. Wogelius. 2012. Chemical mapping of paleontological and archeological artifacts with synchrotron X-Rays. *Annual Review of Analytical Chemistry* 5: 361–389.

Bjonerud, M. 2005. *Reading the rocks: The autobiography of the earth*. Cambridge, MA: Westview Press.

Brochu, C.A., C.D. Sumrall, and J.M. Theodor. 2004. When clocks (And Communities) collide: Estimating divergence time from molecules and the fossil record. *Journal of Paleontology* 78: 1–6.

Bromham, L. (2006). Molecular dates for the cambrian explosion: Is the light at the end of the tunnel an oncoming train? *Palaeontologia Electronica* 9(1.2E). http://palaeo-electronica.org/paleo/toc9_1.htm.

Cleland, C.E. 2002. Methodological and epistemic differences between historical science and experimental science. *Philosophy of Science* 69: 447–451.

Colleary, C., et al. 2015. Chemical, experimental, and morphological evidence for diagenetically altered melanin in exceptionally preserved fossils. *Proceedings of the National Academy of Sciences of the United States of America* 112 (41): 12592–12597.

Conroy, C.J., and M.L. van Tuinen. 2003. Extracting time from phylogenies: Positive interplay between fossil and genetic data. *Journal of Mammalogy* 84: 444–455.

Cuvier, G., and A. Brongniart. 1808. Essai sur la géographie minéralogique des environs de Paris. *Annales du Musée Histoire Naturelle de Paris* 11: 293–326.

D'Alba, L., L. Kieffer, and M.D. Shawkey. 2012. Relative contributions of pigments and biophotonic nanostructures to natural color production: A case study in budgerigar (*Melopsittacus undulatus*) feathers. *The Journal of Experimental Biology* 215 (8): 1272–1277.

Darwin, C. 1859. *On the Origin of Species by Means of Natural Selection*. London: John Murray.

Dietrich, M.R., and R.A. Skipper Jr. 2007. Manipulating underdetermination in scientific controversy: The case of the molecular clock. *Perspectives on Science* 15: 295–326.

Donoghue, P.C. and Benton, M.J., 2007. Rocks and clocks: Calibrating the Tree of Life using fossils and molecules. *Trends in Ecology & Evolution* 22(8): 424–431.

Douglas, H. 2010. Engagement for progress: Applied philosophy of science in context. *Synthese* 177: 317–335.

Easteal, S., C. Collet, and D. Betty. 1995. *The Mammalian Molecular Clock*. Austin: R. G. Lands.

Ebbighausen, R., and D. Korn. 2013. Paleontology as a circumstantial evidence lawsuit. *Historical Biology* 25: 283–295.

Eldredge, N. 1971. The allopatric model and phylogeny in Paleozoic invertebrates. *Evolution* 25: 156–167.

Eldredge, N., and S.J. Gould. 1972. Punctuated equilibria: An alternative to phyletic gradualism. In *Models in paleobiology*, ed. T.J.M. Schopf, 82–115. San Francisco: Freeman, Cooper.

Erwin, D.H. 2014. Temporal acuity and the rate and dynamics of mass extinctions. *Proceedings of the National Academy of Sciences* 111 (9): 3203–3204.

———. 2015. Was the Ediacaran–Cambrian radiation a unique evolutionary event? *Paleobiology* 41: 1–15.

Fisher, D.C., M. Foote, D.L. Fox, and L.R. Leighton. 2002. Stratigraphy in phylogeny reconstruction: Comment on smith. *Journal of Paleontology* 76: 585–586.

Foote, M., J.P. Hunter, C.M. Janis, and J.J. Sepkoski Jr. 1999. Evolutionary and preservational constraints on origins of biologic groups: Divergence times of eutherian mammals. *Science* 283: 1310–1314.

Forber, P. 2009. Spandrels and a pervasive problem of evidence. *Biology and Philosophy* 24: 247–266.

Forber, P., and E. Griffith. 2011. Historical reconstruction: Gaining epistemic access to the deep past. *Philosophy and Theory in Biology* 3: 1–19.

Forster, M.R., and E. Sober. 1994. How to tell when simpler, more unified, or less Ad Hoc theories will provide more accurate predictions. *British Journal for the Philosophy of Science* 45: 1–35.

Fox, D.L., and D.C. Fisher. 2004. Dietary reconstruction of Miocene Gomphotherium (Mammalia, Proboscidea) from the Great Plains region, USA, based on the carbon isotope composition of tusk and molar enamel. *Palaeogeography, Palaeoclimatology, Palaeoecology* 206: 311–335.

Goldschmidt, R. 1940. *The Material Basis of Evolution*. New Haven: Yale University Press.

Gould, S.J. 1970. Evolutionary paleontology and the science of form. *Earth-Science Reviews* 6: 77–119.

———. 1977. The return of hopeful monsters. *Natural History* 86: 22–30.

Grantham, T. 2004. Conceptualizing the (Dis)unity of science. *Philosophy of Science* 71: 133–155.

Hasegawa, M., J.L. Thorne, and H. Kishino. 2003. Time scale of eutherian evolution estimated without assuming a constant rate of molecular evolution. *Genes & Genetic Systems* 78: 267–283.

Huelsenbeck, J.P. 1995. Performance of phylogenetic methods in simulation. *Systematic Biology* 44: 17–48.

Hughes, W.W. 1985. Planetary rotation and invertebrate skeletal patterns: Prospects for extant taxa. *Geophysical Surveys* 7: 169–183.

Huss, J.E. 2009. The shape of evolution: The MBL model and clade shape. In *The paleobiological revolution: Essays on the growth of modern paleontology*, ed. D. Sepkoski and M. Ruse, 326–345. Chicago: University of Chicago Press.

Huttenlocher, J., L. Hedges, and V. Prohaska. 1988. Hierarchical organization in ordered domains: Estimating the dates of events. *Psychological Review* 95: 471–484.

Jeffares, B. 2010. Guessing the future of the past. *Biology and Philosophy* 25: 125–142.

Keller, G. (2011). Defining the Cretaceous-Tertiary boundary: A practical guide and return to first principles. In *The end-Cretaceous mass extinction and the Chicxulub impact in Texas. SEPM Special Publication 100*, eds. Gerta Keller and Thierry Adatte, 23–42.

Kumar, S., and S.B. Hedges. 1998. A molecular timescale for vertebrate evolution. *Nature* 392: 917–920.

Li, Q., K.-Q. Gao, J. Vinther, M.D. Shawkey, J.A. Clarke, L. D'Alba, Q. Meng, D.E.G. Briggs, and R.O. Prum. 2010. Plumage color patterns of an extinct dinosaur. *Science* 327: 1369–1372.

Li, Q., K.Q. Gao, Q. Meng, J.A. Clarke, M.D. Shawkey, L. D'Alba, R. Pei, M. Ellison, M.A. Norell, and J. Vinther. 2012. Reconstruction of Microraptor and the evolution of iridescent plumage. *Science* 335: 1215–1219.

Maynard Smith, J. 1984. Paleontology at the high table. *Nature* 309: 401–402.

Mayr, E. 1963. *Animal species and evolution*. Cambridge, MA: The Belknap Press at Harvard University Press.

———. 1982. Speciation and macroevolution. *Evolution* 36: 1119–1132.

Mazumder, R., and M. Arima. 2005. Tidal rhythmites and their implications. *Earth-Science Reviews* 69: 79–95.

McKellar, R.C., B.D.E. Chatterton, A.P. Wolfe, and P.J. Currie. 2011. A diverse assemblage of late Cretaceous dinosaur and bird feathers from Canadian amber. *Science* 333: 1619–1622.

McPhee, J. 1981. *Basin and Range*. New York: Farrar, Straus, Giroux.

Miall, A.D. 2004. Empiricism and model building in stratigraphy: The historical roots of present-day practices. *Stratigraphy* 1: 3–25.

Morgan, G.J. 1998. Emile Zuckerkandl, Linus Pauling, and the Molecular Evolutionary Clock, 1959–1965. *Journal of the History of Biology* 31: 155–178.

Moyer, A. E., Zheng, W., Johnson, E. A., Lamanna, M. C., Li D., Lacovara, K. J. & Schweitzer, M. H. (2014). Melanosomes or microbes: Testing an alternative hypothesis for the origin of microbodies in fossil feathers. *Science Reports* 4: Article number 4233.

Newell, N.D. 1959. The nature of the fossil record. In *Commemoration of the centennial of the publication of "The Origin of Species" by Charles Darwin. Proceedings of the American Philosophical Society* 103 (2): 264–285.

———. 1966. Problems of geochronology. *Proceedings of the Academy of Natural Sciences of Philadelphia* 118: 63–89.

Peters, S.E. 2007. The problem with the Paleozoic. *Paleobiology* 33: 165–181.

Peters, S. E. 2011. A new view of the sedimentary rock record: Drivers of Earth system evolution are encoded by gaps and rocks. *The Outcrop*, Department of Geoscience, University of Wisconsin-Madison, 12–14.

Peters, S.E., and R.R. Gaines. 2012. Formation of the 'Great Unconformity' as a trigger for the Cambrian explosion. *Nature* 484: 363–366.

Peters, S.E., and N.A. Heim. 2011. Macrostratigraphy and macroevolution in marine environments: Testing the common-cause hypothesis. In *Comparing the geological and fossil records: Implications for biodiversity studies*, ed. A.J. McGowan and A.B. Smith, 95–104. London: Geological Society, Special Publication 358.

Phillips, J. 1829. *Illustrations of the Geology of Yorkshire*. York: T. Wilson.

Pompea, S.M., P.G.K. Kahn, and R.B. Culver. 1979. Paleoastronomy and nautiloid growth: A perspective. *Vistas in Astronomy* 23: 185–205.

Raup, D.M., and J.J. Sepkoski Jr. 1984. Periodicity of extinctions in the geologic past. *Proceedings of the National Academy of Sciences of the United States of America* 81: 801–805.

Resnick, I., Shipley, T. F., Newcombe, N., Massey, C., Wills, T. (2012). Examining the representation and understanding of large magnitudes using the hierarchical alignment model of analogical reasoning. *CSS 2012 Conference Proceedings*, 917–922.

Rudwick, M.J.S. 1985. *The Great Devonian Controversy*. Chicago: The University of Chicago Press.

Runcorn, S.K. 1966a. Corals as paleontological clocks. *Scientific American* 215 (4): 26–33.

———. 1966b. Middle Devonian day and month. *Science* 154: 292.

Scrutton, C.T., and R.G. Hipkin. 1973. Long-term changes in the rotation rate of the earth. *Earth-Science Reviews* 9: 259–274.

Sedgwick, A., and R.I. Murchison. 1839. Classification of the older stratified rocks of Devonshire and Cornwall. *Philosophical Magazine and Journal of Science Series 3.* 14: 241–260.

Sepkoski, D. 2012. *Rereading the fossil record: The growth of paleobiology as an evolutionary discipline*. Chicago: University of Chicago Press.

Shen, S.-Z., et al. 2011. Calibrating the end-permian mass extinction. *Science* 334: 1367–1372.

Sisterna, P.D., and H. Vucetich. 1994. Cosmology, oscillating physics, and oscillating biology. *Physical Review Letters* 72: 454–457.

Smith, W. 1799. *Tabular view of the order of strata in the vicinity of Bath with their respective organic remains.*

Smith, A.B. 2000. Stratigraphy in phylogeny reconstruction. *Journal of Paleontology* 74: 763–766.

Sober, E. 2009. Absence of evidence and evidence of absence: Evidential transitivity in connection with fossils, fishing, fine-tuning, and firing squads. *Philosophical Studies* 143: 63–90.

Sober, E., and M. Steel. 2014. Time and knowability in evolutionary processes. *Philosophy of Science* 81: 537–557.

Stigler, S.M., and M.J. Wagner. 1987. A substantial bias in nonparametric tests for periodicity in geophysical data. *Science* 238: 940–945.

———. 1988. Testing for periodicity of extinction. *Science* 241: 96–99.

Turner, D. 2005a. Local underdetermination in historical science. *Philosophy of Science* 72: 209–230.

———. 2005b. Misleading observable analogues in paleontology. *Studies in History and Philosophy of Science* 36: 175–183.

———. 2007. *Making prehistory: Historical science and the scientific realism debate*. Cambridge: Cambridge University Press.

Vinther, J. 2015. A guide to the field of palaeo colour. *BioEssays* 37(6): 643–656.

Vinther, J., D.E.G. Briggs, R.O. Prum, and V. Saranathan. 2008. The colour of fossil feathers. *Biology Letters* 4 (5): 522–525.

Wacey, D., M.R. Kilburn, M. Saunders, J. Cliff, and M.D. Brasier. 2011. Microfossils of sulphur-metabolizing cells in 3.4-billion-year-old rocks of Western Australia. *Nature Geoscience* 4: 698–702.

Wells, J.W. 1963. Coral growth and geochronometry. *Nature* 197: 948–950.

Whewell, W. 1840. *Aphorisms Concerning Ideas, Science, and the Language of Science*. London: Harrison and Company.

———. 1858. *Novum Organon Renovatum*. London: John W. Parker.

Wilson, A.C., H. Ochman, and E.M. Prager. 1987. Molecular time scale for evolution. *Trends in Genetics* 3: 241–247.

Wimsatt, W.C. 1981. Robustness, reliability, and overdetermination. In *Scientific inquiry in the social sciences* (a festschrift for Donald T. Campbell), ed. M. Brewer & B. Collins, 123–162. San Francisco: Jossey-Bass.

———. 1987. False models as a means to truer theories. In *Neutral models in biology*, ed. M. Nitecki and A. Hoffmann, 23–55. Oxford: Oxford University Press.

Winchester, S. 2002. *The Map that Changed the World: William Smith and the Birth of Modern Geology*. New York: Harper Collins.

Zuckerkandl, E., and L. Pauling. 1962. Molecular disease, evolution and genetic heterogeneity. In *Horizons in biochemistry: Albert Szent-Györgyi dedicatory volume*, ed. M. Kasha and B. Pullman, 189–225. New York: Academic.

———. 1965. Evolutionary divergence and convergence in proteins. In *Evolving genes and proteins*, ed. V. Bryson and H. Vogel, 97–166. New York: Academic.

Chapter 11
The Biologist's Time and Deep Time: Essay on the Psychology of the Paleobiologist

Armand J. de Ricqlès

> *"... Pur di Tempo si breve*
> *voi l'aurora, e non il fine..."*
> Disinganno in
> *Cardinale* Benedetto Pamphilij 's libretto of G.F. Haendel's
> *"Il triumpho del Tempo e del Disinganno"* (1707)

Abstract The paleontologist is familiar with "Deep Time", a concept which emerged during the late eighteenth century, slowly matured, but did not gain general recognition until the end of the twentieth century; although it did create a silent but nevertheless significant revolution in the layman's mind. Being simultaneously a biologist and a geologist, the paleontologist has to deal with issues of Time which, though interrelated, fundamentally differ from those of the current biologist's work. To this extent, the paleontologist may be understood as a paleobiologist: a specialized biologist who studies biological issues within Deep Time. Nevertheless, differences in time scope introduce fundamental discrepancies, especially in the ways that scientific evidence can be demonstrated and, accordingly, in the working and expression of scientific results. All this tends to set the paleobiologist apart psychologically from the current mainstream biologist, whose work is now too often pervaded by urgent biomedical issues. Conversely, the paleobiologist is familiar with general evolutionary issues which now tend to pervade "short time biology", as evidenced by current concerns with, for example, bacterial evolution and the biodiversity crisis.

In Georg Friedrich Haendel's first Oratorio, written during his sojourn in Rome, four allegoric characters: Time, Beauty, Pleasure and Verity, altenatively compete, ally and fight each other. Framed within the typical Christian (and Neoplatonician) sensiblility of the Italian Baroque, its most interesting character, from a modern

A.J. de Ricqlès (✉)
Collège de France, Paris, France
e-mail: armand.de_ricqles@upmc.fr

C. Bouton, P. Huneman (eds.), *Time of Nature and the Nature of Time*, Boston Studies in the Philosophy and History of Science 326,
DOI 10.1007/978-3-319-53725-2_11

perspective, is that of Verity. Called "Disinganno" in Italian, (or disillusion in English, with the meaning of non illusion - or truth), the name strongly connotes the idea of the necessary loss of naivety, which permits the mind to be able to contemplate nude Verity. In the Oratorio, Disinganno thus allies with Time to force Beauty to part with Pleasure - this in order to acknowledge and accept her own fragile nature in this Time dominated world, but her eternity in a timeless Heaven...

It is interesting to note that the same word is used in the title of one of the first books on the true nature of fossils. *La vanna speculazione desigannata dal senso* of Agostino Scilla published in Naples (1670) indeed introduces us to a scientific, realistic interpretation of the nature of fossils and, implicitly, to a new consideration of Time (Zammit Maempel 2007, Romano 2013). This sets the scene for the intimate relationship between the science of Palaeontology and the concept of Deep Time.

11.1 Introduction: The Rise of Deep Time

The concept of Deep Time, put forward by Stephen Jay Gould (1987) and Martin Rudwick (1992, etc.)[1] during the last thirty years or so, has finally made explicit to the layman a so far "silent revolution" in our perception of Time. This revolution, born during the Enlightenment with personnages such as Buffon, matured with Darwin through the nineteenth century and finally blossomed during the first half of the twentieth century, with the development of Palaeontology, Historical geology and Cosmology.

Today, every youngster, everywhere, is familiar with dinosaurs and happily plays with tens of millions of years; but it may be helpful to take a glance backwards, in order to check the magnitude of the Time Revolution – and its psychological consequences – at least at the level of Western Civilisation. Even in our own Western World, millions of people still do not share our current scientific knowledge of the duration of geological and biological time, not to speak of the various situations prevailing in other geographic areas and religious contexts.

For the Western world at least, the main framework from which a psychological perception of Time could first be derived was the Bible. Amalgamated to an expanding Christian faith which, after the fall of classical Antiquity, in turn defined the values, psychology and even the landscape of Western Europe, Genesis remained a matter of undisputable faith for almost two millenia.

Accordingly, our planet is of approximately the actual age announced by biblical texts: about six thousand years. Each plant and animal species is the result of a distinct act of creation, and remains unchanged in its essential character. Man is

[1]According to Martin Rudwick (1992) the expression "Deep Time" appears to have been forged in English by a journalist, John Mc Phee, in a book on the geology of Nevada and Utah : Bassin and Range (1981), before it was popularized by Gould and Rudwick. I thank Dr. Claudine Cohen (EHESS, Paris) to have brought this reference to my knowledge.

practically contemporaneous of the planet that bears him and of all the diverse living beings that exist on it. Finally, the earth experienced a relatively recent and worldwide catastrophic event: the Great Flood.

Hence, it is in accordance with the intellectual framework imposed by Biblical tradition that the actual data, brought by the observation of nature and slowly accumulated through the centuries, were necessarily interpreted and understood. The obligatory literal interpretation of Genesis, with its own short concept of time and its diluvial tradition, constrained the mind for centuries regarding the origin of mankind and of the planet that bears it.

During the Renaissance, despite the realistic interpretation of fossils proposed by Leonardo da Vinci or Bernard Palissy (1580), one still viewed them as "lusus naturae", (perhaps produced by demonic forces to test the good faith of men!). During the next century, the first stratigraphic interpretations of Nicolas Steno in Tuscany were published (1669), but these new data also failed to critically question diluvianism, supposed to account for everything. Strongly constrained by the biblical view of time, *the Sacred theory of the Earth* of Thomas Burnett (1684) remains an astounding example of a first tentative synthesis between "global geology", Genesis and Christian Revelation. Through a grand cyclical vision of Time, our continents are interpreted as mighty ruins left by the Flood.

This book remains for us a moving and fascinating testimony of a human mind still entirely dominated by myth, but nevertheless endeavouring to make sense of the data from nature (Gould 1987).

Buffon's vison of the history of the earth was initially conceived as a prologue to his *Histoire naturelle* (*Théorie de la Terre*, 1749). A decisive methodological step in his work was to set theology apart and separate from natural history. Although Buffon was forced by the Church to renounce (1751) his published views on the earth formation he went on experimenting with the cooling time of large iron balls to get an idea of how old was the earth. In the ultimate form of his view on earth history, *Les époques de la nature* (1778), Buffon published the conclusion of his experiments that the age of our planet should be close to 75,000 years, an unbelievably long amount of time for his contemporaries, although it is known that in his manuscripts he envisaged an age of 2– 3 million years. Buffon conceived a mighty succession of physical (and geographical) changes that could hardly be accomodated within the duration specified by the Bible (Roger 1989).

A great step towards a vision of Deep Time is evident in James Hutton's *Theory of the Earth* (1785) in which he concludes, after consideration of the mighty cycles of production and erosion of rocks, that the evidence suggests "neither a begining nor an end" for geological processes. His vision of an indefinite expansion of Time, although impressive, is nevertheless tempered by a lack of tentative quantification of the duration of geological time, in spite of the trend already foreshadowed by Buffon.

George Cuvier's position towards a vision of Deep Time is sometimes misunderstood. Indeed, his concept of "catastrophes" concentrates enormous amounts of change in a short time, and thus can make the consideration of immense time durations unnecessary, as he notes explicitly, countering Lamarck's views. In his *Discours sur les révolutions du globe*, and the *Recherches sur les ossements fossiles*

(1812–1825) Cuvier tacitly ignores the long duration of geological time, but is mainly impressed by the clear-cut succession of strata, each one with a distinct fossil fauna, a fact well documented by his field work with Alexandre Brongniart in the Paris bassin (1804–1811). This lead him to his famous concept of "catastrophism" to explain the history of the earth as well as the history of life, as epitomized by the extinction of the mammoth (Cohen 2002). In accordance with Cuvier's view, Jameson (1813) interprets the last and most recent of Cuvierian "catastrophes" as the Biblical Flood (diluvium). This explains why the much later recognition of ancient human fossils (1847) became known as "antediluvian".

Nevertheless, and although still generally unrecognized now, Cuvier's intellectual progeny was quite influential towards a better recognition of Deep Time because it focused on the necessity of a precise analysis of the succession of episodes in earth history, as recorded by stratigraphy and the succession of fossils.

Accordingly, in 1822 Conybeare defines the Carboniferous and Brongniart the Jurassic Systems. Alcide d'Orbigny defines 27 geological stages (étages) in 1827, Friedrich von Alberti names the Trias in 1834. Roderick Murchison defines the Silurian in 1839, and the same year, after much controversy, he named the overlaying Devonian System together with Adam Sedgwick. Again, the Permian is named by Murchison in 1841 and finally John Philips defines the great three Paleozoic, Mesozoic and Cenozoic Eras in 1860, although this concept originated as early as 1838. These divisions of geological time (and many others) became so useful that most of them are still accepted and even recognized as "facts of nature" today. Regarding the development of life in Deep Time, the general concept of biostratigraphy, although recognized and used since the end of the eighteenth century (William Smith 1799), was formalized by Louis Dollo as late as 1909.

Charles Lyell 's influential Principles of Geology (1833) not only puts forward the principle of "actualism" (or uniformity of present and ancient causal agents to explain geological data, as opposed to Cuvierian catastrophism) but also refines the methods of assessing the relative age and duration of stratigraphic formations. Lyell used the percentage of extinct to extant species to evaluate the relative age of Tertiary sedimentary rocks. Of course, actualism also opens the mind to the consideration of immensely deep time, provided that the present causal agents are the only ones which ever shaped the geological record. Lyellian concepts did eradicate Cuvier's catastrophism and effectively dominated Geology for the next 150 years, although the accumulation of data over the decades and a sharper epistemological analysis of Lyell's dictum have considerably modulated the situation. Indeed, during the last 30 years or so it became clear that Lyell's Uniformity covers several concepts. While uniformity of law and uniformity of process (actualism) are recognized as sound general principles, Lyell's uniformity of rates (gradualism) and uniformity of state (nonprogressionism) are definitly not warranted by the observation of Deep Time. That rates of processes have not changed and that earth (and life) circle endlessly in a steady state, non progressive way cannot be accepted anymore (Benton 2008).

Charles Darwin, although mainly recognized as a biologist, was also a highly competent geologist (Herbert 2005), who fully grasped and endorsed the concept of Deep Time. Indeed, he had been much influenced by Lyell and had with him *The*

Principles of Geology during his travel on HMS Beagle (1831–1837). Moreover Darwin cogently appreciated that the duration of the "Precambrian" was immensely longer than the later "Phanerozoic", the period in which the plentiful record of fossils makes occurrence of life obvious. In *the Origin of species* (1859) Darwin states that the reader who has not grasped, nor can accept the incredible duration of time necessary to build the tree of life can "close the book at once". Darwin tried to evaluate quantitatively the amount of time necessary to lay down, raise and then erase the cliffs in the geological formations of England, in order to emphasize his concept of Deep Time. During Darwin's lifetime an apparently formidable opposition to Deep Time appeared in the form of Physics, with Lord Kelvin's calculation of the Sun's lifespan and of earth's cooling time. This did not impress Darwin because he was well aware of the constraints imposed by geology. And Deep Time was a prerequisite for his theory of evolution. It was absolutely impossible to accept that earth was six thousand years old, or even a few million years old.

By the end of the nineteenth century a full appreciation of the global stratigraphical scale was reached and the relative order and duration of the main stratigraphic units clearly recognized. It was known that the Cenozoic was shorter than the Mesozoic, itself shorter than the Paleozoic and so on. However the relative datations available left open the exact knowledge of how "deep" was Deep Time: tens and hundreds of millions of years were the orders of magnitude considered, but beyond that, evaluations diverged widely.

It was reserved for the second half of the twentieth century, thanks to radiometric methods using decay curves of various radioisotopes, to create the means of providing absolute datations, rather than only relative ones. In connection with other approaches, notably cosmological, the absolute datations made it possible to firmly set the depth of geological time, while comforting the general patterns already derived from relative datations. So it is now generally agreed that the earth's crust is four billion years old and that life may have started by 3.5 billion years BP (before present). The Neoproterozoic, which is the most recent period of the Precambrian, spans the period from one billion to 545 million years BP. With the Cambrian begins the Phanerozoic (time of visible life), from 545 million years BP to the present. This is in turn divided into the Paleozoic (ancient life), from 545 to 230 million years BP, the Mesozoic (from 230 to 65 million years BP) and by the Cenozoic and Quaternary eras (from 65 million years BP to present). The Quaternary roughly covers the last four million years of earth's history.

11.2 The Psychology of Deep Time

Up to the beginning of the nineteenth century the amount of time elapsed since the period of classical Antiquity was psychologically perceived as enormous. Starting with the Renaissance, the well educated classes generally enjoyed a good historical knowledge of classical Greek and Latin antiquity, and of course of Biblical and Christian traditions. Although this was enhanced by the actual knowledge and practice of the ancient greek, latin and hebraic languages, the time duration represented by the Christian era, principally the Middle Ages, was nevertheless felt as vast. It psychologically pushed back Antiquity to remote times, more or less gently blending, through Homeric or other legendary tales, with those of classical mythology and other myths of the Earth's origins (eg.du Cleuziou 1887). The study of Egyptian antiquity, stirred by Bonapart's "Expédition d'Egypte" (1798–1801) and the subsequent deciphering of hieroglyphs by Jean-François Champollion (1822), extended the duration of Antiquity backwards to more than three thousand years, a length of time almost impossible to grasp by the minds of average persons, although still relevant to biblical chronology. Animal mummies (notably of the sacred Ibis bird) discovered in Egypt were used as "tests" for or against the evolution of species, because it was assumed that, if the views of the transformists (Lamarck, Geoffroy Saint Hilaire) were correct, the "enormous" amount of time elapsed since those animals were alive would have allowed morphological changes (Le Guyader 1998). Retrospectively, the argument suggests how remote from the concept of Deep Time some scientists of the early nineteenth century could remain.

Our current vision is quite different. It is understood now that from Jesus Christ or Julius Caesar to this day, the number of human generations is a mere eighty to one hundred, allowing from four to five generations a century, which makes those personnages roughly our contemporaries! It takes about 25,000 years to produce 1000 human generations, which brings us back to the Upper Paleolithic, close to the origins of artistic expression by *Homo sapiens*. In one million years, one may assume 45,000 generations in the Genus *Homo*, within and beyond or own species. With this duration, one is reaching the scale of Deep Time, where significant evolutionary patterns become clearly noticeable by vertebrate paleontology.

To summarize, the "Deep Time revolution" has entirely altered – even on a subconscious level – the layman's view of the very significance of time. It has schrunk the classic time of History, formerly perceived as enormous, to a close and relatively negligeable proximity. It is a psychological revolution that now, for a person of average education, historical time has been reduced to a mere instant in earth and life history, the last 10,000 years or so since the rise of Neolithic cultures – encompassing all the great human civilisations. Conversely this revolution has extended the perception of time, now replete with non human life and geological events, to a domain within which the mind can move – in imagination, at least – with a feeling of relative comfort and even facility. Has the "Dark abyss of Time" of Buffon (1788) been filled up by an apparent familiarity with it? Yes and no. From a strictly intellectual point of view, yes: it is no longer difficult to play with millions of years, to calculate the motions of wandering continents or biological evolutionary rates. From a subjective point of view, the answer is no: When one observes a fossil in place at the bottom of a formation and, turning up one's head, one sees a hundred meter cliff hanging above, with its monotonous succession of chalk and flint strata, and one realises that its deposition has been a mere instant in earth's history (perhaps half a million years), then this can only be a faint grasp of the meaning of Deep Time for an individual. It took 250 years for "Disinganno" to rid the human mind of Genesis myths in order to reach a scientific vision of Deep Time – and the work is not over!

11.3 Biologist's Time

For our purposes, Time for the biologist may be understood as those time spans within which scientists can directly observe and survey the actual processes of life (reproduction, competition, predation, migration, ...) as they unfold naturally. It also covers the length of time within which scientists can directly experiment on living organisms, in order to decipher various biological processes and functions. These time spans are manifold, going from milliseconds, for some biochemical changes within cells, to years, decades and a few centuries at most, for population changes within species or ecosystemic changes of various scales.

Time is perceived as a separate forth dimension, associated with the three dimensions of Galilean space, which flows continuously and constantly and forms a framework of factual descriptions and causal explanations. It is generally irrelevant, at the level at which the biologist works, whether his "common sense" (Newtonian) view of time is in agreement or not with modern Quantum Physics and Space-Time concept of Relativity (See parts I and II of this book). For instance the biologist is not concerned with time's physical nature, whether it is continuous or discontinuous, whether there is a lower "quantum of time" limit or not, or whether time flow should be "balanced" by something in the universe (entropy growth?) and so on. Obviously the current experimental advances of quantum physics towards the "decoherency limits" which bring quantum effects close to the borders of our "macroworld" may sooner or later meet aspects of the biological realm, but such prospects need not concern our current views of biological work in practice.

Conversely, biologist's time is pervaded by numerous cycles and rhythms. These are easily understood as consequences of the characteristics of our Solar System, which produce daily (diurnal), lunar, seasonal (yearly) and various longer period cycles. These cycles are now understood as external synchronizers which monitor biological activities at various integrative levels : cells, tissues, organs, organisms, populations. It is assumed, from much experimental data, that rythmicity of activities is somehow inbuilt (genetically?) in living organisms, forming the "internal clock", and that external clues (day/night, etc.) do not produce the cyclical activities but rather control them. This reminds one of Bergson's remark (1911, p. 17) that "wherever anything lives, there is, open somewhere, a register in which time is being inscribed." Actually, the synchronizing of biological activities to whatever external cycle that might prove meaningful for survival, appears as a basic adaptation. Cells cannot proceed with all their molecular activities simultaneously and their functions are highly time structured. This holds for higher levels of organisation too, where the synchronism of multicellular activities (eg. as in the secretive epithelium of a gland) may prove highly significant to proper function and ultimately, to survival.

Hence a whole field of chronobiology has blossomed, which tries to decipher the relation of time with life at all and every relevant level, where those relations can be discovered, observed and experimented on (eg. Fraisse et al. 1979). An especially obvious domain of this time-related biology is that of the hard, mineralized tissues of the skeleton; biomineral depositions generally act as excellent time recorders. Examples of daily cycles of deposition are known and well studied among corals, sea shells, fish otoliths (carbonate concretions of the inner ears) or tooth cementum in mammals. Longer period cycles, such as lunar cycles and seasonal (yearly) cycles, are also recorded in the hard tissues of molluscs (shells) and vertebrates (dentine, bone) (e.g. Carter 1990, Ricqlès and Livage 2004). Analogous recordings are also well known in the hard but non mineralized keratinous tissues such as nails, horns or scales, and most prominently among plants in the wood of trees.

Within the framework of time, biology is dominated by a consideration of functions. Functional Biology, as defined by Mayr (1961), deals with the biology of proximal causes, acting within organisms at various levels (from molecular biology to physiology) or between them (population biology, epidemiology, ecology). (See Introduction, this volume). Functional biology is akin to physics and chemistry, as it seeks explanations which are nomological ie. rooted in general laws of nature, and independant of temporal circumstances. Accordingly, for the biologist, to explain is to account for a phenomenon, given known "initial conditions", by using theoretical principles as general as possible.

To sum up, the Time of the biologist usually spans periods ranging from milliseconds to hours, days, years and decades; time spans in which not only observations but actual experimentations can take place. With some emendations, it can be accepted that the biologist's Time may be expanded to include the last few centuries, or even millenia, covering at least a part of historic and prehistoric times, i.e. wherever the actual methodologies of biology can be reasonably used, and conclusions derived from analogy with the present directly drawn. Nevertheless, for such time spans the most powerful tool that biology shares with physics and chemistry – the direct experimental approach – is no longer available. One is left, then, with the methods of the historical sciences which, contrary to Biology *sensu stricto*, cannot generally rely on experimental tests to impose the weight of evidence, as later discussed.

11.4 Paleontologist's Time

Being simultaneously a geologist and a biologist, the paleontologist has to deal with time dimensions (Deep Time: million to billion of years) which differ fundamentally from those in which the ordinary biologist works, although these different time scales may be bridged to some extant by intermediate situations, as mentioned above. As far as he deals with fossils, the remains of past living organisms, and tries to understand them as living beings, the paleontologist may be understood as a paleobiologist: a specialized biologist who studies biological issues within Deep Time. The obvious intellectual framework of such studies is Evolution, both evolution of planet Earth and of the living organisms themselves.

Nevertheless, one aspect of the paleobiolgist's work is functional, roughly the same as the biologist's: namely, trying to decipher the adaptations, functions and life history traits of past organisms, irrespective of the amount of time elapsed since they were alive. This was already George Cuvier's vision of the aim of palaeontology in a non evolutionary context. This approach must rely markedly on what is known about extant organisms and to draw analogies, extending to fossils, *mutatis mutandis*, with the various aspects of current biological knowledge. In this kind of approach, Deep Time is paradoxically irrelevant, almost erased, as if the fossil under study was an extant organism (see also Chap. 10 by John Huss, this volume).

Other aspects of the paleobiologist's work, on the contrary, take Deep Time fully into account: these are the studies of lineage evolution. This includes lineage origination (evolutionary innovations), progressive changes within lineages (anagenesis), splittings (cladogenesis) and extinctions, the combinations of which frame the "Tree of life", as more or less precisely pictured by Systematics. Since the 1970's, Cladistics (or phylognetic analysis) has become the most important method of discovering natural lineages (or clades) and following their evolutionary pathways within Deep Time (see Chap. 12 by Pascal Tassy, this volume). This approach has enabled the retrieval of an unprecendented amount of testable data on the historical development of life through the geological eras. The various contexts in which fossils once were alive are also intimately rooted in Deep Time. The study of past ecosystems, past climates and past geographies, with their slow but ultimately extensive changes in Deep Time are very important components of palaeontology as well as geology. A multi-dimensional "puzzle" is slowly constructed showing how life has spread and evolved in an ever changing scene of wandering oceans and continents, sometimes offering ecological challenges completely different from anything known in our short time world.

Taken all together, the data from the fossils themselves, as well as from their environments, suggest that the rates of change in Deep Time may have varied extensively. There are periods in which one seems to observe slow, progressive changes; there are also episodes when the rate of change (eg. of origination or extinctions of clades) seems to raise significantly, as for the five great "extinction crises" of the Phanerozoic. Ever since Darwin these data have been difficult to assess. The great incompleteness and discontinuity in the stratigraphic columns suggested to Darwin a very imperfect recording of time, in such ways that the "sudden" appearance (or disappearance) of species or higher systematic groupings had to be interpreted as artefacts and not be taken at face value. All we can get are only "occasional scene(s), taken almost at hazard, in an ever slowly changing drama" (Darwin 1872, p. 380). 150 years of progress in the worldwide knowledge of stratigraphic data can leave little doubt, however, that the rates of evolutionary change have tremendously varied, and that there is also an important clade-specific component in rates variation (see Chap. 10, this volume).

A matter of current interest and debate is whether the consideration of evolutionary processes among living (extant) organisms, as studied by evolutionary biologists (genetics, ecology) can be readily extrapolated to Deep Time to explain all aspects of biological evolution, and especially the long time evolutionary patterns revealed by the fossil records. Conversely, are there special causal agents rooted in Deep Time which should be taken into consideration to fully explain the actual historical course taken by life on this planet? A case in point are the dramatic changes in selective regimes which appear to have triggered the great "sudden" phases of mass extinctions, and subsequent expansions of a new biodiversity. There appears to be, in this case, a special, perhaps hazardous surge of "events", which introduce a kind of causality specific to Deep Time (e.g. Gould 2002, Benton 2008, Gayon 2009).

In conclusion, for most paleobiologists, there is a specificity of Deep Time, with its own problems and specific modes of causality which cannot be reached simply by extrapolating causal agents of biological evolution observed in biologist's time.

11.5 Deep Time and Demonstration of the Evidence in Science

A primary consideration is that the paleontologist, as every scientist dealing with historical time, must work on past events and thus cannot readily use the canonical method of the experimental sciences, namely the "hypothetico-deductive method". In this method the experimental test, continually repeated or modulated, plays a fundamental role, answering carefully formulated questions submitted to nature. Since Karl Popper, it is generally accepted that a "successful experiment" (namely one where the expected result occurs) does not prove or demonstrate the ultimate validity of the hypothesis, but only that it has passed the test and remains plausible. Only those hypotheses that are clearly disproved by experimental tests can be known to be wrong. In other words experimental refutation is much more robust than confirmation, and the pathways of experimental sciences should be designed accordingly. Be that as it may, experimental sciences, exemplified by physics and chemistry, are nomological, ie. they attempt to discover "laws of nature", as general as possible, which are often expressed in mathematical terms. These laws are timeless (ie always applicable, irrespective of temporal circumstances). In this context, the data (facts of nature) can always be accounted for (explained), given known initial conditions, by the action of natural laws, as generalized as possible. Also the experimental sciences are, per force, limited to "short time" (ie. compatible with the experimentators life spans). Modern biology, especially in its more functional aspects (molecular biology, biochemistry, populations genetics, physiology…) is very akin to physics and chemistry in its methodology, and intensively relies on experimental tests to bring demonstrative evidence, hence it is rooted in short time.

The modern concept of Deep Time underlines a distinction proposed by Whewell (1840), who defined paletiological sciences apart from nomological sciences. Paletiological – or historical – sciences are "the sciences of what only happened once" and where, accordingly, unique causal explanations must be researched in a more or less remote past (Gohau and Tirard 2005). In this context, to explain is to be able to reconstruct the unique chain of successive causal events and circumstances which necessarily produced the observed result. Although the visions of causality in paletiological and nomological sciences are different, they are not antagonistic, but rather, complementary. Indeed, at each and every step of the historical chain, nomological causality applies. Nevertheless, the historical sciences fully take into account the uniqueness of events surging in the flow of time, raising them as new causal agents (they are changes of "initial conditions" for the nomological sciences). This is why the consideration of Evolution, as an intrinsic consequence of the universal time flow, remains a fundamental dimension of the paletiological sciences.

At the level of Biology, Mayr (1961) has clearly focused on the distinction between the nomological and paletiological approaches. He differentiates Functional biology from Evolutionary biology. The former seeks explanations from within the living organism itself, and relies heavily on experimental evidence, as typically exemplified by physiology. This approach is very close to chemistry and physics, with which biology tends to blend methodologically, apart from the complexity of the objects.

How is it possible to manifest demonstrative evidence in sciences that cannot rely on direct experimentation? This strategy was named Zadig's Method by Thomas Henry Huxley, after the famous character in Voltaire (Cohen 2011). It relies on the comparative use of large quantities of circumstantial evidence. Each manifestation alone does not offer definite proof of the validity of a hypothesis, but merely comforts it. It should also have an heuristical value by giving rise to research of other, potentially comforting circumstantial data which may test the hypothesis. It is the convergence and addition of much circumstantial evidence which, if mutually confirming, may bring, if not a formal proof, at least a firm conclusion that may ultimately raise above reasonable doubt. Note that the concept of test is fully relevant here, but the tests generally cannot be experimental ones. This consiliency method, used in all historical sciences, was consciously used by Darwin, following Whewell (1840), as the epistemological framework of his "long argument" of descent with modification and natural selection (Gayon 1998, 2009). Note that this method, less constraining for the mind than the formal or demonstrative evidence of experimental sciences, is a never ending process of accumulation of data, in which each piece of information reinforces the tentative conclusion. Hence the nature and epistemological value of the methods demonstrating evidence are not the same in experimental and paletiological sciences; this difference should be emphasized, rather than hidden, especially in pedagogical contexts (Ricqlès 2008).

It is noteworthy that one nomological science, once proud of the abstract, timeless generality of its laws, has actually changed its views, to some extant: this is Physics. The highly deterministic classical (Newtonian) physics of the nineteenth century has been forced to include a vision of Time and time changes, since the early twentieth century. This was first evident with the discovery of radioactivity, and generalized later on with the development of modern Astrophysics and Cosmology, which have forcefully introduced a vision of Time-dependant Evolution of the Cosmos itself. Hence the once nomological Physics turns, in parts, into a paletiological science!

11.6 The Functions of "Story Telling"

In their famous paper on "The spandrels of San Marco" (1979) Gould and Lewontin criticized what they called the "adaptationist program" of the Synthesis, and especially the "functional explanations" often put forward within the Synthesis framework, as the corollaries of natural selection. Hence a lot of plausible and ingenious explanations for adaptations were produced, but because they were most often *ad hoc* and not testable, they were criticized as unscientific by Gould and Lewontin and held in contempt as just "story telling".

If I generally agree with Gould and Lewontin's critical analysis concerning non testable and *ad hoc* "evolutionary scenarios", I would like to rescue "story telling" from its Gouldian purgatory (Ricqlès 2008). Clearly, all historical sciences must, sooner or later, use some kind of discursive narrative form, in order to readily communicate the results of their research (Cohen 2002). Evidence for historical explanations must be framed within the natural temporal succession of events, which is also the spontaneous time flow of the Story. Even the inversion of time sequences (flash back) often used in cinema is a paradoxical tribute to the logical power of the story. Indeed, from documented scientific history to fantasy or fiction, and with all the possibilities in between, story telling always uses the same tools of natural language. Of course, it is generally easy to differentiate scientific accounts from fantasy, because the critical apparatus (reference and discussion), obligatorily linked to scientific research, is always prominent in scientific texts. That form of fiction which imitates scientific texts through the use of spurious references can often deceive the naive reader, and is ironically often especially praised by scientists!

To sum up, "story telling" seems to me to be an unescapable tool of all historical sciences (cosmology and palaeontology included). It facilitates the synthesis and communication of state of the art knowledge, as well as new thoughts and discoveries. Again, there is a difference here between the situation of the current biologist and all other practitioners of experimental sciences, for whom "story telling" may be a far less useful tool of communication than statistics and graphs. Because the style of their scientific communications can thus be very different, it is understandable that the psychology and approaches of the current biologist may diverge extensively from those of the paleobiologist.

Another obvious difference, again time governed, is linked to urgency, and consequently to the nature of the scientific questions preferably envisioned by both. The current biologist is overwhelmed by urgent, practical, even vital, issues, such as improving food and water availability and quality, pest control, animal and plant selection, etc. Biomedical applications of fundamental biology currently explode with issues such as the prevention and control of epidemics, rare genetic diseases, preimplantation and prenatal genetic exploration, with all the ethical problems involved; add to this the experimentation of new drugs, functional brain explorations by MNR, etc. In view of all these pressing demands, the harrassed biologist – and society at large – may perhaps consider the Deep Time immersed activities of the paleobiologist and of the "historical biologist" as a waste of time, at best!

Nevertheless, there is now a definite – if still poorly recognized – feedback of Paleobiology to short time biology itself. The current knowledge and understanding of Evolution rooted in Deep Time will make a greater practical contribution to short time biology than was once expected. Paleobiologists (and comparative biologists at large) are familiar with the evolutionary history of life, and this gives them a unique expertise on the study of phylogenetic relationships between organisms, evolutionary processes, rates of sustenable environmental and evolutionary change, patterns and processes of adaptations to various environments, the dynamics of biodiversity changes, etc. This "natural history" knowledge is priceless. If sixty years ago practical biologists and medical doctors had been more familiarized with Biological Evolution at large, massive worldwide mismanagement of pesticides and antibiotics could have been prevented. Current involontary selection of variants of bacteria, parasites and insects resistant to new molecules, resulting from this past (and current) mismanagement, offer "large scale evolutionary experiments"; nicely illustrating evolutionary processes, of which the hazardous results are now fully appreciated. An intimate knowledge of the lessons of the past (remote and recent) is encapsulated in the knowledge and experience of paleobiologists and natural-history trained field biologists. If taken seriously by decision makers, this knowledge will be an invaluable asset, in the decades to come, in taming the current dramatic loss in worldwide biodiversity and other related problems, eg. population displacments, invasive species, incidence of GMO, etc. No scientific knowledge is ever useless.

11.7 Concluding Remarks: On Deep Time, Fossils and Beauty

For the scientist as well as for the layman, there is often a strong aesthetic appeal in fossils. The regularity of shells or ammonites, subtly suggesting their own growth processes in accretional striae or spiral development, the clean, stony nature of the materials, the highly complex, elegant, non developable curvaceous shapes of bones, the intriguing contrasts with neighbouring minerals – all arouse interest and fascination (e.g Halstead and Middelton, 1972). Truly, here one can decipher, frozen into fossils, the dynamics of life itself, at all levels of integration. The almost abstract, beautiful shapes and textures of most fossils often make them praiseworthy ornaments in exhibits, museums and even private homes. This is not the case with the rare fossil that clearly displays its exact origin: a living organism overwhelmed by death; when the non mineralized parts (muscles, skin, hairs or feathers) have left imprints, we witness hard evidence of the post mortem decay processes of a corpse in a sedimentary environment. This far less sympathetic situation (which often, however, provides unique information, as for "feathered dinosaurs", see also Chap. 10 by John Huss, Sect. 10.6, this volume) emphasizes that, rather than a contemplation of death, what scientists and laymen alike really seek in fossils and their aesthetics is to capture life itself, especially a life unfamiliar and long past.

Thus, returning to the allegories we started with, indeed Beauty in living beings is frail and vanishes quickly, within the short time of individual human experience, which is also the kind of time familiar to the current Biologist. This has been the inspiration of countless poems and artworks, which adorn every human language and culture. The situation is quite different for the Paleobiologist. Here, Beauty in fossils indeed wins over Time, and the Paleobiologist has the privilege of contemplating and understanding her, sometimes down to the last minute life-induced detail and its significance; as if he were himself granted some sort of eternity, the ultimate illusion he is so well aware of.

Acknowledgments I thank Christophe Bouton and Philippe Huneman who offered me the possibility to submit an essay in the framework of the written version of the Symposium on Time they organized at the Bordeaux Montaigne University in September 2012. Discussions, over the years, on Time and Time-related matters with Jean Gayon (Paris 1 University), Philippe Taquet and Pascal Tassy (MNHN, Paris), Hervé Le Guyader (UPMC, Paris), the members of the GEPS (groupe d'étude pratique de squelettochronologie, Paris 7 University), Kevin Padian (U of C, Berkeley), Claudine Cohen (EHESS, Paris) and the late Stephen Jay Gould (Harvard University) have been illuminating experiences. I thank Arlène Thiel (Paris) and Adam Hocker for suggestions and revision of the English text.

References

Benton, M.J. 2003. (second edt. 2008) *When life nearly died – The greatest mass extinction of all time*. London: Thames and Hudson.

Bergson, H. 1911. *Creative evolution*. Trans. A. Mitchell. New York: Henry Holt and Company.

Carter, J.G., ed. 1990. *Skeletal biomineralisation: Patterns, processes and evolutionary trends*. Vol. 2 Vol. New York: Van Nostrand and Reinhold.

Cohen, C.L. 2002. *The fate of the mammoth – Fossils, myth and history*. Chicago: The University of Chicago Press.

———. 2011. *La méthode de Zadig*. Paris: Le Seuil.

de Ricqlès, A., Livage J. (Eds.) 2004. Biomineralisation: Diversity and unity. Comptes Rendus Palevol, 3 (6–7), 435–628.

———. 2008. L'Evolution , nouveau « récit de création » ou synthèse de toute la Biologie ? In *Le Récit, Actes de savoir* 4, ed. W. Marx, 13–18. Paris: PUF.

Du Cleuziou, H. 1887 *La création de l'Homme et les premiers âges de l'humanité*. Paris : C. Mapron et E. Flammarion.

Fraisse, P., F. Halberg, H. Lejeune, J.A. Michon, J. Montangero, J. Nuttin, and M. Richelle, eds. 1979. *Du Temps biologique au Temps psychologique*. Paris: PUF.

Gayon, J. 1998. *Darwinism's struggle for survival : Heredity and the hypothesis of natural selection*. Cambridge: Cambridge University Press.

———. 2009. Death or persistance of Darwinism? A philosopher's point of view. *Comptes Rendus Palevol* 8 (2–3): 321–340.

Gohau, G., and S. Tirard (eds.). 2005. Les sciences des causes passées. *Cahiers François Viète* 9–10. Université de Nantes.

Gould, S.J. 1987. *Time's arrow, time's cycles : Myths and metaphor in the discovery of geological time*. Cambridge, MA: Harvard Univ. Press.

———. 2002. *The structure of evolutionary theory*. Cambridge, MA: Harvard University Press.

Gould, S.J., and R.C. Lewontin 1979. The spandrels of San Marco and the Panglossian paradigm: A critique of the adaptationist program. *Proceeding of the Royal Society of London B* 205: 581–598.

Halstead, L.B., and J. Middelton. 1972. *Bare bones, an exploration in art and science*. Edimburgh: Oliver and Boyd.

Herbert, S. (2005). Charles Darwin, *Geologist*. Ithaca: Cornell University Press.

Jameson, R. 1813. *Essay on the theory of the earth, with mineralogical notes and an account of Cuvier's geological discoveries*, Edinburgh.

Le Guyader, H. 1998. *Geoffroy Saint-Hilaire – un naturaliste visionnaire*. Paris: Belin.

Mayr, E. 1961. Causes and effects in biology. *Science* 134: 1501–1506.

Mc Phee, J. 1981. *Basin and range. annals of the former world*. New York: Farrar, Straus and Giroux.

Roger, J. 1989. *Buffon, un philosophe au jardin du Roi*. Paris: Fayard.

Romano, M. 2013. "The vain speculation disillusioned by the sense": The Italian painter Agostino Scilla (1629–1700) called "the discoloured", and the correct interpretation of fossils as "lithified organisms" that once lived in the sea. *Historical Biology* 26 (5): 1–21.

Rudwick, M. 1992. *Scenes from deep time - Early pictorial representation of the prehistoric world*. Chicago/London: The University of Chicago Press.

Scilla, A. 1670. *La vana speculazione disingannata dal senso (etc.)*. Napoli: Andrea Colichia.

Whewell, W. 1840. *The philosophy of inductive sciences, founded upon their history*. London : J.W. Parker.

Zammit, M.G. 2007. *Illustrators and their illustrations of maltese fossils and geology – A historical and biographical account*. Malta: P.E.G.

Chapter 12
The Making of Paleontological Time

Pascal Tassy

Time passes slowly up here in the daylight We stare straight ahead and try so hard to stay right

(Bob Dylan, *Time passes slowly*)

Abstract Usually, the notion of time shared by paleontologists is an arrow. Hence, time is a synonym of duration, either when discussing questions relevant to one thousand years or to a billion years. Geological time, or “deep time”, is a simple concept. This concept of time has been built from lithology and from the study of faunas and floras inside these lithologic layers, or, differently said, from fossils. Stratigraphy, as a science, rests on lithology and fossils. Modern methods which produce radiometric dates (absolute ages) confirmed two centuries of stratigraphers'observations and speculations. Nevertheless, comparison of the linear time of geology and the branching time of living beings is a source of severe difficulties. Although they seem familiar, the relationships between life and time, between differentiation and duration, or between evolution and chronology are rather complex. Time and fossils are less secure allies than usually thought. The deployment of living beings in the temporal dimension creates multiple parallel durations and independent developments. Hence, the historical reading of the tree of life and the interpretation of geological time are not immediate. The notions of hierarchical time and of temporal paralogy developed in this paper, demonstrate these difficulties, often underestimated.

P. Tassy (✉)
Muséum National d'Histoire Naturelle, Paris, France
e-mail: pascal.tassy@mnhn.fr

C. Bouton, P. Huneman (eds.), *Time of Nature and the Nature of Time*, Boston Studies in the Philosophy and History of Science 326,
DOI 10.1007/978-3-319-53725-2_12

12.1 An Introduction with Google

The eleventh of January 2012, Sténon's 374th anniversary was celebrated by Google: that day websurfers discovered a geological logo (Google letters were drawn in the form of geological layers). This dramatic cultural signal pleased geologists and paleontologists who all know what they inherited from the Danish bishop. Nicolas Sténon (1663–1686) or Nicolai Stenonis, aka Niels Stensen, is the founder of that part of geology which studies strata (rock layers) called stratigraphy, at least of its major concept (Stenonis1669). He acknowledged the relation between the superposition of strata and the irreversibility of time. Hence he has been named "une sorte de Champollion de la géologie" (a kind of Champollion in geology) (Gohau1987: 67).

Sténon's discovery is the starting point of the notion of geological time. The stratigraphical and paleontological background of deep time (geological time) is discussed in a first section.

In the second section, it is acknowledged that the graduation of the geological time (the stratigraphic time table) fits with the arrow of time (a linear concept), and the deployment of the tree of life during the geological time (phylogeny) fits with a mathematical hierarchy. Phylogeny is the history of living beings and phylogenetics is its study. Phylogenetics produces phylogenetic trees. Their pattern is structured by common ancestry, or by propinquity of descent, to use Darwinian words. With its degrees of common ancestry this pattern is a hierarchy.

The contradictory relations between the arrow and the hierarchy are discussed in the third section. The phylogenetic hierarchy is a branching pattern which displays an internested set of nodes (nodes are common ancestry, or, better said, are hypotheses of common ascent). It does not display real biological ancestors (that is populations – or varieties in Darwinian words) which are unknowable in deep time. Regardless, the pattern implies that there are degrees of ancestry and as a consequence an implicit chronology. Because phylogeny is not a one-dimensional tree-trunk (linear), but involves instead a multidimensional branching (non linear) of independent histories that derive from these various nodes (ancestries). For example, among amniotes (the branch of tetrapod vertebrates comprising the amphibians, reptiles, birds, and mammals), the timing of the interrelationships of mammals is independant of the timing of the interrelationships of birds, as among mammals the timing of interrelationships of primates is independent of the timing of interrelationhisps of carnivores. The consequence of this pattern is that the phylogenetic time is at odds with the arrow of time. The conclusion is that the quest for a measured fit between stratigraphy (arrow) and phylogeny (hierarchy) is somehow doomed to fail. This situation is explained by the concept of paralogy. Paralogy applies to any structure that combines independent parallel logics and exemplifies the error of mixing two independent logics. Temporal paralogy explains why the more complete a phylogenetic tree with multiple branchings and chronologies is, the less informative it is with respect to its fit with stratigraphy.

12.2 Stratigraphic Succession

One's perception of time in daily activities has nothing to do with the paradoxical nature of time in physics. Despite this, I would say that the notion of geological time falls in the middle of these two views. As with physics, deep time for the paleontologist is an abstraction, and as such, is a difficult notion to grasp. On the other hand, when I evaluate how much time I need to correct the pile of examinations on my desk, and when I touch a series of lithologic layers that make an outcrop, I feel the same, and therefore feel that I have a direct experience of geological time.

Reiterating, stratigraphy is the prior key to understand geological time. Stratigraphy involves knowledge of layers and of their faunas and floras. The other key is the evolution of fossils, and its sequels, biostratigraphy and phylogenetics. These two disciplines involve more than the study of the succession of faunas and floras. The first is datation according to the evolutionary level of species, and the second is the study of relationships between species or taxa (a taxon is a species or a group of species). We know that early stratigraphers and paleontologists were creationists and fixists, and as such did not have a clear idea of deep time. In France it is not a surprise that the first theoretician of evolution, Jean-Baptiste Monet de Lamarck, was also a promoter of deep time, that is of an infinite time. Among many activities, Lamarck studied the invertebrate fossils of the Paris Basin.

In the second chapter (called "the first synthesis") of his book on Lamarck, Pietro Corsi (2001) rightly compares Lamarck and Jean-Claude Delamétherie. Both inherited from Buffon the idea that the antiquity of the Earth is beyond human experience and cultural traditions. Delamétherie (1797: 142–143) supported the idea that geological layers such as coal and limestone were due to living beings and their fossilisation during "des siècles innombrables" (numerous centuries) and concluded "Le temps est tout pour nous, et rien pour la nature" (time is all for us and nothing for nature). Lamarck (1802: 89) clearly connected deep time and life: "Combien cette antiquité du globe terrestre s'agrandira encore aux yeux de l'homme, lorsqu'il se sera formé une juste idée de l'origine des corps vivants, ainsi que des causes du développement et du perfectionnement graduels de ces corps" (The antiquity of the planet earth will be much extended to man's eyes when one gets a better understanding of the origin of living beings, and of their gradual development and progression). In other words (Tassy 1994: 58), for Lamarck, "le démiurge, c'est le temps" (time is the creator).

Yet, despite these pioneering insights, geological succession and the evolution of organisms were largely thought to be separate. For many, evolution of organisms was no more than an unsound theory. Before discussing some aspects of Georges Cuvier's work, the prominent palaeontologist of his time, let us anticipate and conclude on the stratigraphic side of the question. In the nineteenth century, William Smith in England and Alcide d'Orbigny in France conceived the basis of the geological timetable. They did it without assuming continuity between members of the different faunas and floras, or in other words, species evolution. In 1816, Smith published *A geological table of organized fossils* where he recognized

17 levels, and a couple of years later added a further 10 (Smith2010). D'Orbigny (1849) organized the chart in 28 stages as well as 28 different faunas due to 28 creations. D'Orbigny's work is the basis of the present stratigraphic chart developed by the International Commission on Stratigraphy (ICS), which includes 100 stages (http://www.stratigraphy.org ; Gradstein et al. eds. (2012)).

As a geologist, Smith was interested in drawing the geological map of Great Britain. It can be noticed that the standard colors of the timetable are the same as those of the geological map. The orange, pink and yellow of the Tertiary layers of the map are the same as the colors of time. There is no better way to demonstrate that lithology and time are the same, and that time and space are united. This connection, or unification, was soon made accessible to the geologist by the fate of erosion and exposition of outcrops.

George Cuvier deserves a few more words. Cuvier together with Alexandre Brongniart inspected in detail the succession of geological layers and their fossils around Paris (Cuvier and Brongniart 1808). Their purpose was to establish the first geological map of the Paris Basin, published 3 years later. Cuvier became extraordinarily famous, not as a geologist, but as a descriptor of extinct vertebrate species. He wrote an introduction that can be described as the first textbook on paleontology, and as an "exposé" (presentation) of his methodology and ideas on the theory on earth (ideas later recalled as "catastrophism"), *Recherches sur les ossemens fossiles* (Cuvier 1812, 1834). This introduction, first called *Discours préliminaire*, was later published as a book (Cuvier 1825). The ultimate version includes a stratigraphical chart (Fig. 12.1). This pioneering chart made by Alexander von Humboldt shows several fossils associated with distinct layers, such as *Ananchites* (Fig. 12.1). The layer, called "Gypse à ossemens", that made Cuvier famous is also seen in Fig. 12.1.

Cuvier, as Brongniart, Smith, D'Orbigny, and others, grasped the concept of succession, for both rocks and fossils. But they all failed to grasp that of evolution. Still, deep time was difficult to conceive (see de Ricqlès' Chap. 11 in this volume). The best example of this difficulty is Cuvier's well known study of Egyptian mummies, some of them collected during Bonaparte's expedition to Egypt. In Cuvier's description of the mummified ibis, we found a nice comparison with the living ibis (today known as *Threskiornis aethiopicus*), and a taxonomy of the different ibis species. Cuvier concluded rightly that the mummy and the extant species were the same, and that time did not change anything. Delamétherie and Lamarck were apparently refuted. The problem is that Cuvier's concept of time did not refer to the proper duration.

Stratigraphy is empirical. Evolutionary thinking is deductive. In other words, the superposition of lithological layers is directly read, but by contrast evolutionary relationships can only be inferred. In both cases duration is not measured. The conceptual dichotomy between stratigraphy and evolution explains why, in the beginning, the construction of geological time by the means of the stratigraphic chart is not connected to a proper understanding of time, that is of deep time. Lamarck, as a biologist, supported the idea of an infinite time, a prerequisite to accept continuous anatomical transformations. On the contrary, stratigraphical

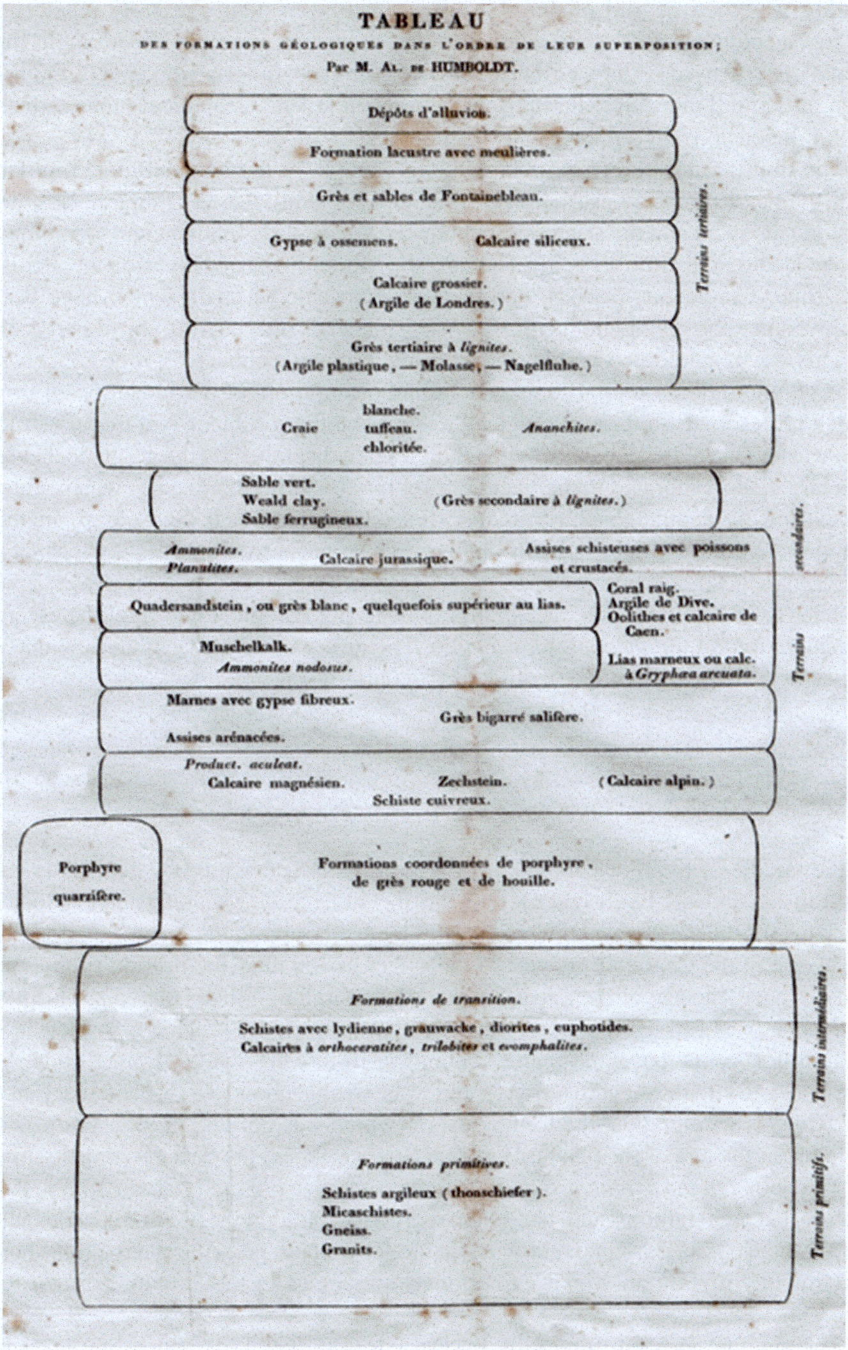

Fig. 12.1 The stratigraphic chart designed by Alexander von Humboldt and published by Georges Cuvier (Taken from the fourth edition of *Recherches sur les ossemens fossiles* (Cuvier 1834, plate face p.355))

scaling was then devoid of duration, of deep time. Yet, in the early nineteenth century geological time is already conceived of as an arrow, and according to this model the arrow does not go backwards. Geological time is irreversible, and as such, it is simple to follow through the geological layers. Geological time seems to lack a dimension of complexity.

The stratigraphic chart says that graduation allows us to travel in time. Empirism is at the heart of this construction of time. The analogy between superposition of geological layers and the succession of temporal periods is fundamental to geology. Hence geological time is vertical as are the white cliffs of Dover of the British Coastline (Cretaceous period), natural wonders made of chalk which have been studied since the eighteenth century: what is under is old, what is above is recent.

This is why geologists built a sensitive and pragmatic concept of time conceived of as chronological order. The understanding of duration, of the 'quantity' of time which corresponds to these geological layers is another problem. Yet this construction was remarkably successful. The relative scale displayed by the geological chart was largely corroborated in the twentieth century by radiometric dates which added a measure of depth (absolute dating) (See Huss' Chap. 10, Sect. 10.6, in this volume). We know, rather precisely, how long it takes to climb the vertical scale of geological time.

This brief comment on the stratigraphical side of the topic anticipates any consideration on the paleontological side. The arrow of an irreversible time determines first the history of the making of paleontological time.

12.3 Ages

In the middle of the nineteenth century deep time is understood by geologists and naturalists, such as Lyell and Darwin, in the same way we understand it today. Darwin (1859:287) estimated from measurements of sedimentation and erosion that what we call the Cenozoic era was "a far longer period than 300 million years. As a consequence, the age of the Earth should be measured in billions of years. This was criticized by Lord Kelvin who claimed that our planet cannot be older than 100 million years, based on thermodynamics. Both were wrong. Lord Kelvin did not know the natural radioactivity, and was wrong about the age of the Earth. Darwin was wrong about the Cenozoic era (which was only 65 million years), but right about the age of the planet. However, both accepted the concept of deep time.

The book *On the origin of species* includes only one picture (Darwin 1859). This picture became famous and has been reproduced hundreds of times, especially during the Year of Darwin in 2009. It is a model which represents and explains both patterns of differentiation and evolutionary processes in the temporal dimension (see Fig. 12.2a).

The genuine idea put forcefully by Darwin (1859, p. 124) is that the intervals between the horizontal lines in his diagram represent "a thousand generations (...), a million or hundred million generations, and likewise a section of the successive

a

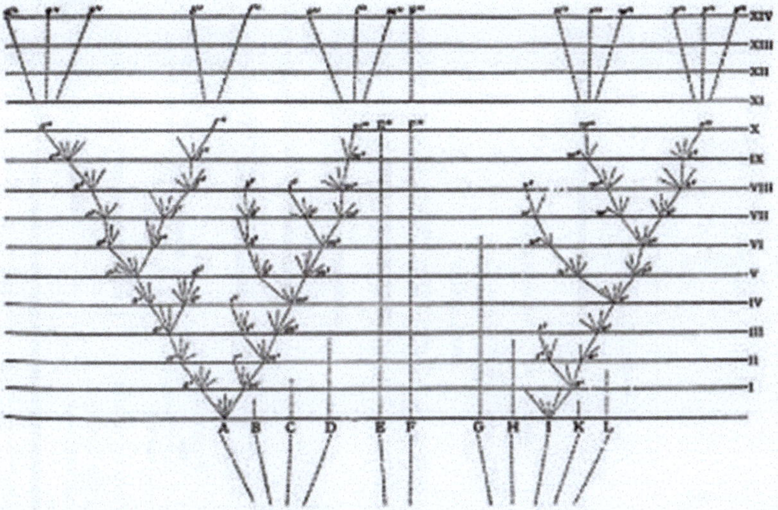

Fig. 12.2a Darwin's famous illustration of the origin of species: diversification of species *A–L* through generations and geological time

strata of the earth's crust including extinct remains". Hence, for Darwin the notion deep time of geology and that of deep time in biology (including paleontology) are the same time. Nevertheless, what Darwin also shows in his diagram, is that biological time has the shape of a tree not of an arrow. Today the phylogenetic tree – or the diagram of relationships – of any group (Fig. 12.2b) resembles Darwin's illustration.

This revolutionary view of evolution in the time dimension, that is, phylogeny (a word coined in 1866 by Ernst Haeckel), was far ahead of its time. Moreover it was doomed to be reduced by many a palaeontologist as a display of the succession of fossils; a crude way to answer Sober's (1988, p. 1) question: "when is history knowable?". In fact, during most of the twentieth century, the basic method adopted by paleontologists to reconstruct lineages in the time-dimension was a direct reading of evolution in the strata. Gingerich (1979) describes this approach under the term "stratophenetics" which produces "gradual phyletic evolution", and roots the method (Gingerich 1976, p. 22) in a paper published by the American palaeontologist William Diller Matthew 50 years before: "One could actually trace in the succession of strata the progressive evolution of the different races, verifying in specimen after specimen the primitive characters of those from the lower layers, the progressive characters of those from the upper layers" (Matthew 1926, p. 454). Matthew's assertion is acceptable at the level of a gross view of evolution but not on the minute level of phylogenetic relationships. Simple plants such as mosses appear in the stratigraphical timetable before flowered plants. Yet at a minute level (genera, species), the imperfect stratigraphical record does not display a harmonious succession of primitive and derived characters.

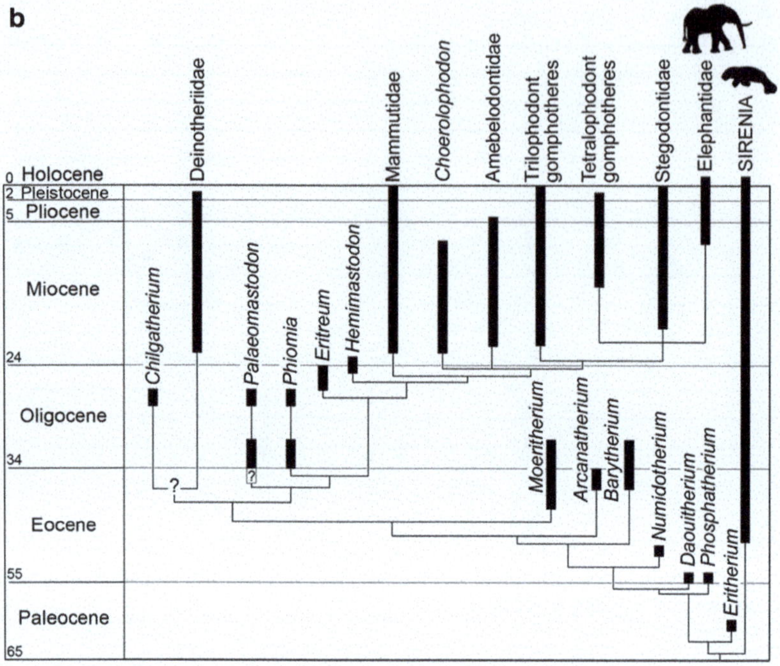

Fig. 12.2b An illustration of phylogeny through time, here that of proboscideans and their sister group, the sirenians (manatees and dugongs). Among proboscideans only elephants survive. *Thin lines*: hypothesized relationships; *thick lines*: known stratigraphical range (Modified from Gheerbrant and Tassy 2009)

Contemporaneous cladistic approaches following Hennig (1966) focus on the branching pattern. It is based on character analysis (either anatomical or molecular) with distinction of primitive and derived character states. Although the original spelling is "phylogenetic systematics", the word cladistics better emphasizes the concept of clade (from greek *klados*), that is a branch. Cladistics ruins the direct reading of evolution: geological time does not give relationships. Let us choose a crude example taken among fishes (see Fig. 12.3).

In the living lamprey *Petromyzon marinus* the anterior branchial arch is identical with its follower arches (a primitive character state). In the extinct species *Bothriolepis canadensis* (Frasnian stage, circa 380 million years before present) of the Miguasha National Park in Quebec, the anterior branchial arch is transformed into dorsal and ventral parts, that is maxilla and mandible, making a mouth (a derived character state). Moreover, at Miguasha numerous fossils with non transformed anterior branchial arch, such as that of living lampreys, coexist with jawed *Bothriolepis canadensis*, making the picture even more complex. This means that the orientation (called "polarity") of evolutionary transformations is not directly seen in the stratigraphical succession. The equations, ancient = primitive, and, recent = evolved, (i.e, respectively, the presence of a maxilla and a mandible

Fig. 12.3 Two aquatic vertebrates. *Above*, living lamprey (*Petromyzon marinus*). The anterior slit visible on the lateral side corresponds to the first branchial arch not transformed into jaws. *Below*, the extinct (380 million years old) placoderm (*Bothriolepis canadensis*), a jawed animal (*Petromyzon* copyright Emily Damstra. http://emilydamstra.com/. *Bothriolepis* copyright François Miville-Deschêne http://www.virtualmuseum.ca/)

in species dated 380 millions years ago, and the presence of an anterior branchial arch in forms that live today), fall into "the trap of common sense" ("le piège du bon sens") as said by Dupuis (1986: 233). This aspect of phylogenetics will not be discussed here.

Two conclusions can be drawn from this. On the one hand, radiometric dating confirms without a doubt the historical approximations of the geologists' deep time: the Earth is older than four billion years. The graduation of time has benefited from the concept of *Mega annum* (Ma), (i.e. million years (My) before present), due to Van Couvering and Berggren (1977: 285), now universally adopted. On the other hand, biological evolution is not understood as a simple lineage, like an arrow, but as a tree (Fig. 12.2b). This tree, with its degrees of recency of common ancestors or common ascent (i.e. the nodes of the picture), is a hierarchy. Hence, speaking about the nodes, this hierarchy has a time dimension (Fig. 12.4 below). Of course in Fig. 12.4 the age t1 is older than the age t3 and this fits with the concept of arrow; but the age of X is independent of the age of t3 or of Y and of Z : it can be older than t3 or more recent than t3, the same in relation to Y and Z. Said differently, the logic of time duration of X and of Y and Z in taxon A are independent.

Hence the fact that a tree conceived of as a hierarchy is imprinted on the arrow of time (such as Figs. 12.2a, b) yields several paradoxical situations only recently explored. These are discussed in Sect. 12.4 below.

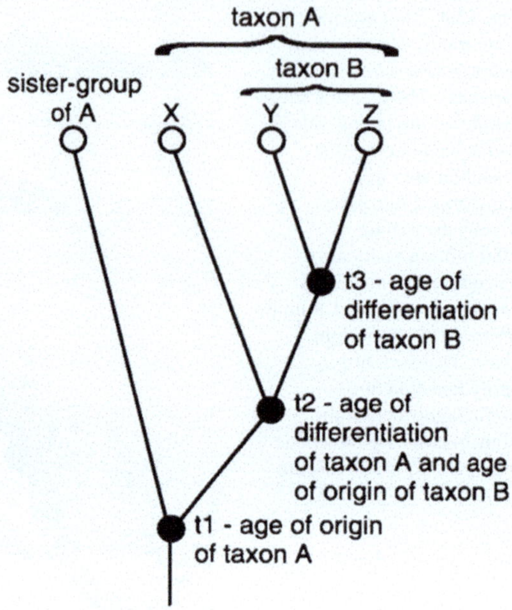

Fig. 12.4 The age of taxa according to Hennig's concepts of age of origin and age of differentiation (From Hennig's (1966) modified by Tassy (1999, fig. 2))

12.4 The Hierarchy of Time

12.4.1 Hierarchy and Arrow

For the French essayist Philippe Sollers (2003: 228) « [l]e temps, lui, ne peut être, à chaque instant, que vertical, étagé, feuilleté, poudroyant, ouvert ». (Time, at any one moment, will be vertical, stratified, multilayered, shimmering, open).

Vertical time, stratified, and multilayered is the time of stratigraphers. The aspect of being multilayered nicely evokes layered rocks, shales and slates, the sculpting of which produces plates and sheets. On the contrary, shimmering rather evokes a bursting time, a burst that, as such, may be opened. If one thinks of something opened, one can think about filling it. Time is doomed to be filled. But I will not venture further on Soller's metaphors.

Yet there is another way to look for opened and filled time: a hierarchical time which is not, as such, vertical, or stratified nor multilayered.

The notion of a time hierarchy contrasts with the notion of the arrow of time. Hierarchy is here understood as a mathematical hierarchy, that is, a classificatory structure where the intersection between two classes is empty or is one of the two classes. The following section illustrates this distinction.

Contrary to the time hierarchy, the geological scale understood by geologists and paleontologists as the arrow of time, can be considered as a different mathematical structure, which consists of several ordered partitions of relative time (either eras, periods, epochs, stages, etc. ; periods in Fig. 12.5a, b).

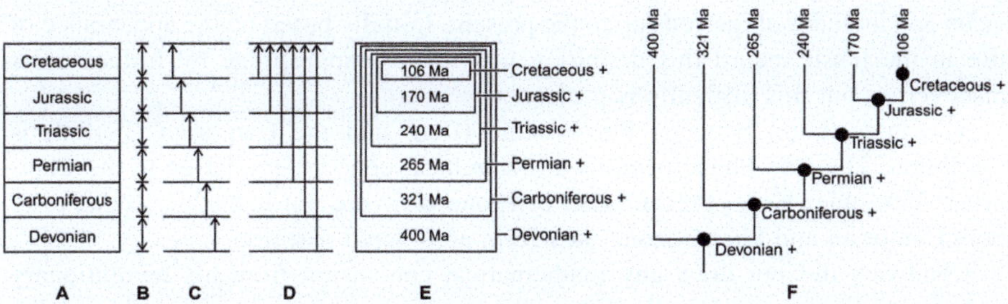

Fig. 12.5 The hierarchy of time. (**a**) the geological time scale with periods defined as an ordered partition. (**b**) linear time or the "arrow of time": the same event defines the beginning of a period and the end of the previous one. (**c**) geological periods are defined by their temporal extensions, (**d**) temporal extensions up to the present, only the beginning of a period defines periods viewed hierarchically. (**e**) temporal hierarchy as a Venn diagram derived from D: particular ages are shown as individuals belonging to each class. (**f**), temporal hierarchy as a tree derived from D: its shape is that of a pectinate tree, the geological periods are nodes (Modified from Zaragüeta Bagils et al. (2004, fig. 2))

Two properties define partitions: each individual belongs to one class; and there is no intersection between classes. Here, each age (defined in absolute time, e.g. 360 Ma, 80 Ma) belongs to one temporal class (360 Ma belong to Devonian, 80 Ma belong to Cretaceous etc.). There is no intersection between Devonian and Cretaceous. What we need to know when we manipulate the geological scale is what is the oldest or the more recent period.

Interestingly, geologists and stratigraphers who generally produce partitions also introduced hierarchies in the geological scale, but without conceptual generalizations. As soon as the nineteenth century, the scale conceived by Alcide d'Orbigny (D'Orbigny 1849) contains the stage Cenomanian (between 100.5 and 93.9 Ma) which is included in the Cretaceous period (between 145 and 66 Ma). This subordination is seen up to the latest version of the chart endorsed by the ICS. Hence, one can also see the scale as a juxtaposition of independent hierarchies, each used according to the desired chronological precision: a cretaceous age is less precise than a cenomanian age. But when Cenomanian is compared to Turonian (between 93.9 and 89.8 Ma) one follows the logic of the arrow, and it is in this way that the scale has been manipulated, following the succession of classes, according to the partition. Combining arrow and hierarchy but relying only on the arrow is a long-time paradox that remains the same in the modern chart.

12.4.2 Phanerozoic as an Example

Introducing the Phanerozoic in 1930, the American stratigrapher George Chadwick defines a hierarchical concept of time (Chadwick 1930). The Phanerozoic (*Phaneros*, appearance, and *Zoon*, life) is a vast section of time, called an Eon, according to the stratigraphical nomenclature. This Eon contains the last 542 million years and includes all periods up to the present. Initially based on the appearance of life in the fossil record this definition was made inappropriate by further older discoveries, but this difficulty is irrelevant.

The intersection between Phanerozoic (Eon) and the Cambrian (Period) is Cambrian. The Eon Phanerozoic is a hierarchical concept in relation to the Cambrian. Yet, older Eons, Precambrian or Proterozoic are not hierachies, relative to both Cambrian and Phanerozoic; they form an ordered partition.

Chadwick did not draw any mathematical conclusion from his revolutionary concept of Phanerozoic, and neither did any of his followers. The idea of the Phanerozoic is to exemplify duration, the duration of life. This duration involves a constant renewing up to the present, just as does life. The Phanerozoic is 85 years longer than it was in 1930: The Phanerozoic includes the whole future duration. In the same way, because of descent with modification, present life is the consequence of past life and the cause of future life. Said differently, future life is both the continuation and differentiation of present life.

The concept of the Phanerozoic became popular and useful during the last 30 years, mainly due to the work of the American palaeontologist Jack Sepkoski. Sepkoski's interest was the variation of biodiversity through the ages as given by the fossil record, especially the succession of mass extinctions. This purpose made Sepkoski and followers use the Eon Phanerozoic (i.e., Sepkoski1984). Each successive system, series or stage – that is, each element of the partition – is not informative in itself; only duration is, that is, the Phanerozoic. Diversity curves with massive extinction events make sense only in continuity (see Huneman's Chap. 14, Sects. 14.4.2.2 and 14.5, this volume). A geologic period brings information only in relation to the preceding period. Moreover, what precedes makes sense to what follows. One can conclude that a hierarchical concept – here Phanerozoic – gives to the arrow its sense, or its direction.

12.4.3 Mathematical Hierarchy

In a temporal hierarchy if two temporal periods intersect, the intersection is always one of these periods. Hence, if the beginning of any geological period is an event, one is able to understand the set of events as a hierarchy (Fig. 12.5c–e). The geological time, as a hierarchical structure, can be shaped as a tree (Fig. 12.5f). Individuals which belong to each class are ages, i.e. 400 Ma (Fig. 12.5f). Terminals (the leaves of the tree) are these ages (Fig. 12.5f). The temporal hierarchy specifies that each temporal period has differentiated from a longer, more general or inclusive one. In this view, geological time is still understood as a duration, a long duration which includes shorter durations; and it contains an operational instant, i.e. terminal ages. On Fig. 12.5f, the classical names of the periods of the geological scale are maintained to qualify subordinated periods, with a simple suffix (+) to stress their new meaning. The nodes of the tree are these geological periods +.

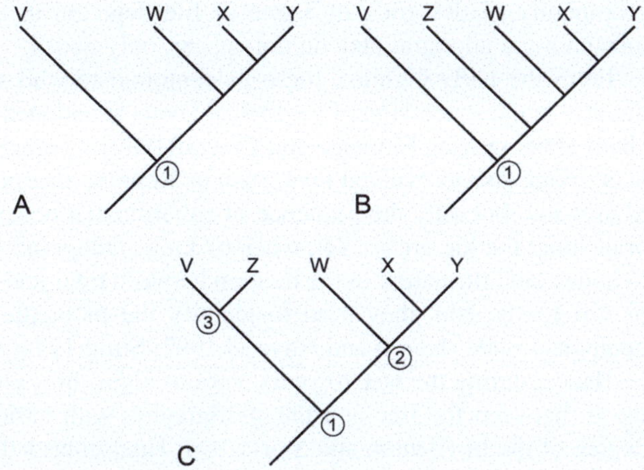

Fig. 12.6 Two ways to incorporate a new taxon (*Z*) among four taxa (*V*, *W*, *X*, *Y*). On (**b**), the position of Z (sister group of WXY) does not change the previous hierarchy (**a**). The tree is called a pectinate tree. On (**c**), the position of Z (sister group of *V*) creates a new hierarchy (node 3, independent of node 2). The tree is called a balanced tree

The "time tree" is a pectinate diagram, an unbalanced tree or "comb-like" clearly reminiscent of a phylogenetic tree with its subordinated taxa (Figs. 12.2b, 12.4 and 12.6). In such a tree only one descendant branch continues to diversify after a split (or node). This is why traditional classification bears some kind of hierarchy. For instance, the Mammalia includes the Primates which includes the Catarrhini, which includes the Hominoidea, which includes the Hominidae, which, finally, includes *Homo sapiens*.

Since what has been called the "cladistic revolution" in the seventies, the search for relationships between species and groups of species has been done through a cladistic analysis, the result of which is called a cladogram (Fig. 12.6), which is a hierarchy (today, extremely popular non-cladistic programs, such as likelihood and Bayesian programs, also yield hierarchical trees). In a cladogram, taxa (i.e. Mammalia) differentiate into subtaxa (i.e. Primates) in an unending movement. Taxinomists (Dupuis1992; Minelli et al. 1991) discovered that the fractal structure of such a taxonomic diagram was surmised by Mandelbrot as soon as 1956, although for Mandelbrot (1956) the starting point was more the hierarchy of categories than of the tree.

12.4.4 A Time for Living

Viewing time as a hierarchy is not only a formal matter. Any evolutionary biologist (neontologist or paleontologist) sees life as a deployement, a diversification, a differentiation, that occurred during more than three billions years ago. The tree of life roots in the deep time and the two million extant tips of its branches are the living species known so far. The connection between these species, are the nodes of the tree which display their common ascent, or, using Darwinian words, "the propinquity of descent". In a tree of life depicted in the form of a cladogram, metaphorical ancestral taxa do not appear, only nodes.

From the early eighties, these paleontologists who stopped to read evolution directly in the succession of geological layers abandoned what was left of the great chain of beings (i.e. Eldredge and Cracraft1980). In other words, no more ancestral taxa giving rise to evolved taxa, such as Reptilia ancestors of Aves (birds) and of Mammalia, but only the sequence of subordinated nodes, that is, of the historical branching: the hierarchy. Yet, when making cladograms paleontologists soon tried to understand the nature of the relation between trees and the stratigraphical record; or differently said, they tried to identify the fit of the phylogenetic tree to the geological scale (Norell and Novacek1992; Smith1994; Tassy 1990).

Hence, during the last 20 years, various algorithms were conceived to calculate the fit between the tree and the stratigraphy, with various indices to qualify the degree of the fit (Benton and Storrs1994; Huelsenbeck 1994; Norell and Novacek 1992; Wagner 1995). The problem is that these indices appear to be biased due to the shape of the tree: pectinate trees always display better indices than balanced trees (such as Fig. 12.6b, c), even when the chronology of the taxa is the same (Pol and Norell 2001; Pol et al. 2004; Siddall1996, 1997, 1998; Wills1999).

These indices indeed express something, but unfortunately not the fit, precisely what they are supposed to express. They rather express the degree of adequacy of an empirical tree compared to an ideal tree, that is, an entirely pectinate tree. The reason lies in the fact that the succession of the nodes of a pectinate tree is linear (see Fig. 12.5f and 12.6a–b), as the arrow of time. The explanation of the distortion is the superimposition of two independent logics: the hierarchy of taxa and the linear structure of the geological scale (Siddall1998; Zaragüeta Bagils and Lelièvre2001). The hierarchical representation of time allows for the problem to be solved (Zaragüeta Bagils et al. 2004).

Fossils integrate the time dimension in a peculiar way. When a fossil is taxonomically identified, it represents a species (and the whole classificatory hierarchy which includes the specie: genus, tribe, family etc.). Usually the age allocated to the specimen is the age of the geological formation where it is found. Sometimes a radiometric date is available (such as 125 $\pm$ 3,2 Ma). Hence a species seen as a terminal in a cladogram is also an age, as we will see in the next section (Fig. 12.5f and12.7a, ages 1–6).

12.4.5 Temporal Paralogy

The way all terminal ages are used to draw an evolutionary scenario is not devoid of ambiguity, even if paleontology has long been understood as a narrative during the deep time, sometimes dramatic with such episodes as "the conquest of land" or the "hominization" (see Lecointre 2015).

This narrative ambiguity is enlightened when time is seen as a hierarchy. This ambiguity has a name: "temporal paralogy" (Zaragüeta Bagils et al. 2004; Lelièvre et al. 2008) or, more accurately, "taxic temporal paralogy" as Zaragüeta Bagils (pers. comm. 2014) prefers. The concept of paralogy was first put forward in evolutionary studies by Walter Fitch in 1970 relative to molecular biology, followed by Gareth Nelson and Pauline Ladiges in the field of biogeography (Nelson and Ladiges 1996).

Biochemist Walter Fitch (1929–2011) pioneered the field of molecular biology. His outstanding statement of paralogy played a major role in the domain of molecular phylogenies up to the present. The comparative studies of proteins (such as hemoglobin) made him identify convergent processes that give mistaken results. Fitch used the example of α and β hemoglobins to stress the difference of two genes present in the same organism due to gene duplication (α and β hemoglobins) and two genes present in two different organisms due to common ancestry (α hemoglobin in humans and in mice). "Where the homology[1] is the result of gene duplication so that both copies have descended side by side during the history of an organism, (for example α and β hemoglobin) the genes should be called *paralogous* (para = in parallel). Where the homology[2] is the result of speciation so that the history of the gene reflects the history of the species (for example α hemoglobin in man and mouse) the genes should be called *orthologous* (ortho = exact)" (Fitch 1970 :113). Distinguishing between paralogy and orthology is still a challenge in modern transcriptome analyses (analyses of messenger RNA molecules of gene expression) (Kocot et al. 2013).

The key is the idea of independant histories. Paralogy is responsible for the histories of the duplicated genes. The result is that hemoglobins of different organisms appear more closely related than were α and β hemoglobins of the same organism, ensuing a repetition of the taxa under study in the resulting tree.

Temporal data associated with a cladogram can be represented by a tree where the terminal taxa are the ages of fossil organisms and the nodes are inferred ages of origin and differentiation of taxa (Fig. 12.4). When a temporal period (a node) appears several times in such a temporal cladogram (called "horogram" by Zaragueta Bagils et al. 2004: 385), it defines an instance of a class of paralogy named "temporal paralogy" (e.g. age 2 of taxa E and X in Fig. 12.7b).

Balanced nodes lead to two other nodes: such nodes are named paralogous (black circle in Fig. 12.7b). Pectinate nodes lead to a terminal (particular age) and a period (a node): these nodes are named orthologous (white circles in Fig. 12.7a–b). A paralogous node leads to independent histories in each sister-taxon (in Fig. 12.7b: (C, X) on the one hand, and (D,E,F) on the other). Temporal ambiguity is yielded by nodes which result in either orthologous or paralogous nodes (grey circles in Fig. 12.7a–b).

This independence explains why measurement fits of trees to stratigraphy are shape-dependent, and thereby provide ambiguous or irrelevant results for cladograms which are not strictly pectinate.

[1]"Homology" here is to be understood as similarity due to resemblance.

[2]"Homology" here is to be understood as similarity due to ancestry.

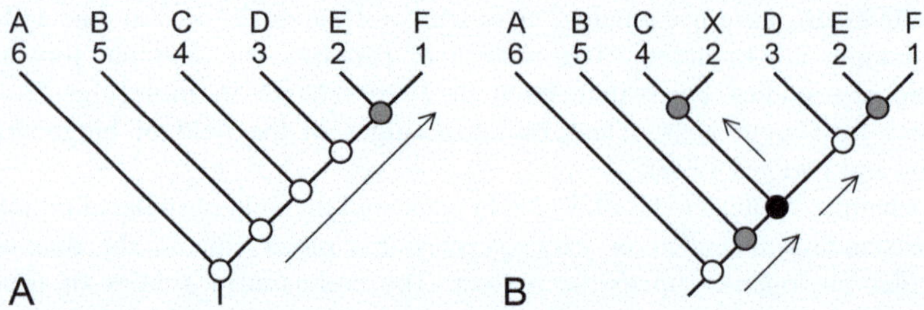

Fig. 12.7 Cladograms and temporal information. (**a**), a maximally informative pectinate cladogram of taxa (*A–F*) and their ages (6–1): a relation of orthology defines all nodes (*white color*; *grey circle* introduces temporal ambiguity. (**b**), a supplementary taxon (*X* with age 2) introduces paralogy and decrease resolution. Only *white circles* are connected by a relation of orthology. The *black circle* is a node which introduces paralogy without information on the respective inclusiveness (relative ages) of the sister taxa (*C, X*) and (*D, E, F*) (Modified from Zaragüeta Bagils et al. (2004, fig. 3))

This question is not only relevant for a restricted question, that is, the fit of trees on stratigraphy. Paralogy is a more comprehensive phenomenon and forces us to think about congruent hierarchies: in the present example, hierarchy of time and hierarchy of taxa. This congruence implies a specific treatment of the data. It is not sufficient, as it is regularly done, to put a tree (a cladogram) in the geological chart and estimate the amount of lineages without fossils (usually named "ghost lineages") due to the hypotheses of relationships. It is necessary to extract from the tree what is orthologous and what is not, with the purpose to make equivalent both hierarchy of taxa and hierarchy of their ages. The more complete the tree of life is (which is balanced), the more numerous are paralogous nodes. Hence, the quest for a fit may be meaningless. What is left from the traditional view of phylogeny given by the rocks may be abandoned. Or differently said, one might abandon the idea that a given phylogenetic hypothesis should be, in some way, congruent with the fossil record, a largely accepted opinion, regularly endorsed up to the present: "Paleontologists have long understood that phylogenies should be congruent with the order of appearance of fossils in the rock record… and that information about phylogeny is contained in the distribution of fossiliferous horizons for fossil species" (Huelsenbeck and Rannala1997: 174). These words are reminiscent of Matthew's words discussed above, and the conclusion is the same: this statement is true for gross evolution, largely speaking, not for phylogenetic details.

12.5 Conclusion

Time is not a secure ally for fossils. It plays tricky things, as do fossils relative to time. The topic discussed here is the necessity of a connection between the two hierarchical structures, time and life. This may be a minor topic. More important is to recognize that when time is viewed as a hierarchy – still a marginal approach – it enlightens the hierarchical structure of the history of life.

Today microbiologists emphasize that the first two-thirds of the history of living beings is a network, not a tree (i.e. Doolittle and Bapteste 2007). Horizontal gene transfers among bacteria, archaea and simple eucaryotes blur the dichotomies, the nodes, and, as a consequence, blur the irreversibility. By definition, the structure of a network is not a hierarchy. The skeleton of this network shares something with that of a tree: descent with modification implies vertical relations, blurred, or not (i.e. Theobald 2010).

Yet, for one interested by the phylogeny of Metazoa (animals), tree reconstruction is as new as it was when Lamarck (1809) published the first tree that can be understood as a phylogenetic tree (Tassy 1991). The tree of animals is a minor biological problem compared to the tree of life. The interrelationship of animals applies to more than one million and two hundred thousand extant species (exactly 1,211,612, according to Lecointre and Le Guyader (2006)), still a formidable task.

Yet, the intimate connection between time and the tree of life is less than evident. The deployment of living beings in the temporal dimension creates multiple parallel durations and independent developments. The historical reading of the tree of life and the interpretation of geological time are not immediate. As counterintuitive as it can be, the notion of taxic temporal paralogy forces us to accept the idea that the arrow of time does not help the paleontologist anymore. As phylogeneticists, palaeontologists can rely on the accumulation of empirical data but, more than ever, they are doomed to propose hypotheses, hypotheses of internal nodes, that is, hypotheses of relationships, and the epistemological status of which remains an open debate.

Acknowledgements I thank Christophe Bouton and Philippe Huneman for their invitation to participate in this volume. This article is derived from a presentation given on the 27th of September 2012 at the conference "The Time of Nature, the Nature of Time" organized by C. Bouton, Ph. Huneman and P. Duris, in Bordeaux, France. A former version of the manuscript benefited from René Zaragüeta's criticisms and suggestions. Two careful reviewers helped greatly to improve the text. I greatly appreciated the artistic help of Alexandre Lethiers and Sophie Fernandez. I am also grateful to Andrew McFarland for a thorough language check of the manuscript.

References

Benton, M.J., and G.W. Storrs. 1994. Testing the quality of the fossil record: Paleontological knowledge is improving. *Geology* 22: 111–114.

Chadwick, G.H. 1930. Subdivision of geological time. *Geological Society of America Bulletin* 41: 47–48.

Corsi, P. 2001. *Lamarck. Genèse et enjeux du transformisme*. Paris: CNRS Editions.

Cuvier, G. 1812. *Recherches sur les ossemens fossiles*. Vol. 4. 1st ed. Paris: Deterville.

———. 1825. *Discours sur les révolutions de la surface du globe*. Paris: Dufour et D'Ocagne.

———. 1834. *Recherches sur les ossemens fossiles, tome premier*. 4th ed. Paris: Edmond d'Ocagne.

Cuvier, G., and A. Brongniart. 1808. Essai sur la géographie minéralogique des environs de Paris. *Annales du Muséum d'Histoire Naturelle* 11: 293–316.

Darwin, C. 1859. *On the origin of species by means of natural selection*. London: John Murray.

Delamétherie, J.-C. 1797. *Théorie de la Terre* (2e édition, 5 vols.). Paris: Maradan.

Doolittle, W.F., and E. Bapteste. 2007. Pattern pluralism and the tree of life hypothesis. *Proceedings of the National Academy of Sciences* 104: 2043–2049.

D'Orbigny, A. 1849. *Cours élémentaire de paléontologie et de géologie stratigraphiques*. Vol. 1. Paris: Masson.

Dupuis, C. 1986. Darwin et les taxinomies d'aujourd'hui. In *L'Ordre et la diversité du vivant*, ed. P. Tassy, 215–240. Paris: Arthème Fayard/Fondation Diderot.

———. 1992. Permanence et actualité de la Systématique – III – Regards épistémologiques sur la taxinomie cladiste. Adresse à la XIe session de la Willi Hennig Society (Paris 1992). *Cahiers des Naturalistes* 48: 29–53.

Eldredge, N., and J. Cracraft. 1980. *Phylogenetic patterns and the evolutionary process. Method and theory in comparative biology*. New York: Columbia University Press.

Fitch, W. 1970. Distinguish homologous from analogous proteins. *Systematic Zoology* 19: 99–113.

Gheerbrant, E., and P. Tassy. 2009. L'origine et l'évolution des éléphants. *Comptes Rendus Palevol* 8: 281–294.

Gingerich, P.D. 1976. Paleontology and phylogeny: Patterns of evolution at the species level in early Tertiary mammals. *American Journal of Science* 276: 1–28.

———. 1979. The stratophenetic approach to phylogeny reconstruction in vertebrate paleontology. In *Phylogenetic analysis and paleontology*, ed. J. Cracraft and N. Eldredge, 41–77. New York: Columbia University Press.

Gohau, G. 1987. *Histoire de la géologie*. Paris: La Découverte.

Gradstein, F., J. Ogg, M. Schmitz, G. Ogg (eds.) (2012). *The geologic time scale 2012*. Amsterdam: Elsevier.

Hennig, W. 1966. *Phylogenetic systematics*. Urbana: University of Illinois Press.

Huelsenbeck, J.P. 1994. Comparing the stratigraphic record to estimates of phylogeny. *Paleobiology* 20: 470–483.

Huelsenbeck, J.P., and B. Rannala. 1997. Maximum likelihood estimation of phylogeny using stratigraphic data. *Paleobiology* 23: 174–180.

Kocot, K.M., M.R. Citarella, L.L. Moroz, and K.M. Halanych. 2013. PhyloTreePruner: A phylogenetic tree-based approach for selection of orthologous sequences for phylogenomics. *Evolutionary Bioinformatics* 9: 429–435.

Lamarck, J.-B. Monet de. 1802. *Hydrogéologie ou Recherches sur l'influence qu'ont les eaux sur la surface du globe terrestre*. Paris: L'auteur.

Lamarck, J.-B Monet de. 1809. *Philosophie zoologique*. Paris: Dentu.

Lecointre, G. 2015. Telling the story of life: On the use of narrative. In *Handbook of evolutionary thinking in the sciences*, ed. T. Heams, P. Huneman, G. Lecointre, and M. Silberstein, 387–410. Dordrecht/Heidelberg/New York/ London: Springer.

Lecointre, G., and H. Le Guyader. 2006. *The tree of life*. Cambridge: The Belknap Press of Harvard University Press.

Lelièvre, H., Bagils R. Zaragüeta, and I. Rouget. 2008. Temporal information, fossil record and phylogeny. *C. R. Palevol* 7: 27–36.

Mandelbrot, B. 1956. La distribution de Willis-Yule, relative aux nombres d'espèces dans les genres biologiques. *Comptes rendus hebdomadaires de l'Académie des Sciences* 242: 2223–2226.

Matthew, W.D. 1926. Early days of fossil hunting in the high plains. *Natural History* 26: 449–454.

Minelli, A., G. Fusco, and S. Sartori. 1991. Self-similarity in biological classification. *Biosystems* 26: 89–97.

Nelson, G.J., and P.Y. Ladiges. 1996. Paralogy in cladistic biogeography and analysis of paralogy-free subtrees. *American Museum Novitates* 3167: 1–58.

Norell, M.A., and M.J. Novacek. 1992. The fossil record an evolution: Comparing cladistic paleontologic evidence for vertebrate history. *Science* 255: 1690–1693.

Pol, D., and M.A. Norell. 2001. Comments on the Manhattan stratigraphic measure. *Cladistics* 17: 285–289.

Pol, D., M.A. Norell, and M.E. Siddall. 2004. Measures of stratigraphic fit to phylogeny and their sensitivity to tree size, tree shapes, and scale. *Cladistics* 20: 64–75.

Sepkoski, J.J. Jr. 1984. A Kinetic model of Phanerozoic taxonomic diversity. III. Post-Paleozoic families and mass extinctions. *Paleobiology* 10: 246–267.

Siddall, M.E. 1996. Stratigraphic consistency and the shape of things. *Systematic Biology* 45: 111–115.

———. 1997. Stratigraphic indices in the balance: A reply to Hitchin and Benton. *Systematic Biology* 46: 569–573.

———. 1998. Stratigraphic fit to phylogenies: A proposed solution. *Cladistics* 14: 201–208.

Smith, A.B. 1994. *Systematics and the fossil records. Documenting evolutionary patterns*. Oxford: Oxford Blackwell Scientific Publications.

Smith, W. 2010 [first edition 1817]. *A stratigraphical system of organized fossils*. Cambridge: Cambridge University Press.

Sober, E. 1988. *Reconstructing the past*. Cambridge, MA: The MIT Press.

Sollers, P. 2003. *Eloge de l'infini*. Paris: Gallimard, Folio.

Stenonis, N. 1669. *De solido intra solidum naturaliter contento. Dissertationis prodromus*. Florence.

Tassy, P. 1990. Phylogénie et classification des Proboscidea (Mammalia): Historique et actualité. *Annales de Paléontologie* 76: 159–224.

———. 1991. *L'Arbre à remonter le temps*. Paris: Christian Bourgois.

———. 1994. Le temps entre stabilité et rupture: La dimension paléontologique. *L'Inactuel* 2: 57–64.

———. 1999. Willi Hennig et l'objet paléontologique. *Geodiversitas* 21: 5–23.

Theobald, D.L. 2010. A formal test of the theory of universal common ancestry. *Nature* 465: 219–222.

Van Couvering, J., and W.A. Berggren. 1977. Biostratigraphic basis of the Neogene time scale. In *Concepts and methods of biostratigraphy*, ed. E.G. Kauffman and K.J.F. Hazel, 282–306. Stroudsburg: Dowden, Hutchinson & Ross.

Wagner, P.J. 1995. Stratigraphic tests of cladistic hypotheses. *Paleobiology* 21: 153–178.

Wills, M.A. 1999. Congruence between phylogeny and stratigraphy: Randomization tests and the gap excess ratio. *Systematic Biology* 48: 559–580.

Zaragüeta Bagils, R., and H. Lelièvre. 2001. Ages et noms versus caractères et phylogénie: le problème de l'ajustement entre la stratigraphie et la phylogénie. In *Systématique et Palé ontologie*, Biosystema, ed. P. Tassy and A. de Ricqlès, vol. 19, 169–185.

Zaragüeta Bagils, R., H. Lelièvre, and P. Tassy. 2004. Temporal paralogy, cladograms, and the quality of the fossil record. *Geodiversitas* 26: 381–389.

Part IV
Time of the Evolution (Evolutionary and Developmental Biology)

Chapter 13
Repetition and Reversibility in Evolution: Theoretical Population Genetics

Jean Gayon and Maël Montévil

Abstract Repetitiveness and reversibility have long been considered as characteristic features of scientific knowledge. In theoretical population genetics, repetitiveness is illustrated by a number of genetic equilibria realized under specific conditions. Since these equilibria are maintained despite a continual flux of changes in the course of generations (reshuffling of genes, reproduction...), it can legitimately be said that population genetics reveals important properties of invariance through transformation. Time-reversibility is a more controversial subject. Here, the parallel with classical mechanics is much weaker. Time-reversibility is unquestionable in some stochastic models, but at the cost of a special, probabilistic concept of reversibility. But it does not seem to be a property of the most basic deterministic models describing the dynamics of evolutionary change at the level of populations and genes. Furthermore, various meanings of 'reversibility' are distinguished. In particular, time-reversibility should not be confused with retrodictability.

13.1 Introduction

Evolutionary biologists commonly assume that "evolution is unique and irreversible". In contemporary literature, this claim is often closely related to the claim that evolution is historically contingent from top to bottom with no laws and no genuine theories. Although the authors share John Beatty's assertion that all (or almost all) biological generalizations are ultimately historically contingent[1], they believe that

[1]On the 'historical turn', see also Williams (1992), and Griffiths (1996). For a criticism of Beatty (1995), see Sober (1997).

J. Gayon (✉)
Institut d'Histoire et de Philosophie des Sciences et des Techniques (IHPST) - UMR 8590, Université Paris 1 Panthéon-Sorbonne, 13, rue du Four, Paris 75006, France
e-mail: jean.gayon@gmail.com

M. Montévil
Laboratoire "Matière et Systèmes Complexes" (MSC), UMR 7057, Université Paris Diderot, Paris, France
e-mail: mael.montevil@gmail.com

C. Bouton, P. Huneman (eds.), *Time of Nature and the Nature of Time*, Boston Studies in the Philosophy and History of Science 326,
DOI 10.1007/978-3-319-53725-2_13

the phrase "evolution is unique and irreversible" is far too general and too vague to be plausible. In reality, contemporary biology offers significant examples of repetition, invariance, and reversibility, both at the theoretical and the experimental level. Such examples may help to get out of the too-narrow alternative between "historical contingency" and "lawfulness" in biology, and particularly in evolutionary biology. In a sense, this alternative suffers from its excessive philosophical radicalness. The issues of repeatability *vs.* non-repeatability and reversibility *vs.* irreversibility of evolutionary phenomena offer a useful tool to make the debate more nuanced. It may be the case that repeatability and reversibility in evolution are marginal; nevertheless there are clear cases, both at the theoretical and the experimental level. The present paper will concentrate exclusively on theoretical population genetics.[2]

What exactly do the terms "repeatability" and "reversibility" mean? The definition is a delicate issue here, especially for the second notion. Does reversibility in evolution mean that an evolving entity (e.g. a population or a species) can return to a previous state (whatever the trajectory), or that the reverse trajectory should be strictly symmetrical with the direct trajectory? In his papers on the irreversibility of evolution, the Belgian Palaeontologist Louis Dollo was particularly concerned by this latter sense of "reversibility": "In order for [evolution] to be reversible, we would have to admit the intervention of causes exactly inverse to those which gave rise to the individual variations which were the source of the first transformation and also to their fixation in an exactly inverse order" (Dollo, quoted in Gould 1970, p. 199). Another difficulty arises from the technical notions of reversibility used in mathematics and physics. Do these notions apply to evolutionary biology? One of the main objectives of this chapter is to clarify the varying meanings of repetition and reversibility applicable to evolution. Repetition is a simpler matter, but this notion also involves a certain amount of ambiguity. Indeed, the two notions of repetition and reversibility should not be conflated as George Gaylord Simpson did when he commented on Dollo: "... evolution is a special case of the fact that history does not repeat itself. The fossil record and the evolutionary sequences that it illustrates are historical in nature, and history is inherently irreversible" (Simpson 1964, p. 196; quoted in Gould 1970, pp. 208–209). Simpson's statement seems a little too obvious. Although evolutionary reversibility is most often related to the kind of repetition that is so important for living beings—reproduction—repetition and reversibility should be clearly distinguished from one another.

When Louis Dollo introduced his famous "law of irreversibility in evolution", he defined it in the following terms: "... an organism cannot return, even partially, to a former state already realized in the series of its ancestors." (Dollo 1893, translated by Gould 1970, p. 211). As suggested by this formula, Dollo was interested in the problem of reversibility be it at the level of the organism or at least of a complex organ. Furthermore, as a palaeontologist, he conceived his "law" as applying to a large temporal scale (macroevolution in modern terms). Dollo did not deny that reversion could occur at more elementary levels. Moreover, neither genetics nor

[2]Another paper, devoted to experimental biology, will be published separately.

even less population genetics, existed when Dollo proposed his law of irreversibility in evolution. Therefore, and in view of present knowledge, it seems appropriate to reassess the problems of invariance and reversibility at a microevolutionary level.

Repetitiveness and reversibility in evolution can be assessed at two different levels, empirical and theoretical. At an empirical level, living objects exhibit properties of invariance that are crucial for evolutionary change. Current examples of invariance include gene replication and constancy of the number of chromosomes in the process of cell reproduction; and reproduction and alternate generations at organism level. In such cases, invariance is not absolute, indeed the replication of genetic material is not always perfect; correlatively, hereditary material exhibits an ability to change (genic mutations, recombination, chromosomal accidents…). Similarly, reproduction can encounter accidents (*e.g.* developmental anomalies), and exists under various modes (*e.g.* asexual *vs.* sexual reproduction, diverse schemas of alternate generations). Replication and reproduction are very general properties of living beings, and provide a basis for evolutionary models. They objectively exist throughout the living world. Of course, they result from a historical process, and for that reason, they cannot be thought of in terms of "laws of nature" in the sense of universal statements of unlimited scope, applying everywhere and at any time in the universe. One of us advocates the use of the concept of constraints in order to discuss limited invariance in the context of biological historicity (Longo and Montévil 2014, Montévil and Mossio 2015).

Population geneticists also share an intuitive notion of reversibility. Some biological processes make the return of a population to a previous state possible. Obvious examples include reverse mutation, especially if repeated; backwards selection (*i.e.* inverted selection coefficients); and chance (random drift). What "reversibility" precisely means in these examples is open to question, however the idea that populations can return to a previous state is perfectly plausible, given the nature of the basic biological processes involved in genetic evolution. There is another manner of formulating the intuitive notion of reversible evolution, which is more precise and better adapted to present genetic knowledge, namely: "for a given individual, consider the set of all its possible genetic states. One can move from one state to another thanks to the ordinary sources of genetic change (substitution of nucleotides, deletion, insertion, recombination, etc.). It is obvious that any sequence of states E_1, E_2… E_i… E_k that an individual can follow can also be followed in the reverse direction." (Goux 1979, p. 568, our translation).

These intuitive notions of repetitiveness and reversibility come *prior* to the construction of models in population genetics. They should be carefully distinguished from the properties discovered through the development of theoretical models describing the genetic evolution of populations. At that theoretical level, non-trivial notions of repetitiveness and reversibility occur; they result from modelling itself. Section 13.2 shows how population genetics models satisfies a characteristic feature of scientific knowledge currently found in the physical sciences, namely the discovery of formal properties of invariance through transformation. The next section examines whether theoretical genetics has also the capacity of discovering properties of reversibility or not. This is a more difficult issue. After defining several possible meanings of reversibility, Sect. 13.3 shows that time reversibility in the mathematical sense is illustrated by some stochastic models, whereas basic deterministic models do not exhibit the property of time-reversibility. The concluding section raises serious doubts about the traditional comparison made between classical mechanics and the deterministic models of population genetics.

13.2 Repetitiveness in Theoretical Population Genetics

Jean-Michel Goux observes that the source of a number of equilibria in population genetics is the endless repetitiveness of the life cycle (1979, p. 567). We will here freely expand on this proposal. In spite of its sophisticated use in mathematics, the notion of invariance under transformation can be defined in a simple and general way, and can be applied to many different areas of knowledge, not only in mathematics and theoretical physics. For a given class of objects, an invariant is a property that remains unchanged when a specified type of transformation is applied to the objects. This concept is especially fruitful when the objects and their relations are described by mathematical formulae; in such cases, a precise sense can be given to what is said to be invariant.

One of the most famous examples of invariance to transformation in physics is the Galilean transformation. In its traditional formulation in classical mechanics, Galileo's principle of invariance (also called Galileo's principle of relativity) states that the laws of motion are the same in all inertial frames. Based on the postulate of absolute time and absolute metric of space, this principle makes the transformation of spatial and temporal coordinates from one inertial referential system to another possible. For instance, if the speed of material point in S is v, its speed in S' will be[3]:

$$V_x{}' i = dx'/dt = d(x - vt)/dt = v_x - v$$

Invariants through transformation may be of many kinds. In classical and relativist mechanics they are relative to motion. However they can also be relative to structures (*i.e.* the composition of a particular class of objects). This section considers the case of invariance relative to the genetic structure of a population in specified conditions. Some classic examples of genetic equilibria are given below, which all belong to what could be called *evolutionary statics*, as opposed to evolutionary dynamics. Time-reversibility is also an extreme case of invariance through transformation. This notion will be considered in Sect. 13.3, devoted to evolutionary dynamics in population genetics.

[3]In the simple case where the spatial coordinates are chosen so that the origins O and O' of the two referential systems coincide for space and time. Then the three axes move along a line Ox.

The Hardy and Weinberg equilibrium is certainly the best-known example of a structural invariant. Consider a single locus with two alleles *A* and *a* with frequencies *p* and *q* (with $p + q = 1$).[4] The Hardy and Weinberg law states that, irrespective of the initial gene frequencies and the initial genotype frequencies, if (1) all crosses occur within the same generation (no overlapping generations), if there is (2) no selection, (3) no migration, (4) no mutation, if (5) mating is random, and if (6) if the population size *N* is big enough to consider that $1/N \approx 0$, then the genotypic frequencies are constant and depend only on the gene frequency of the initial generation (for a precise formulation of these conditions, see Jacquard 1971, pp. 48–58, and Hartl 1980, pp. 93–94). Under such conditions, the expected genotypic ratios are *AA*: p^2; *Aa*: 2*pq*; *aa*: q^2. In a discrete-generation population, the population immediately achieves these proportions from the first generation of mating, and the expected ratios remain constant as long as the six conditions mentioned are satisfied. The Hardy and Weinberg [HW] law derives its name from the British mathematician G. H. Hardy[5] (1877–1947) and the German physician and obstetrician Wilhelm Weinberg (1835–1937), who independently and simultaneously demonstrated it in 1908 (Hardy 1908; Weinberg 1908). Some authors call it a 'principle' (Crow and Kimura 1970). But it is more accurately characterized as a theorem, because it can be demonstrated on the mere basis of the Mendelian law of segregation and the six conditions stated above. It is also commonly referred to as "the Hardy-Weinberg equilibrium", where "equilibrium [refers] to the fact that there is no tendency for the variation caused by the co-existence of different genotypes to disappear" (Edwards 1977, p. 7). The reason why this law is so important is that it purely expresses the effect of Mendelian inheritance in the absence of any factor able to change the genetic frequencies (*i.e.* gene frequencies and genotypic frequencies).[6] As Edwards puts it, "this ability to maintain genetic variation is one of the most important aspects of Mendelian genetics" (*ibid.*). The constancy of genetic frequencies under Mendelian inheritance provides a reference model for describing the effects of evolutionary "forces", especially mutation, migration, selection, population size, inbreeding, and the mating system (homogamy vs. heterogamy), which may modify the genotypic structure of the population.[7] Returning to the problem of repetitiveness, the HW

[4]In genetics, a locus is a particular position on a chromosome, occupied by a gene, which can itself exist under several alternative versions, named 'alleles'. The Hardy-Weinberg equilibrium applies to sexually reproducing and diploid species, where all chromosomes (except fot the sexual chromosome) exist in pairs.

[5]Godfrey Harold Hardy did not use his first Christian name with his friends, but rather 'Harold' (Anthony Edwards, personal communication).

[6]This is why Sober calls the HW principle the "zero force law of population genetics" (Sober 1984). Gayon (1998) qualifies the Hardy-Weinberg equilibrium as an equivalent of the principle of inertia in classical mechanics (see however the conclusion of the present paper), which challenges this view.

[7]Some of these factors modify both the gene and the genotypic composition of the population. Others (homogamy) modify only the genotypic structure.

equilibrium is typically an invariant under transformation, because it identifies something (the distribution of gene and genotypic frequencies) that persists in spite of the indefinite reshuffling of the alleles that meiosis dissociates at each generation. Of course, the HW law is an idealization, because no real population ever strictly satisfies the conditions that permit its derivation.

Another classic example of structural invariance under transformation in population genetics is "Wright's principle", also called "Wright's law of equilibrium". This gives the frequency distribution of genotypes in an infinite population, for a diallelic locus[8]:

$$\left(p^2 + \mathrm{F}pq\right) + 2pq(1 - \mathrm{F}) + \left(q^2 + \mathrm{F}pq\right) = 1$$

where F is the coefficient of inbreeding. This law expresses the zygotic proportions[9] expected in a population with a certain amount of inbreeding, in other words a population where mates are more closely related than if they were chosen at random. The Hardy-Weinberg equilibrium is in fact a particular case of Wright's equilibrium, corresponding to $\mathrm{F} = 0$. Therefore, Wright's law of equilibrium takes into account one of the major causes of departure from the HW equilibrium (the other one being assortative mating). The F coefficient may of course change. Nevertheless, Wright's formula states that for a given F, and if no other factor is allowed to modify the genotypic frequencies, the genetic structure of the population is invariant from generation to generation. Apart from the Hardy and Weinberg law and Fisher's 1918 paper on the correlation between relatives under Mendelian Inheritance (Fisher 1918), this is one of the oldest results in theoretical population genetics. It has been demonstrated several times, and improved and generalized (multi-allelism) since Wright's original paper in 1921 (Wright 1921; Malécot 1948; Li 1955).

As said earlier, the Hardy and Weinberg equilibrium is established as early as the first generation of crossing (the first zygotes made from the previous generation). But this is true only if generations do not overlap (see above: condition 1). If generations overlap, it takes more time for the HW equilibrium (as well as Wright's equilibrium) to be established, but the population does converge towards this equilibrium. Therefore, the equilibrium does not emerge immediately, as in the discrete case, but gradually, as what could be described as a "trend".

This notion of "trend", which is very ubiquitous in theoretical population genetics, leads to an important remark about equilibria in this discipline. The search for equilibria is an important part of theoretical population genetics. Developing an idea suggested by J. B. S. Haldane, Crow and Kimura speak of "evolutionary statics", as opposed to "the dynamics of evolution". In a paper entitled "The Statics of Evolution", Haldane (1954) declared that, in spite of evolution being a "dynamical process", a good deal of it is better understood in terms of "statics". Haldane

---

[8]Diallelic locus: refers to a locus with two alleles.

[9]A zygote is a diploid cell (two stocks of chromosomes) resulting from the fusion of two haploid cells (spermatozoon and ovum), which have only one stock of chromosomes.

stated that the reason for this was that evolution is usually an extremely slow process, which may, nevertheless, rely upon strong forces (esp. selection). Whence the idea that an important part of evolutionary processes should be thought of in terms of "approximate equilibria" resulting from a balance of 'forces' that quite often conflict with each other (e.g. various kinds of selection, selection and migration, selection and mutation, etc.), and that result in the persistence genetic polymorphism. The 'statics' of evolution is indeed one of the most spectacular parts of theoretical population genetics. Quite often, the results are simple, elegant and easily found, in contrast to the difficulty associated with the mathematical treatment of the 'dynamics' (which will be evoked in the next section, when coming to 'reversibility'). For this reason, equilibrium formulae play an important role in the elementary teaching of population genetics.

In their *Introduction to Population Genetics Theory* (1970), Crow and Kimura devote an entire chapter to "Populations in approximate equilibrium" (Chap. 7). They examine an impressive list of factors maintaining gene frequency equilibria. All these factors "depend on some kind of balance between opposing forces" (Crow and Kimura 1970, p. 256). A partial list of the types of such equilibria is given below (from Crow and Kimura 1970, p. 256–196). Quite often, the models obtained are simple. Illustrative examples are given for the first two categories in the list.

- *Equilibrium between selection and mutation*. For instance, in the case of a single locus with complete dominance where the recessive homozygous-mutant genotype is disadvantaged, and where mating is random, the equilibrium is reached when $q = \sqrt{\frac{u}{s}}$ with q: frequency of the mutant gene; u: mutation rate; s: selection coefficient.
- *Equilibrium under mutation pressure* (infinite population, no random drift). For instance, in the simple case of a two-way recurrent mutation, the equilibrium is reached when$p = \frac{v}{u+v}$, with u and v being the mutation rates from and to allele A, and p the frequency of A.
- *Equilibrium between migration and random drift*.
- *Equilibrium under selection*: stabilizing selection (selection directed towards the elimination of deviants),[10] advantage to the heterozygote,[11] frequency dependent selection,[12] disruptive selection (selection in varying directions), multi-niche polymorphism...
- Selective models accounting for the *constancy of sex ratio* (most often 1:1 at the age of reproduction).

[10]This kind of selection favours the mean type.

[11]This kind of selection favours the individuals with genotype *Aa*. A classic example is the better resistance to malaria of individuals who are heterozygotes for the gene responsible for sickle cell anemia. Double recessives *aa* suffer from severe anemia and most often die at an early age; double dominant *AA* are much less resistant to malaria than heterozygotes *Aa* in areas infected by *Plamodium falciparum*. See Fig. 13.7.

[12]In this kind of selection, the selective values of the genotypes depend on the allelic frequencies. This results in an intermediate equilibrium.

All these equilibria—and this list is not exhaustive—isolate an invariant under transformation. For instance, the model for a two-way recurrent mutation tells us how the genetic structure of a population (both gene frequencies and genotypic frequencies) is preserved as long as the same conditions hold. In spite of the endless reshuffling of genes and of changes recurrently caused by two mutation pressures, an equilibrium is attained. Despite the obvious complexity and historicity of evolutionary phenomena, equilibrium models reveal invariant relations between parameters (gene and genotypic frequencies, mutation rates, selection rates, etc.) under idealized conditions.

The discovery of formal properties of invariance through transformation is an important component of scientific knowledge, whether in biology, physics or economics. In his *Models of Discovery*, Herbert Simon once wrote that "the notion of invariance under transformation as a necessary condition for a 'real' property of a physical system has provided a leading motivation for the development of relativistic mechanics and other branches of physics" (Simon 1977, p. 79, n. 8). We should not be surprised to find such invariants in evolutionary theory. As previously stated, repetitiveness is a massive empirical property of living beings: repetitiveness of life cycles, repetitiveness of cellular reproduction, repetitiveness of gene replication, and also repetitiveness of occasional phenomena such as recurrent mutation at population level. Given that repetitive phenomena are so widely observed at an elementary level, it is reasonable to expect that more formal invariants emerge when population genetics extrapolates from these empirical cases of repetitiveness to the behaviour of populations. As suggested in the introduction, the huge degree of historicity and contingency in evolution should not dismiss a certain amount of lawfulness, at least at the microevolutionary level.

13.3 Time-Reversibility

Time reversibility is a less obvious notion than invariance through transformation, for two reasons. First, it refers to problems that may become highly technical, and counter-intuitive in terms of their mathematical treatment. Secondly, evolutionary biologists use different notions of reversibility, and very often they do this unconsciously. Discussing time reversibility with several population geneticists, we have been struck by the combination of spontaneous certainty and doubt manifested in their spontaneous reactions to this subject. One of the most common reactions was: "yes, obviously, the deterministic models (esp. models of selection) are reversible, but, if random drift is taken into account, the opposite is true... Markovian processes are deprived of memory...". Another rarer response was: "Most probably, the equations describing the effect of deterministic processes do not describe reversible processes, but it seems evident that there is a large amount of reversibility in the models describing stochastic events". Such contradictory statements, made by some renowned population geneticists, triggered our curiosity. But the most common thought was of the following type: "obviously" the biological processes involved in the evolution of a population may "in principle" cause the return of the population to a previous state. For instance, if a mutant allele gets fixed, there is always the possibility that a reverse mutation will trigger a reverse evolution (through the accumulation of such mutations, or appropriate selective pressure, or random drift). Similarly, if the hierarchy of selective pressures is inverted, then reverse evolution will follow.[13] In fact, it seems that population geneticists, although excited by the problem, do not have a clear and articulate position. They use the word to have various meanings. This section intends to clarify the several possible meanings of 'reversibility' with respect to population genetics.

Three different senses of 'reversibility' can be found in the current—mainly mathematical and physical—scientific literature. After giving definitions for these, the level of their applicability to population genetics will be considered and the less conventional and specifically biological meanings of 'reversibility' among population geneticists will be discussed.

13.3.1 Three Senses of 'Reversibility'

To correctly treat reversibility as an operational concept in mathematics, physics and other exact sciences, would require a more formal and detailed analysis. The remarks that follow will only sketch out some distinctions that may help clarify the problem of reversibility in population genetics.[14]

13.3.1.1 Retrodictability

In classical mechanics, prediction and retrodiction are symmetrical: knowing the law (s) governing the development of a certain phenomenon through time and the state of the system at a given time *t*, it is possible to infer the state of the system at any other time, past or future. For instance, given Kepler's laws for the motion of the planets in the solar system, and given appropriate information about the state of a planet a time *t* (that is the position and instantaneous speed of the planet considered, as well as of other planets that interact with it), the position and speed of this planet at any past or future time can be inferred. What is required for retrodictability is the possibility of deriving a backward equation from the forward equation that describes the past trajectory as precisely as the forward equation describes the normal motion. For such an inference, the astronomer's theoretical framework does not need to be perfect. It may have, and certainly has its own limits (for instance Poincaré's three bodies problem). We just assume a certain more or less sophisticated theoretical system,

---

[13]Here is an example: in the nineteenth century, the proliferation of melanic forms of moths in industrial regions resulted from the darkening of the bark of trees by soot: the dark forms were better protected against predation by birds. With desindustrialization, light forms replaced dark forms. This is a typical case of inversion of selective pressure.

[14]We are very much indebted to Jean-Philippe Gayon, Anthony Edwards, Pierre-Henri Gouyon, and Michel Veuille, for their helpful interaction on this subject.

describing a deterministic process. If the process is deterministic, prediction and retrodiction are expected to be equally possible; note also that, in discrete time, retrodictibility may sometimes be impossible for deterministic systems (see Appendix A.1). Retrodictability is often associated and identified with time reversibility in the mathematical sense (see below (13.2)), and the two notions may indeed be closely related in particular cases. But as will be seen shortly, they are distinct. Because of the confusions resulting from equating retrodictability and reversibility, speaking of retrodiction (inference to the past) as a case of 'reversibility' should definitely be avoided. Although not common usage, the rest of this paper will repeatedly distinguish 'time-reversibility' (reversibility *sensu stricto*) and retrodictability.

13.3.1.2 Time Reversibility in the Conventional 'Mathematical' Sense

The notion of time-reversibility is based on a comparison between normal trajectories and trajectories after time reversal, that is to say where the past becomes the future and the future becomes the past. Reversibility occurs when these two trajectories follow the same law. Conversely, when the dynamics are irreversible, the law provides an orientation to time (the 'arrow of time'). The question of time-reversibility in this sense is commonly discussed in theoretical physics, see for example Chaps. 7 and 8 in this volume.

From a technical point of view, a time-reversible process is such that the equations describing its dynamics are invariant if the sign of time is reversed. In other words, if '$-t$' is substituted for 't', the law(s) governing the phenomenon is unaffected. In the case of classical mechanics, this will usually be checked by looking at the second order derivative of the equation describing the trajectory of the system. Consider for instance Galileo's law of falling bodies, which states that the distance x travelled by a free-falling body is directly proportional to the square of the time t for which it falls: $x = 1/2gt^2$. The first order derivative, $dx/dt = gt$ gives the speed. The second order derivative, $d^2x/dt^2 = g$ gives the acceleration, which is the key element from a dynamic point of view. It is easily seen that substituting $-t$ for t in this second order derivative does not change anything. The same 'law' holds in both time-directions. This means that, if a ball is thrown up, the law governing the motion of the rising ball is identical to the law that describes the motion of the falling ball (not similar, but identical). The direction of the trajectory will be inverted, of course, and the speed will diminish instead of increasing, but the rate of the decrease will be strictly the same as the rate of the increase: the dynamics governing the process is the same. This is exactly what the second order derivative says: since t exists only as a square number in $d^2x/dt^2 = g$, this equation is not affected by an inversion of time. The transition from t to $t + 1$ and the transition from $t + 1$ to t are said to follow the same 'law'. Indeed, this insensitivity can be demonstrated for both a discretized formulation of Galileo's law (where the evolution of the system is described at successive discrete steps, t, $t + 1$, etc.) and for continuous time (that is using a differential equation). In this precise sense, time reversibility is traditionally seen as an almost universal property of the laws of classical mechanics. In this respect, the best examples of time-reversibility (T-symmetry) are to be found in the fundamental laws of mechanics, which give the basic dynamics underlying mechanical processes. Newton's law expressing the relationship between force and acceleration (F = ma) would certainly be a better example than Galileo's law, an empirical law describing a trajectory rather than genuine dynamics. It is easy to understand that $d^2x/dt^2 = F(x)/m$ holds also after time reversal. With Newton's law, there is no need to be concerned with the counter-intuitive representations associated with a falling body governed by the same law both when it falls and when it is 'thrown up', or 'reverts' (when, how, what initial conditions, etc.—all elements that are unclear in the example). It should be noted, however, that Newton's law, $F = ma$, is time-reversible if and only if F is symmetrical by time-reversal (see Appendix A.1.1 for a more formal definition), for example when F depends only on x. This is the case for all classical fundamental forces, gravitation and electromagnetism. However, in other cases such as friction, where $F = -fdx/dt$ (where f is the friction coefficient), the law is no longer reversible.

The relation between reversibility and retrodictability is a delicate problem. The two notions may be closely related in particular cases. For instance, Galileo's law discussed above, allows for both retrodictability and reversibility. Knowing the position and the speed of a material point at any instant enables the prediction of the future and of the past position and speed of this material point at any time (within the limits of the actual trajectory). But Galileo's law also satisfies the condition of reversibility: it describes a transformation in both possible directions. However, the two notions are not necessarily associated. Consider the case of discrete time equations, which are particularly important in population genetics. The function g allowing a 'prediction of the past' may be totally different from the function that describes how the system passes from t to $t + 1$. Now consider a recurrence equation of the form $p(t + 1) = g[p(t)]$. Reversibility means:

$$p(t+1) = g[p(t)] \rightarrow p(t) = g[p(t+1)] \tag{13.1}$$

with p designating some function of time, and g the function defining the dynamic of p (*i.e.* what future state follows from the current state).

This formula means that the transition from t to $t + 1$ and the transition from $t + 1$ to t follow the same law.

We can similarly express the idea of retrodictability:

$$p(t+1) = g[p(t)] \rightarrow p(t) = h[p(t+1)] \tag{13.2}$$

where h is a function derived from the recurrence formula, which allows 'retrodiction'. The functions h and g may be totally different. For more precise definitions of reversibility and retrodictability, see Appendix A.1.1 (continuous time) and A.1.2 (discrete time).

Finally, there is another, important reason why retrodictability and reversibility should not be confused. So far, time-reversibility has been discussed only in the context of deterministic processes. However, time-reversibility can also be a property of stochastic processes: if the stochastic properties of a process depend on the direction of time, this process is said to be irreversible; if they are the same for either direction of time, the process is called reversible. Let $X(t)$ be a stochastic process, and τ a time increment. A standard definition of time-reversibility for a stochastic process is:

$$\text{A stochastic process } X(t) \text{ is } \textit{reversible} \text{ if } (X(t_1), X(t_2), \ldots, X(t_n)) \\ \text{has the same distribution as } (X(\tau - t_1), X(\tau - t_2), \ldots, X(\tau - t_n)) \\ \text{for all } t_1, t_2, ..., t_n\text{" (Kelly 2011, } p.5) \tag{13.3}$$

This definition means that the joint probabilities of the forward and reverse state sequences are the same for all sets of time increments. This notion can be applied to a Markovian process (*i.e.* in a state of equilibrium), that is to say a random process that undergoes transition from one state to another with no memory of the past. In a state of equilibrium, the condition for reversibility is:

$$\Pi_i P_{i,j} = \Pi_j P_{j,i} \tag{13.4}$$

where $P_{i,j}$ is the transition probability from state i to state j, and $\Pi_i P_{i,j}$ the probability flux from state i to state j. Π_i is the proportion of the population in state i. This formula does not explicitly include the time parameter t, but time is implicit through the notion of transition.

This statistical notion of reversibility has been fruitfully applied to a number of subjects, such as queuing networks, migration processes, clustering processes (esp. the equilibrium distribution polymerization process), and also population genetics, where Markovian processes are tremendously important for the treatment of random genetic drift (Kelly 2011). In contrast with reversibility in deterministic systems, stochastic time-reversibility is hardly compatible with retrodictability. Retrodictability is ordinarily understood as the possibility to reconstitute the actual trajectory that led to the present state, and is strongly associated with determinism, or at least to the idea of a causal sequence that has driven the trajectory. The notion of retrodictability could perhaps be extended to stochastic processes, but this is not the usual way of thinking about it. Furthermore, this would probably be a strange way of thinking in the case of time-reversible stochastic evolution, because stochastic reversibility is a typical case of an invariant and stationary state.

To sum up, although simple in principle (insensitivity of a given law to time reversal), the 'mathematical' notion of reversibility is delicate. It is not synonymous with retrodictability, and it does not only apply in deterministic situations.

13.3.1.3 'Physical' or 'Thermodynamic' Notion of Reversibility

The 'physical' notion of reversibility is closely related to thermodynamics. Both denominations are unsatisfying, because the former arbitrarily restricts the notion of the 'physical',[15] while in fact the latter extends beyond thermodynamics. 'Physical' reversibility means that a physical system can spontaneously return to a prior

[15] As seen in the previous paragraph, the insensitivity of the laws of classical mechanics to an inversion of time is as 'physical' as thermodynamic irreversibility is 'physical'.

physical state. A classical (although not perfect) example is given by a spring, which returns to its prior state after being elongated. By contrast, we do not expect that a broken glass will spontaneously find again its original shape. This is quite different from the property of insensitivity to time reversal. Consider a glass that breaks into pieces after falling. The glass would not be expected to spontaneously reacquire its original, ordered structure. Another example is that of a marble in a bowl. If there were absolutely no friction, the ball could go up and down the sides of the bowl indefinitely. But there is always some degree of friction. Therefore the oscillations of the marble will steadily decrease in amplitude, and in the absence of an external force, the marble will finally stop moving. These two examples (broken glass and rolling ball) illustrate the notion of physical irreversibility.

The traditional physical notion of reversibility is closely associated with thermodynamics: reversible evolution of a system is a type of evolution where no entropy is produced. Conversely, the greater the entropy, the more irreversible the transformation of the system. In a closed system, entropy is a quantity that can only increase. A precise definition of what this quantity precisely means is not needed here, nor are the classical debates on entropy as a property of macroscopic systems, as opposed to the reversible phenomena that underlie irreversible behaviour at a macroscopic level. For the needs of the present paper, suffice to retain that the physical (or thermodynamic) notions of reversibility and irreversibility are closely related to those of conservative *vs.* dissipative systems. A conservative system allows for reversible transformation; a dissipative system implies a dissipation of energy (*e.g.* diffusion of heat and friction), which renders the transformation irreversible (see Chap. 4, Sect. 9 in this volume for short discussion on the origin of irreversibility in thermodynamics).

From a thermodynamic point of view, biological phenomena are widely thought to be far-from-equilibrium processes: they maintain a relatively low entropy thanks to flows of energy and matter. Because they produce entropy, they are irreversible from the thermodynamic perspective. However, this aspect concerns energy and energy dispersal in a space of positions and momenta (measured in physical units), which is different from the space of gene populations that we discuss in this paper (gene frequencies are not measured in physical units). As a result, thermodynamic irreversibility is analytically independent from the question of the intrinsic time-reversibility or time-irreversibility of population genetics models.

13.3.2 Retrodictability and Reversibility in Theoretical Population Genetics

The various notions of reversibility mentioned in the previous paragraph will now be applied to theoretical population biology. No attempt is made here to be exhaustive. The present analysis will be limited to a few typical cases. Stochastic and deterministic processes will be successively examined, and a few remarks will be made about cases of 'reversibility' in population genetics that do not fit with the proposed categorization, cases where the notion of evolutionary reversibility relies on specific biological processes.

13.3.2.1 Stochastic Processes

Stochastic processes probably offer the best and most spectacular cases of time-reversibility in the strictest mathematical sense, that is to say insensitivity of a given model to reversal of time. This is most explicit in G. A. Watterson's two articles on "Reversibility and the Age of an Allele" (1976, 1977). These papers consider the probability distribution of the age of a mutant allele, whose present frequency is known. The 'age of an allele' is defined as the time that has elapsed between the introduction of the allele by mutation and the present. In the first paper (1976), the author assumes that there is no mutation and no selection. The model is set in discrete time, and admits that there is an infinite number of possible neutral alleles and thus discuss genetic drift. The method consists in taking y, the present frequency of the mutant gene, as "the initial state of a stochastic process, and studying how long diffusion would take to reach state x (or state 0) for the first time" (x being the frequency at a time t units prior to the present). Watterson is very explicit about the role of time-reversibility in his model, the general spirit of which is presented in the following terms:

> This interpretation seems surprising on two counts: first, because it means that the published results are simply moment of extinction times for diffusions, and second, there are stochastic processes for which this reversibility is valid, that is, processes whose behaviour looking into the past is statistically identical with their future behaviour. (Watterson 1976, p. 240)

This declaration is accompanied by a no less explicit figure (Fig. 13.1), illustrating the symmetry between the 'age of an allele' and 'extinction time'.

The model itself relies on an estimation of the probabilities of all the possible transitions from one state to another, in both directions of time (β → β' and β' → β). Since the Markov chain considered has a stationary distribution, which means that the process described is time reversible (Kelly 2011), Watterson states:

> The consequence of reversibility for the stationary configuration process is that we can discuss its past evolution with equal ease as its future evolution; $\beta(t)$, $\beta(t-1)$, $\beta(t-2)$, $\beta(t-3)$,... have the same joint [*sic*. 'probability' ?] as have $\beta(t)$, $\beta(t+1)$, $\beta(t+2)$, $\beta(t+3)$,... (Watterson 1976, p. 243).

The main conclusion of the paper is that "...the age distribution $\{P_i(a)\}$ for an allele now represented by i genes in the population is the same as the extinction time distribution for such an allele. Moreover, $\{P_i(a)\}$ is independent of the frequencies of the other alleles in the population." (*Ibid.*, p. 246). This is a remarkable result, illustrating the usefulness of time-reversibility for the elaboration of models in population genetics: "By reversibility we mean that given the present state of a stochastic process, the statistical properties of its future behaviour are the same as those for its past history treated as a stochastic process with time running in the reverse direction." (Watterson 1977, p. 179) Watterson's notion of reversibility corresponds to that defined in Eqs. (13.3 and 13.4).

In the same spirit, time-reversibility has been extensively used in coalescent theory, which is probably the major innovation in population genetics since the 1980s (Kingman 2000). Coalescent theory is concerned with gene genealogies

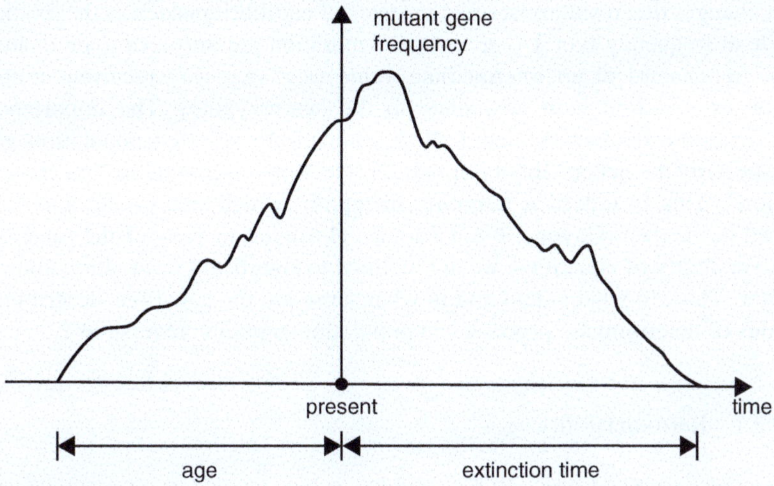

Fig. 13.1 Symmetry between 'age of an allele' and 'extinction time' (Watterson 1976)

within species. Relying on neutral mutations and the assumption of randomness of reproduction, its basic idea is to estimate the average time at which several genes share their most recent common ancestor. Time reversibility of the genealogical process is crucial in this case.

13.3.2.2 Deterministic Processes

Population genetics theory is commonly divided into two main branches. The first is stochastic theory, which focuses on the effect of random changes, especially "random drift", in allelic and genotypic frequencies. The second focuses on "deterministic" effects of factors such as mutation, migration, selection, and mating system. The deterministic theory of population genetics ignores the random changes, and is therefore less complete than the stochastic theory (Ewens 2012). In fact, all real processes in nature include a stochastic aspect, not least because real populations are finite and inescapably subject to random drift. Consequently, in the real world, deterministic factors always interact with stochastic factors. Furthermore, when population geneticists speak of the evolutionary factors in terms of 'forces', it is only a metaphor. Som philosophers defend that population genetics does not deal with physical Population genetics does not deal with physical forces but with statistical effects (Walsh et al. 2002; Matthen and Ariew 2009; Huneman 2013). Nevertheless, the notion of 'deterministic' factors in population genetics is acceptable in the sense of factors that have a directional effect, and tend to push gene and genotypic frequencies in one direction, up, down, or intermediate. Recurrent mutation, selection, migration and non-random mating are the commonest cases of such deterministic factors. Just as forces in mechanics, they produce either secular changes that result most often in states of equilibrium such as the fixation of an allele at frequency 0 or 1 (*e.g.* one-way mutation pressure), or maintenance of genetic frequencies at an intermediate value (*e.g.* two way recurrent mutation pressure, or selection with advantage to the heterozygote). The importance of these "deterministic factors" led J. B. S. Haldane to say that population genetics—especially the genetic theory of natural selection—is part of the "mechanics of evolution". This is indeed a tempting metaphor, which one of the authors has endorsed in the past (Gayon 1998, Chap. 8). However, in view of the issue of the time-reversibility of equations, we are inclined to significant reservation about this metaphor. Thus, to what extent and in what sense are the equations describing the dynamics of deterministic processes in population genetics 'reversible'?

13.3.2.2.1 Retrodictability

The issue of retrodictability will be considered first. Intuitively, if the existence of deterministic models in population genetics is accepted, the answer seems obvious. Richard Lewontin is particularly clear on this issue:

> It is usual for population geneticists, who claim that they are studying the dynamics of evolution, to divide the kinds of models that they deal with in two sorts. They talk about deterministic models and about stochastic models? What is meant by deterministic models is that, given the initial conditions of the population, which I will simply specify by X_0 (although that will be some sort of a vector of conditions of the population), and some set of parameters, then it is possible to predict exactly the condition of the population at some other time, τ. (Lewontin 1967, p. 81)

By "some other time", Lewontin means any other time, past or future, as explicited a few pages later in the famous 1967 paper on "historicity in evolution": "If I just give you one history of a deterministic population because it is deterministic I can say everything about its past" (Lewontin 1967, p. 87). This is exactly what retrodictability means.

Population genetics textbooks are replete with deterministic models describing the effects on gene and/or genotypic frequencies, of factors such as recurrent mutation, migration, selection, and mating system. In these models, evolutionary time can be treated in two ways, discrete or continuous.[16] With discrete time, the time-unit is a generation, and the evolutionary dynamics of the population is described by recurrence equations. Continuous time means that generations overlap and that continuous change is assumed (with no discrete time intervals whatever). In fact, the typology of models with respect to time is a bit more complicated (See Crow and Kimura 1970, Chapter 1, "Models of Population Growth", 1.1–1.4), but this will not be enlarged upon here. The type of model chosen will depend on the organisms being considered.

[16]Strictly speaking, this observation applies to all dynamic models in population genetics, stochastic or deterministic. It is introduced here because it will be useful for a proper understanding of the examples taken in this section.

For instance, models with discrete and non-overlapping generations are realistic with annual plants; models with overlapping generations with discrete time intervals are appropriate for animals or plants with a specific breeding season but which survive for several successive seasons. In other cases, birth and deaths occur at all times, and the most realistic models are based on the assumption of overlapping generations and continuous change. Thus, the basic method relies on differential equations. Ronald Fisher favoured this method, which is typically appropriate for the human species. Nonetheless, models relying on discrete time and recurrence equations quite often offer an acceptable first approximation, and are then preferred because of their mathematical simplicity.

Retrodictability seems to be a general property of deterministic models in population genetics. There may be special cases, where retrodiction is rendered impossible because of mathematical intractability. In other cases, recurrence equations may not be inverted because a state has several preceding states, generating ambiguity in the retrodiction (see Appendix A.1); these will not be discussed here. Consider now two examples of retrodictability.

First the case of a population subjected to a recurrent one-way mutation. Since mutation rates are usually very small (10^{-5} – 10^{-9}), the time taken to reduce p from a value close to 1 to a value close to 0 is extremely long. For instance, if the mutation rate is 10^{-6}, and if the initial frequency p is 0.9, it will take 2.2 million generations to reduce the frequency of p to 0.1.[17] Therefore, mutation pressure alone is not likely to be a major cause of evolution, because other factors, particularly random drift and selection, will very probably supersede the effect of recurrent mutation. Nevertheless, if isolated from any other consideration, it can be precisely described. The effect of recurrent mutation on the frequency of the mutated gene is:

$$p(t+1) = (1-u)p(t) \tag{13.5}$$

where u is the probability that a 'normal' allele A of frequency p mutates to a.

Since (1-u) is the probability that A does not mutate, (13.5) says that p at time t + 1 is the fraction of A alleles that do not mutate (Roughgarden 1979, p. 43–45). Equation (13.5) can be iterated analytically:

$$\begin{aligned}
p(1) &= (1-u)p(0) \\
p(2) &= (1-u)p(1) = (1-u)^2 p(0) \\
p(3) &= (1-u)p(2) = (1-u)^3 p(0), \text{etc.}
\end{aligned}$$

Whence the generalization:

$$p(t) = (1-u)^t p(0) \tag{13.6}$$

[17] $t = log(p_t/p_0)/\log(1-10^{-6})$, with $p_t = 0.1$, and $p_0 = 0{,}9$. This formula can be directly derived from the equation.

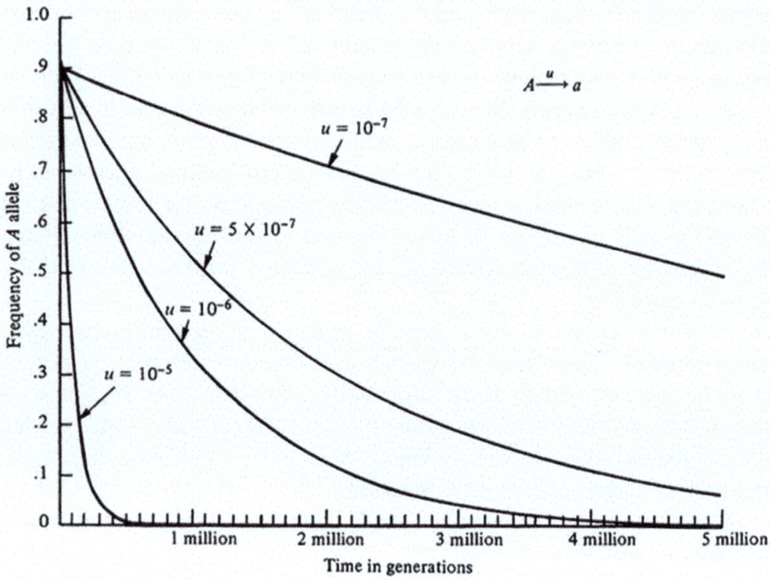

Fig. 13.2 Elimination of an allele by recurrent one-way mutation for various values of u (mutation rate), all other evolutionary factors being ignored. Observe the extreme slowness of the process. (From Roughgarden 1979)

Each generation raises $(1 - u)$ to the new high power. Since $(1 - u)$ is <1, the frequency of a gene reduces at a rate that itself decreases, because the quantity of alleles A in the population is reduced at each generation. This deterministic process is represented in Fig. 13.2 for $p(0) = 0.9$, and various values for the mutation pressure u. It is easy to see why this 'law' makes retrodiction possible. At any generation t, the frequency at previous generation $t - 1$ can be obtained by dividing $p(t) = (1 - u)^t\, p(0)$ by the quantity $(1 - u)$. Figure 13.2 illustrates the evolution of the frequency of the mutated gene for $p(0) = 0.9$ and various values of u. The very possibility of drawing such a curve strongly suggests that the phenomenon is both predictable and retrodictable.

A similar operation can be done for selection at a diallelic locus with constant selective values W_{11}, W_{12}, W_{22}, one of the simplest cases of selection. In the case of discrete generations, all textbooks give the same basic recurrence equation for this process[18] (p: frequency of allele A; q: frequency of allele a; $p + q = 1$):

$$p(t+1) = (p(t)W_{11} + q(t)W_{12})p(t) / \left(p(t)^2 W_{11} + 2p(t)q(t)W_{12} + p^2 W_{22}\right) \tag{13.7}$$

[18]Roughgarden (1979, p. 29); Jacquard (1971, p. 241); Crow and Kimura (1970, p. 179).

The expression ($p(t)W_{11} + q(t)W_{12}$) in the numerator is the selective value W_A of allele A (average selective value of all genotypes including allele A). The denominator is the mean selective value of population W (mean of the selective values of the different genotypes weighted by the frequencies of these genotypes). Equation (13.7) can thus be written:

$$p(t+1) = p(t)W_A/W$$

In fact this formula applies to all alleles. Therefore, if p_i is the frequency of allele i, and p_i' the frequency of i in the next generation:

$$p_i\text{'} = p_i W_i/W \tag{13.8}$$

This is a non-linear equation. Except in very special cases, there is no general method to iterate it analytically, but this can be done manually or using a computer—an easy operation for anyone today with an application like *Populus.*[19] Figure 13.6 (Appendix A.3) gives the result of computer iterations from (13.7) for selection against a recessive allele with strong selection. This is an illustration of a typically deterministic process.

Although analytical iteration is not accessible, it is possible to iterate backwards generation-by-generation. Appendix A.2.2 provides a demonstration of the possibility of retrodiction in the case of a model of selection at a diallelic locus with $W_{11} > W_{12} > W_{22}$. This result could reasonably be extended to all possible fitness ratios, because the backward equation is quadratic and its coefficients show that it has only one positive solution (Anthony Edwards, personal communication). Therefore, at least in this case, retrodiction is possible. Note that the backward equation is *derived* from the forward equation. Appendix A.2.2 also shows why reversibility is not satisfied for the same model. This brings us to the issue of time-reversibility in population genetics.

13.3.2.2.2 Time-Reversibility (Mathematical Sense)

The question of whether the deterministic models of population genetics are reversible in the usual mathematical sense is the most delicate problem. If yes, the 'laws' expressed by them should be unaffected by time reversal, that is substituting $-t$ for t in the equations (see Sect. 13.3.1.2). Since the notion of 'law' is important here, it is worth recalling Elliott Sober's proposal. According to him, the traditional logical empiricist concept of law requires that laws be statements that are both truly universal (that is to say not referring to any place, time, or individual) and empirical. This

[19]Freely accessible on http://cbs.umn.edu/populus/download-populus. This application has been developed by the College of Biological Sciences of the University of Minnesota for pedagogical purposes. We thank Michel Veuille for giving us this information. For some (very) particular cases where Eq. (13.7) is analytically tractable, see Hartl (1980, pp. 209–210).

concept is inappropriate for various domains of science, especially evolution, where the second condition (being an empirical statement) is questionable. For Sober (1997), the process of evolution as studied in population genetics "is governed by models that can be known to be *a priori* true". These models are mathematical truths, which describe how systems of specified type develop through time, whence Sober's expression "process law". For instance, given Mendel's laws, and an operational definition of notions such as recurrent mutation, selective value, etc., the process laws of population genetics describe how these factors determine the trajectory of a population in the space of gene frequencies. This viewpoint is endorsed here (Sober 1984, 1997; see also Brandon 1997, and Gayon 2014).

The reason why the notion of law is important here is that time-reversibility is not so much a property of a *trajectory* as a property of the *law* that governs the dynamics of the system, and therefore underlies the trajectory. Strictly speaking, the property of time-reversibility tells us nothing about the capacity of a given system to 'return' to its former state. This is not excluded, but will depend on the actual conditions imposed on the system. Time-reversibility is a property of a law of transformation that applies to transformations in both directions, whatever the actual fate of the system. For instance, the reversibility of Newton's law (F = ma) does not mean that a given body submitted to force F could effectively come back to a prior state, but that, if it did, the law would still apply. Thus, mathematical reversibility evokes counter-factual prediction, not the capacity of a *given* system to spontaneously return to a prior state (thermodynamic reversibility, see below). Of course, the two notions are often closely related in actual scientific practice, just as retrodictability and mathematical time-reversibility are often connected. However, the concepts are distinct.

Returning now to the examples of recurrent mutation and selection, using exactly the same models as in the previous paragraphs leads to the question "Are the 'laws' expressed in the recurrence equations time-reversible?"

Case 1: Recurrent Mutation

The position defended here is that the recurrence equation describing this process *is not* time reversible. This thesis goes against a common intuition. Since it is one of the simplest dynamic models in population genetics, it will be treated in detail.[20]

To begin with, let us formally define the concept of time reversibility in a dynamic system, the evolution of which can be described by a recurrence equation. We assume that the state of a system $x(t)$ at time t is a function of the p previous states:

$$x(t) = f(x(t-1), x(t-2), \ldots, x(t-p)) \tag{13.9}$$

Given that (13.9) is time-reversible for any sequence $(x(0), x(1), \ldots, x(T))$ solution of (13.9), so the reverse sequence $(x(T), x(T-1), \ldots, x(0))$ is also a solution of (13.9).

[20]The reasoning that follows should be credited to Jean-Philippe Gayon, who is warmly thanked for his help.

Before applying this definition to the one-way recurrent mutation model, an example of a recurrence equation that satisfies our definition of time-reversibility will first be given.

Consider the following recurrence equation:

$$x(t+2) = 2x(t+1) - x(t) + 1 \tag{13.10}$$

It can easily be checked that the sequence (0, 1, 3, 6, 10, 15) is a solution of (13.10).

$x(0) = 0$ (initial condition)
$x(1) = 1$ (initial condition)
$x(2) = (2 \times 1) - 0 + 1 = 3$
$x(3) = (2 \times 3) - 1 + 1 = 6$
$x(4) = (2 \times 6) - 3 + 1 = 10$
$x(5) = (2 \times 10) - 6 + 1 = 15$

Surprisingly, the reverse sequence (15, 10, 6, 3, 1, 0) is also a solution of (13.10):

$x(0) = 15$ (initial condition)
$x(1) = 10$ (initial condition)
$x(2) = (2 \times 10) - 15 + 1 = 6$
$x(3) = (2 \times 6) - 10 + 1 = 3$
$x(4) = (2 \times 3) - 6 + 1 = 1$
$x(5) = (2 \times 1) - 3 + 1 = 0$

More generally, for any sequence ($x(0)$, $x(1)$,. . ., $x(T)$) solution of (13.10), it is possible to prove that the reverse sequence ($x(T)$, $x(T-1)$,. . ., $x(0)$) is also a solution of (13.10). Therefore (13.10) is time-reversible.

Consider now the recurrence equation describing the evolution of the frequency p of an allele A subject to recurrent mutation. As seen above (13.5), this equation is:

$$p(t+1) = (1-u)p(t)$$

The sequence $(1, (1-u), (1-u)^2)$ is solution of (13.5):

$p(0) = 1$ (initial condition)
$p(1) = (1-u) \times 1 = (1-u)$
$p(2) = (1-u) \times (1-u) = (1-u)^2$

On the other hand, the reverse sequence $((1-u)^2, (1-u), 1)$ is obviously not a solution of (13.5) when $u \neq 0$. If the initial condition is $(1-u)^2$, then the sequence obtained by applying (13.5) is

$p(0) = (1-u)^2$ (initial condition)
$p(1) = (1-u) \times (1-u)^2 = (1-u)^3$
$p(2) = (1-u) \times (1-u)^3 = (1-u)^4$

We conclude that Eq. (13.5) describing the evolution of a population under recurrent mutation pressure is not time-reversible when $u \neq 0$ (in the case $u = 0$, the population never changes which is time reversible).

More generally, if any solution of (13.9) is strictly decreasing (respectively increasing), then (13.9) is not time reversible.

However, we anticipate an objection. The practicing population geneticist might say that reversing the direction of mutation ($a \rightarrow A$ instead *of* $A \rightarrow a$) would amount to 'reversing the process'. To make things as symmetrical as possible, a rate of reverse mutation equal to the rate of direct mutation could be taken, and begin with a frequency of say 0.9 in both directions. A law of the *same general form* would describe the 'reverse' dynamics and trajectory, with v (mutation rate for $a \rightarrow A$) replacing u (mutation rate for $A \rightarrow a$). But this would be *another* process; because a crucial parameter, with a different biological meaning (reverse mutation) has been introduced, it is not the *same* law. Changing the direction of mutation does not amount to changing the direction of time. 'Reverse mutation' is a biological concept, which should not be confused with the question of whether the process law describing the diffusion of a 'recurrent' mutation is 'reversible' or not.

Case 2: Basic Models of Selection

The basic equation to predict the evolution of gene frequencies for a diallelic selection submitted to selection has been given in (13.7). This recurrence equation is a rather general one. It ignores the various possible relations between the selective values, for instance: selection against a recessive gene ($W_{22} < W_{12} = W_{11}$), selection against the dominant allele ($W_{11} < W_{22} = W_{12}$), incomplete dominance ($W_{11} < W_{12} < W_{22}$ *or* $W_{22} < W_{12} < W_{11}$), or advantage to the heterozygote (W_{12} superior to the two other genotypic selective values). In all cases, selective values are assumed to be constant.

Now, given (13.8), Δp_i can be calculated simply,[21] giving:

$$\Delta p_i = \frac{(W_i - W)}{W} \tag{13.11}$$

with: p_i: frequency of allele A_i; W_i: average selective value (or "fitness") of the A_i allele (= average selective value of all genotypes containing A_i); W: average fitness of the population (average fitness of all genes at this locus, or average fitness of all the genotypes at the same locus in the population).

Equation (13.11) is of fundamental importance for the genetical theory of natural selection. As underlined by Crow and Kimura (1970, p. 180), it shows that the rate of change of gene frequency is proportional to: (1) the gene frequencies, $p_i\,(1 - p_i)$, which means that a very rare or very common gene will change slowly, regardless of how strongly it is selected; (2) the average excess in fitness of the A_i allele over the population average ($W_i - W$), which can be either positive or negative. If $W_i - W$ is positive, the frequency of the allele will increase; if negative, it will decrease.

[21] See Crow and Kimura (1970, pp. 179–180).

Whether Δp_i always increases, or always decreases, or sometimes increases and sometimes decreases will depend on the relations between the genotypic selective values $W_{11,}$ W_{12}, and W_{22}.

Before commenting on reversibility, another crucial notion must be introduced. If the selective values are kept constant, Δp_i can be rewritten as:

$$\Delta p_i = \frac{p_i(1-p_i)}{2W}\frac{dW}{dp_i} \tag{13.12}$$

This famous equation is known as "Wright's formula" (Wright 1937, 1940). It is fundamental because it connects the change in gene frequency Δp, with the slope of the function W (average fitness). It shows that if W is at a maximum with respect to p, then Δp is zero, or in other words the population is at equilibrium. W is classically interpreted as the "fitness function" or the "adaptive topography". Since W maximizes, a population submitted to selection is seen as "going uphill". Figures 13.3, 13.4, 13.5, 13.6, 13.7, (Appendix A.3) borrowed from Albert Jacquard (1971), give a clear graphic expression of the connection between Δp and W in various cases; these are actually the most common situations taught in an elementary course of population genetics. On these graphs, W_{11}, W_{12}, W_{22} are the selective values (or 'fitnesses'). By convention, Jacquard has systematically taken $W_{11} \leq W_{22}$, and p is the frequency of the unfavoured gene (numbered "1"). This is why Δp is most often negative. (It would be equivalent to observe the increase of the favoured gene; Δp would then be positive; but of course the order of alleles on the graph should be inverted). The arrows in the figures show that W is always $\geq$0 and is systematically maximized. Equilibrium is attained when W is maximal; this maximal value of W corresponds to $\Delta p = 0$. In Figs. 13.3, 13.4 and 13.5, the equilibrium is realized when $p = 0$ (the unfavoured gene disappears, the other gets fixed). In Fig. 13.6 (heterozygote inferiority), there are two stable equilibriums, $p = 0$ and $p = 1$, the outcome depending on the initial conditions). In Fig. 13.7 (heterozygote superiority), the population stabilizes for an intermediate value of p: selection maintains variation. Figures 13.8 and 13.9 are examples of the trajectories of populations submitted to two kinds of selective pressure abundantly realized in nature: selection against a recessive allele (Fig. 13.8), and selection with heterozygote superiority (Fig. 13.9). In each case, several curves are given, corresponding to selective pressures of varying intensity. The *trajectory* given in Fig. 13.8 corresponds to the *dynamics* represented in Fig. 13.5 (selection against a recessive). The *trajectory* given in Fig. 13.9 corresponds to the *dynamics* represented in Fig. 13.7 (heterozygote superiority). The trajectories represented in Figs. 13.8 and 13.9 are the result of computer iterations from the basic recurrence equation. As observed by Roughgarden (1979), it is remarkable that Eq. (13.12) despite its important restrictive condition (constancy of selective values), can generate so many different trajectories.

Are these models time-reversible? Formally, it would be useful here to provide a detailed proof that the basic recurrence Eq. (13.7), or its special applications (sometimes with simplifying assumption) to particular cases (*e.g.* Crow and Kimura 1970, Eqs. 5.2.2, and 5.2.14 to 5.2.17), or the treatment of the same problems with continuous time (*e.g.* Crow and Kimura 1970, Eqs. 5.2.8–5.2.11) are not time-reversible. This will be treated in a future publication. For the time being, Appendix A.2.2 provides a proof in the special case of a diallelic model with $W_{11} > W_{12} > W_{22}$. It shows that the dynamics of this system are retrodictable but not reversible. In fact the situation is identical to that of recurrent mutation. In both cases, it is possible to derive a rule for retrodiction, but this rule is obviously incompatible with the idea that the dynamic equation is reversible.

Furthermore, it should be observed that all standard models of selection with constant selective values describe dynamics that are driven by a maximizing function (maximization W of the average selective value of the population). Therefore it seems hard to imagine that such models could be used to describe a reverse transformation obeying exactly the same law: what would it mean for a population to accomplish the reverse trajectory with W minimized throughout? This would contradict the models. To sum up, all elementary models of selection evoked in the present chapter are deterministic and retrodictable, but they do not seem to describe a time-reversible process.

One final observation on this subject. We mentioned earlier that in selective models with constant genotypic selective values, the average selective value always increases from generation to generation, until it becomes equal to zero when equilibrium is attained, therefore $\Delta W \geq 0$. However, should this be understood strictly or approximately? Is there room for oscillation, as was often suggested in elementary courses of population genetics some decades ago, especially in the case of heterozygote superiority (see Fig. 13.7)? The idea was that the population rises through the adaptive topography a little like a ball which runs down a bowl, goes up the other side of the bowl, comes back, runs up again, etc. However, it has been demonstrated that there is no room for oscillation (Roughgarden 1979). This comes as no surprise if selection takes a population close to the fixation of one of the alleles: once the gene is fixed, no variation is left, and therefore there is no room for further selection. But in the case of heterozygote advantage, this is somewhat more surprising. Roughgarden (1979, pp. 51–53) notes that there is no possibility of "overshoots that are so large as to prevent the convergence to $\widehat{p}$". Therefore W is ≥ 0. No oscillation, no bounce. Although this is only an intuitive comparison, this behaviour can be contrasted with the situation in classical mechanics. If a moving body finds an obstacle on its way, one expects that it will communicate a fraction of its movement to another body, and bounce. Nothing like this is observed in standard models of selection: when the equilibrium point is reached, the movement just stops. This is typical of highly directional and maximizing dynamics, where reversion is hardly conceivable as long as the conditions remain the same.

13.3.2.2.3 'Physical' or 'Thermodynamic' Reversibility

Physical reversibility is the possibility for a system to return spontaneously to a previous state. Does this apply to population genetics? This subject will not be treated here in detail. A few glimpses will suffice. Since Ronald Fisher, the directionality of evolution under natural selection has been regularly compared with the directionality implied by the Second Law of Thermodynamics. The Second Law asserts that irreversible physical processes imply a unidirectional increase of entropy. In his

Genetical Theory of Natural Selection (1930), Fisher 'immediately recognized certain formal analogies between the mechanistic models introduced by Boltzmann (1896) to analyse physical systems, and the selection models proposed by Darwin (1859) to explain adaptation in biological systems' (Demetrius 2000). According to Demetrius, Fisher's fundamental theorem of natural selection is indeed a directionality theorem. This theorem states that 'The rate of increase of fitness of any species is equal to the genetic variance in fitness' (Fisher [1930] 1958, p. 50). By this formula, Fisher meant that the speed of action of selection is a function of the additive genetic variance.[22]

It is worth quoting Fisher here who compares his fundamental theorem with the Second Law of Thermodynamics:

> Both are properties of populations, or aggregates, true irrespective of the nature of the units which compose them; both are statistical laws; each requires the constant increase of a measurable quantity, in the one case the entropy of a physical system and in the other the fitness, measured by m, of a biological population. As in the physical world we can conceive of theoretical systems in which dissipative forces are wholly absent, and in which the entropy consequently remains constant, so we can conceive, though we need not expect to find, biological populations in which the genetic variance is absolutely zero, and in which fitness does not increase. (Fisher [1930] 1958, p. 39)

In spite of these resemblances,[23] Fisher's objective was in fact to emphasize the differences between the Second Law of Thermodynamics and his theorem. Among the five differences that he mentions, one is of special interest for our subject: "Entropy changes are exceptional in the physical word in being irreversible, while irreversible evolutionary changes form no exception among biological phenomena" (Fisher, *ibid*., p. 40).

---

[22]The additive genetic variance is the fraction of genetic variance attributable to the additive effects of genes, ignoring the inter-allelic and inter-genotypic interactions (For detailed comments on Fisher's fundamental theorem, see Price (1972); from a historical point of view, see Gayon (1998), Chap. 9).

[23]It is worth adding that in 1941, Fisher had a controversy with Wright about the meaning of his 'fitness function' or 'adaptive topography'. Commenting on this equation, Fisher wrote: 'Wright's conception embodied in equation (6) (Eq. (13.12) in the present paper) that selective intensities are derivable, like forces in a conservative system, from a simple potential function dependent on the gene ratios of the species as a whole, has led him to extensive but untenable speculations.' (Fisher 1941). Thus Fisher denies that (13.12) is a potential function describing a conservative system—a position also assumed by Crow and Kimura in their 1970 textbook. However Fisher's objection is formulated in a context where the adaptive topography is supposed to describe the behaviour of the species for all gene ratios. In reality, Wright's equation is valid only for a diallelic locus with constant selective values (Edwards 2000). Wright acknowledged this when he gave the equation for the first time in 1937, but he eventually extended it to multi-allelic loci without providing proof. He also suggested that this equation could be related to his notion of 'adaptive landscape', a notion that takes into account the entire genome. Populations are then seen as pushed towards a 'peak' by W, the average selective value, an ever-increasing quantity (For more on this controversy, see Gayon (1998), Chap. 9). However, as noted by Edwards (2000), it is not clear at all that, even in the simplest case of a diallelic locus, W is a potential surface: 'Wright's mistake, repeated by Crow and Kimura, was to interpret his non-standard partial derivative $\partial w/\partial q_i$ as "the slope of w in the direction where the relative frequencies of the other alleles do not change" (Crow and Kimura 1970), whereas in fact it is the rate of change of w in that direction, *but with respect* to *change in* q_i *alone*. This is not a gradient on the w-'surface' at all, and the analogy of a potential function, already tenuous because of the factor $q_i(1-q_i)$, is thus in reality even more remote.' (Edwards 2000, pp. 68–69).

In fact, modern population biologists, or at least some of them, compare entropy and fitness more literally than Fisher used to. A stimulating example is Lloyd Demetrius, who proposes an adaptation of the concept of entropy to evolutionary genetics and ecology, and puts forward the concept of "evolutionary entropy": a measure of the dispersion of the age of the ancestors of a randomly chosen newborn. Demetrius' concept of evolutionary entropy is an explicit attempt to overcome the obvious difference between statistical thermodynamics and population biology; thermodynamics deals with the properties of populations of inert particles, whereas population biology treats the properties of populations of living objects which reproduce, and therefore grow. To overcome this difficulty, Demetrius points to a "formal correspondence" between the thermodynamic variables and the population parameters intervening in population biology: "growth rate" corresponds to free energy; "generation time" corresponds to inverse temperature; "evolutionary entropy" corresponds to thermodynamic entropy. Then, both situations follow the same equations with the appropriate variables (Demetrius 2000). Demetrius' basic intention is to offer an alternative to Fisher's conception of directional change in mean fitness: Fisher's fundamental theorem is based on the relative viability of individuals within the population; Demetrius takes into account the absolute survival and reproduction of populations. In Demetrius' terms, "Darwinian fitness is completely described by evolutionary entropy, a demographic variable which measures the heterogeneity in the age of reproducing individuals in the population." (Demetrius 2000, p. 2) Moreover, "[evolutionary] entropy characterizes Darwinian fitness, the efficiency with which a population acquires and converts resources into viable offspring. Accordingly, entropy predicts the outcome of natural selection in populations subject to different classes of ecological constraints." (Demetrius et al. 2009) We acknowledge that we feel rather uncertain about the precise sense in which "evolutionary entropy" and entropy in the ordinary physical sense are correlated. As Demetrius & al. recognize, evolutionary entropy is "analogue of the Gibbs–Boltzmann entropy in statistical thermodynamics". This analogy is grounded both in a formal similarity of equations, and in the distinction between a microscopic[24] and a macroscopic description.[25] Suffice to say that Demetrius's stimulating proposal represents just one way among others to bridge the gap between population genetics and population ecology, with its emphasis on absolute fitness, population density, and bounded *vs.* unbounded population growth.[26]

---

[24]The uncertainty in the age of the mother of a randomly chosen newborn.

[25]Darwinian fitness, the efficiency with which a population acquires and converts resources into viable offspring.

[26]Just to give an example: "In populations subject to bounded growth constraints, demographically stable states are described by the condition of maximal entropy; in populations with unbounded growth, demographically stable states are described by the condition of minimal entropy. Under bounded growth conditions, entropy increases; under unbounded growth conditions, *large* population size, entropy decreases; under unbounded growth conditions, *small* population size, the change in entropy is random and non-directional " (Demetrius 2000, p. 7; see also Demetrius 1992, with its explicit title, " Growth rate, population entropy and evolutionary dynamics").

Demetrius' proposal will not be analysed further here. What matters for the present paper is a consequence of his approach to the question of whether the evolutionary dynamics are reversible or not in a thermodynamic (or - thermodynamic-like) sense. Rather than emphasizing the kind of unlimited evolutionary irreversibility that Fisher favoured, Demetrius offers a tool that leaves room for both evolutionary irreversible processes (characterized by unidirectional increase of evolutionary entropy), and evolutionary reversible (or virtually stationary) processes. In such a theoretical framework, the reversible/non reversible distinction is not an all or nothing alternative, but rather a tool for describing the dynamics of evolutionary change (Demetrius 2000. For a comprehensive description of this field of research, see Weber, Depew and Smith 1988).

This area of research is fascinating and innovative, but remains quite speculative today. Are there real analogies with friction and heat dissipation in the models of population genetics and ecology, or are these comparisons no more than suggestive metaphors or mathematical analogies? This question will be left open (For a survey of this field of research, see Weber, Depew and Smith 1988).

13.3.2.2.4 Other Senses of Reversibility in Population Genetics

The intuitive notion of reversibility (see above, 2.1, introduction) sometimes encountered among population geneticists will now be reconsidered. For a population geneticist, who usually works at a small evolutionary temporal scale, it seems obvious that most often a given population can return to a prior state. For instance, if a recurrent mutation spreads on its own and is fixed, in principle a reverse recurrent mutation can do the same. If natural selection causes the expansion of an allele in the population, it is always possible that selective pressure in the other direction returns the population to its initial state. If random drift leads to the fixation of a mutated allele, a reverse mutation followed by drift or selection may cause 'reverse' evolution. And so on.

There is no doubt about the plausibility and effectiveness of such processes. But they mean something different from the various senses of reversibility analysed above. Recurrent reverse mutation has been discussed, consider now the case of selection. In the haploid case, how far would an inversion of the selection coefficient amount to a 'reversible process' in a technical sense close to time-reversibility? Appendix A.2.1 provides a formal treatment of this problem in the simple case of haploid populations. The method consists in replacing the selective values W_i (haploid populations) with the inverse quantities $1/W_i$ in the recurrence equation. This inversion preserves the relative values of the *Ws*, but in reverse order. The question then, is whether the inversion of the fitness values preserves the 'law' expressed in the recurrence equation or not. The calculation shows that the dynamics of the initial equation is preserved with the new parameters $1/W_i$. Therefore, the 'law' is preserved by time reversal, except for the constants. This is not 'time-reversibility' in the strict sense, but could be named weaker time reversibility (for a definition, see Appendix

A.1, A.1.2, and example in A.2.1). Incidentally, this demonstration applies to as many alleles and corresponding selective values as required.

Would the same methodology produce a similar case in the case of selection on sexually reproducing diploid population? Certainly not. Suppose for instance that the heterozygote is advantaged, *ie*. W_{12} greater than both W_{11} and $W_{22.}$ The dynamics of such selection is represented in Fig. 13.7. If the values of the W_{ij} were inverted, *i.e*. if the W_{ij} were replaced by expressions of the form $1/W_{ij,}$ then we would have W_{12} smaller than both W_{11} and $W_{22.}$ This would correspond to the dynamics represented in Fig. 13.6 (underdominance), with an unstable equilibrium, and two possible equilibria ($p = 0$ and $p = 1$). This is indeed a totally different dynamics. The same could be said of other possible ratios for the Ws, except when $W_{12} = \sqrt{W_{22}W_{11}}$ (see Appendix A.2.2).[27] In this case, the heterozyogote has an intermediate value, where the term 'weak time reversibility' could be used. In the general case, selection on a diallelic locus is not reversible in any sense, although it is retrodictable (see demonstration for $W_{11} > W_{12} > W_{22}$ in Appendix A.2.2).

13.3.3 Conclusions on Reversibility in Theoretical Population Biology

Although the analysis above remains somewhat sketchy and incomplete, the conclusions that it suggests are quite straightforward:

- Reversibility in the sense of retrodictability (in fact an improper use of the word) seems obvious for deterministic models, with possible exceptions in discrete time. It does not apply to stochastic models.
- Time-reversibility is clearly instantiated in some important stochastic models; it seems inadequate for deterministic models. In one case (selection in a haploid population), one comes close to time-reversibility since a weaker definition than the usual one is met. However, this subject would require more extensive and detailed examination.
- Thermodynamic reversibility, which is a fascinating subject, remains mainly a matter of speculation.
- In classical models of selection, the average selective W can hardly be interpreted as a potential function (see n. 11).

[27]Hartl (1980, Figure 25, p. 212) gives a nice graphical representation of this fact, although his intention is not to comment on reversibility. The figure comprises two graphs showing the effect of selection when allele A is favoured and when allele a is favoured. In both cases $W_{12} = \sqrt{W_{22}W_{11}}$, and several initial frequencies for A are considered. The trajectories are symmetrical.

13.4 Conclusions

Two broad conclusions can be drawn. The first conclusion bears on lawfulness in evolutionary biology. Laws do not need to be empirically true statements of unlimited universal scope; they can also be "models that can be known to be true *a priori*" (Sober 1997). Not all mathematical truths are laws, only those that elaborate on the basis of some empirically plausible conditions. For instance, if Mendelian inheritance is assumed, and if such factors as mutation, migration, mating system and selection are plausible evolutionary factors, then it is possible to derive the typical behaviour of a population under specified conditions. The abstract descriptions obtained are idealizations, no more and no less than those developed by the physical sciences. The question of whether they are useful to predict the actual behaviour of populations or not is irrelevant to their nomological status. They just say that, if conditions C_1, C_2,..., C_n are met, then a specific behaviour should be observed. The Hardy and Weinberg equilibrium is a good example: provided that the six conditions stated in 2.2 are satisfied, then the genetic structure of a population is constant over time. Of course no biologist believes that actual populations are infinite, but the model opens the possibility of making other models that will explain the deviations as a function of population size. Therefore the problem is not whether the HW equilibrium is strictly true from an empirical point of view, but how useful it is for explaining and predicting the behaviour of real populations, in combination with other models that will take into account the fact that this or that condition is not met, and to which degree it is not met. Nevertheless, the HW law remains a mathematical truth *if* the standard conditions are met. Some philosophers might argue here that it would be wiser to give up the term 'law' and just speak of 'models'. This is maybe the case. In practice, population geneticists use the terms 'law' and 'model' indifferently to qualify the HW equilibrium. In other cases (*e.g.* treatment of the effects of mutation, selection or population size), they tend to prefer 'model'. We do not think that this is a crucial issue. Whatever the outcome of such a discussion, one important conclusion of this chapter is that an impressive number of results of theoretical population genetics consist in equilibria, which qualifies population genetics as a genuine science able to identify properties of invariance through transformation. The discovery of such properties has been a distinctive feature of modern science since its inception.

In population genetics, invariance to transformation is mainly defined with reference to the genetic structure of a population. Under a number of conditions, this structure may stabilize. One of the most characteristic features of population genetics as a field of research is indeed the search for "trends towards equilibria". When browsing through text-books, it is amazing to observe in population genetics the definition of the equilibrium, and the estimation of the time required to come close to it, is declared to be more important than the precise description of the trajectory. This is what Haldane meant when he stated that a good deal of evolutionary theory deals with the 'statics' rather than the 'dynamics' of evolution (Haldane 1954).

The second conclusion of this enquiry comes to temper the first one. Time-reversibility has also been said to be a major characteristic of modern science, with special reference to classical mechanics. In the theory of population genetics, time-reversibility is obviously and abundantly present in the treatment of stochastic evolutionary processes. But this is a rather particular sense of reversibility (symmetry of probability distributions in both directions of time). As far as the authors have been able to judge from this first exploratory enquiry, time-reversibility in its mathematical sense does not seem to be a property of the models describing deterministic phenomena. But this problem will have to be examined in a more systematic way. Deterministic models in population genetics are highly retrodictable but this property is not identical to time-reversibility.

Time reversibility is a very impressive property of the equations of classical mechanics in the case of fundamental forces. In classical mechanics, objects follow the principle of inertia so that a homogeneous rectilinear movement does not require an external cause. The structure of selection models in population genetics is different. The changes in a population are directly driven by differences in fitness. If these differences cease to be, then the changes in the population stop immediately. In this sense, selection in population genetics is closer to a mathematized Aristotelian mechanics than to classical mechanics. In classical mechanics energy is conserved and is transferred from potential energy to kinetic energy and vice versa, which ensures the time reversibility of the dynamics. In population genetics, Wright's equation describes a gradient ascent akin to classical potentials, but there is no equivalent to the kinetic energy that would enable the system to descend the gradient, which would be required for time reversibility.

To end up, the present publication is the first part of a broader enquiry on repetition and reversibility in evolution, both at the theoretical and the experimental level. A further paper will examine the rather different problems raised by the issues of the repeatability of experiments in experimental population biology, and of the capacity of populations or species to return to a prior evolutionary state.

Acknowledgements Maël Montévil's work is supported by a grant from the labex "Who am I?". This paper has benefited from extensive discussion and collaboration with Jean-Philippe Gayon, who has been enormously helpful in clarifying the notion of time reversibility from a mathematical point of view. We also thank Anthony Edwards, Pierre-Henri Gouyon, and Michel Veuille. Their great competence in population genetics has been very fruitful to our understanding of the issue of reversibility in population genetics models. We also thank Edwards and Veuille or their careful reading of the text. We thank Elliott Sober for his careful reading of the paper and for his fruitful comments. Véronique Charrière is warmly thanked for her careful linguistic corrections and suggestions.

Appendices

Appendix A.1 provides a formal definition of symmetry by time reversal and of retrodictability. Appendix A.2 applies these definitions to some population genetics models. Appendix A.3 contains the Figs. 13.3, 13.4, 13.5, 13.6 and 13.7.

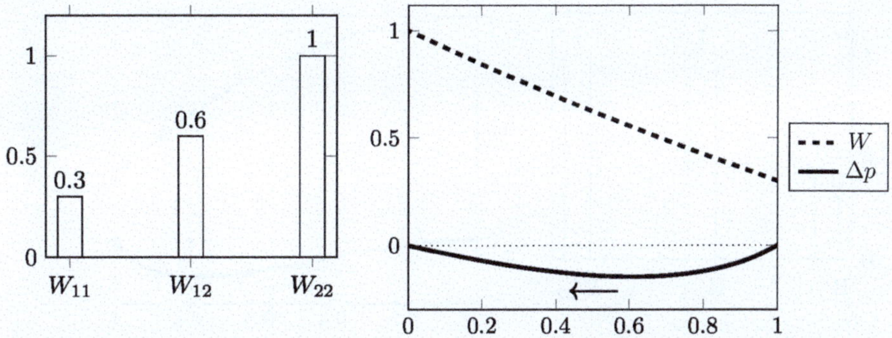

Fig. 13.3 Effect of selection when $W_{11} < W_{12} < W_{22}$ (After Jacquard 1971). The W_{ij} designate the selective values of the genotypes. Allele 1, with frequency p, is disadvantaged. The heterozygote has an intermediate selective value. The average selective value W of the population always increases. It goes from $W = W_{11}$ when all individuals are AA to W W_{22} when all individuals are aa. Δp is always negative, meaning that the unfavoured allele 1 is progressively eliminated. W always increases

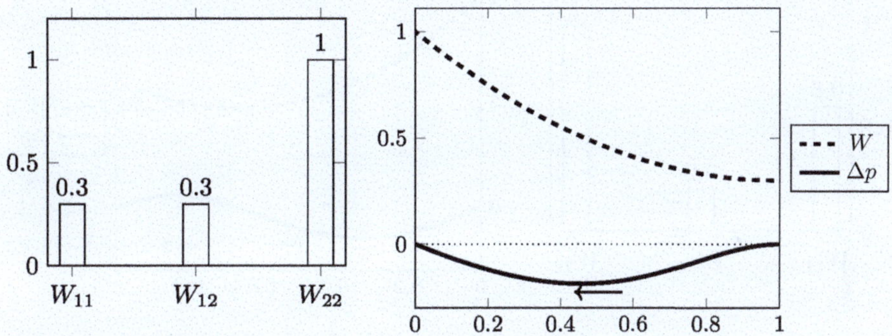

Fig. 13.4 Effect of selection when $W_{11} = W_{12} < W_{22}$ (After Jacquard 1971). Allele 2 is favoured only as homozygote, and spreads very slowly when rare. W always increases

Time Reversibility

As discussed in the body of the text, time reversibility corresponds intuitively to a situation where the dynamics follow the same law before and after time reversal. Only deterministic dynamics are considered here.

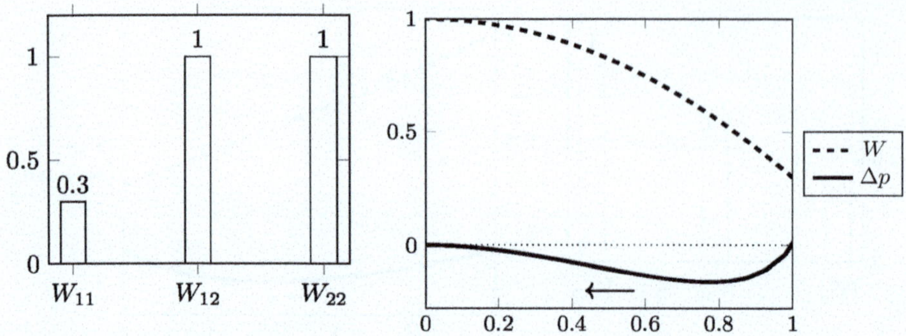

Fig. 13.5 Effect of selection when $W_{11} < W_{12} = W_{22}$ (After Jacquard 1971). This is the typical case of selection against a recessive gene. Allele 1 is disadvantaged only as homozygote, and is eliminated very slowly when rare. W always increases. Selection against a recessive gene is the most frequent case among the standard situations represented in Figs. 13.3 through 13.7. Figure 13.8 represents the shape of the trajectory of gene frequencies for this case

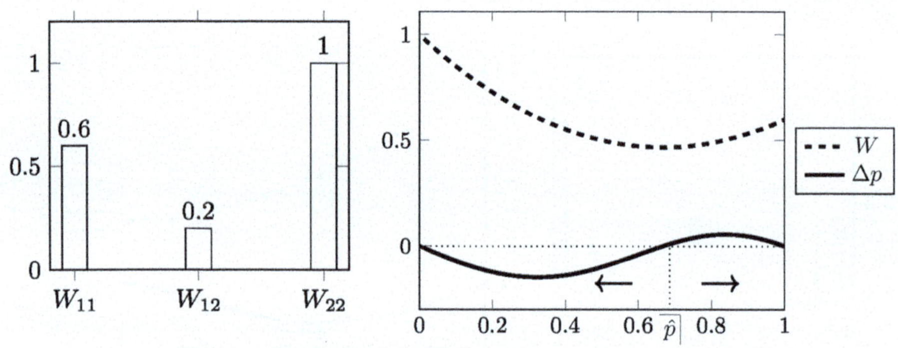

Fig. 13.6 Effect of selection when $W_{12} < W_{11} < W_{22}$ (After Jacquard 1971). The heterozygote is disadvantaged (underdominance). There are two equilibria. Δp is either negative or positive. W always increases, but moves towards two possible stable equilibria, corresponding to $p = 0$ and $p = 1$; $\widehat{p}$ is an unstable equilibrium, corresponding to a minimal value for W. The question is debated whether this situation is realized in nature or not, and what its biological meaning is. But it is probably a very rare case

Continuous Time

In continuous time, a dynamic is typically defined by differential equations. A fairly general definition of differential equations is admitted here:

$$0 = \mathrm{F}\big(\alpha_1, \ldots, \mathrm{x}, \mathrm{dx}/\mathrm{dt}, \mathrm{d}^2\mathrm{x}/\mathrm{dt}^2, \ldots, \mathrm{d}^n\mathrm{x}/\mathrm{dt}^n\big) \qquad (13.13)$$

where the α_i are parameters.

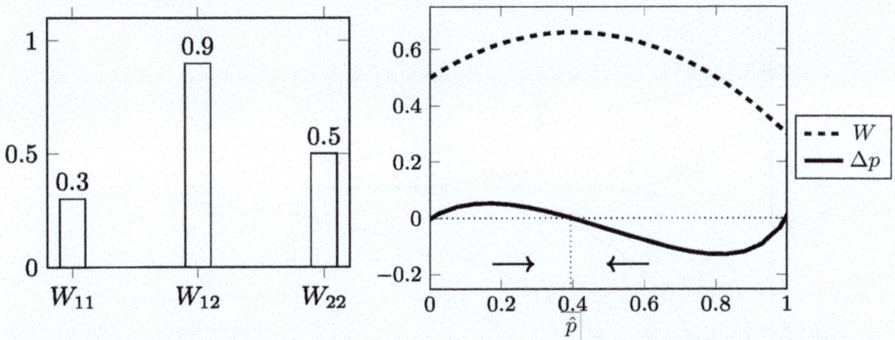

Fig. 13.7 Effect of selection when $W_{11} < W_{22} = W_{12}$ (After Jacquard 1971). Known as over-dominance or 'advantage of the heterozygote'. There are two unstable equilibria, $p = 0$ and $p = 1$, and one stable equilibrium, $\widehat{p}$, corresponding to a maximal value for W, which always increases. Δp is either negative or positive. This is the best-known polymorphic equilibrium, with ample proof of its existence in nature

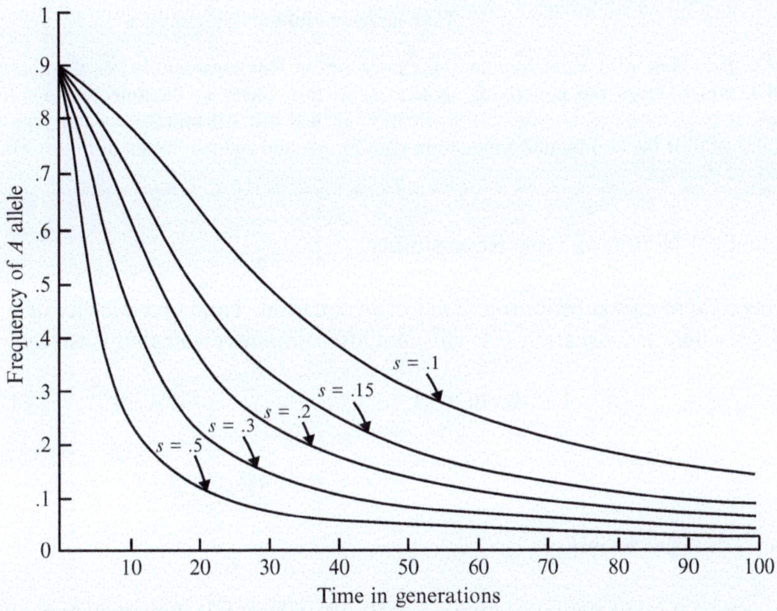

Fig. 13.8 Selection against a recessive allele for various selective coefficients (From Roughgarden 1979). Compare with Fig. 13.5, which gives the underlying dynamics. As almost all trajectories in models of selection with discrete generations, this trajectory is obtained by computer iteration. s is the 'selection coefficient'. The greater s, the stronger the selection against the genotype. The 'coefficient of selection' is a convenient way of treating the relative selective values. If the selective value of the favoured genotype(s) is 1, the selective value of the unfavoured genotype (here the recessive homozygote) is $1 - s$

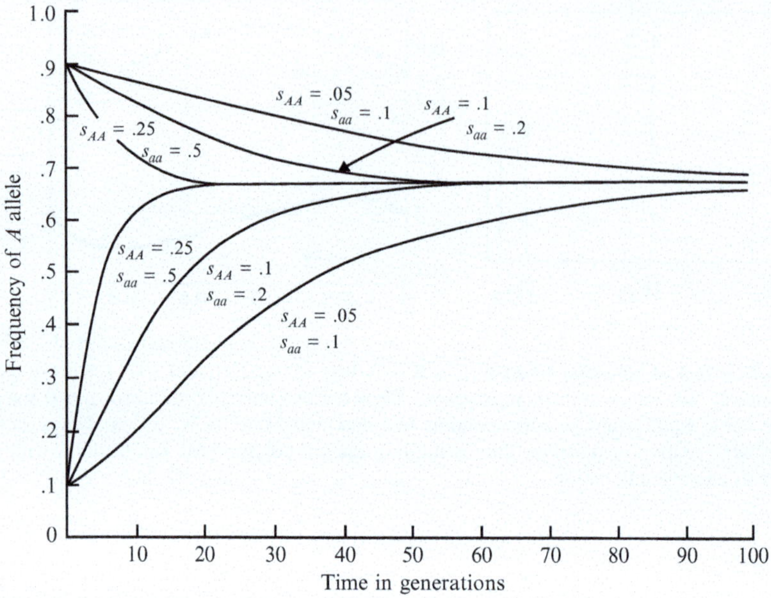

Fig. 13.9 Selection with heterozygote superiority (From Roughgarden 1979). Compare with Fig. 13.7, which gives the underlying dynamics. In this example, the selection against the homozygote *aa* is twice as intense as the selection against *AA*. All trajectories converge to the same value of $\widehat{p}$. If the two homozygotes were equally selected against, the equilibrium would be obtained for $\widehat{p} = 0.5$

Mathematical Notion of Time Reversibility

Time reversal means substituting $-t$ to t in an equation. Time reversibility describes situations where the equation F is still met after time reversal; that is to say:

$$0 = \mathrm{F}\left(\alpha_1, \ldots, \mathrm{x}, -1 \times \mathrm{dx/dt}, (-1)^2\mathrm{d^2x/dt^2}, \ldots, (-1)^\mathrm{n}\mathrm{d^nx/dt^n}\right) \tag{13.14}$$

Weaker Time Reversibility

We propose a weaker notion of time reversibility, where F is preserved even though the parameters α_i are modified through time reversal. Thus, the criterion becomes:

$$0 = \mathrm{F}\left(\alpha_1', \ldots, \mathrm{x}, -1 \times \mathrm{dx/dt}, (-1)^2\mathrm{d^2x/dt^2}, \ldots, (-1)^\mathrm{n}\mathrm{d^nx/dt^n}\right) \tag{13.15}$$

A supplementary condition is required for this notion to have a theoretical meaning. The α_i' have to be possible values for the α_i. For example if α_1 is a mass α_1' has to be a positive number, but it may be different from α_1.

Retrodictability

The Cauchy-Lipschitz theorem also called Picard–Lindelöf theorem, states that under very general hypotheses on the regularity of F, there is only one trajectory which goes through one initial condition. This theorem is the basis of the ability of the differential equation to define a deterministic process, but it also ensures that we can retrodict the past. Increasing or decreasing the time parameter makes no difference in this theorem. Note, however, that solutions to differential equations may be valid only for a limited time interval, either because a variable becomes infinite or takes a value that has no physical or biological meaning.

As mentioned in the main text, the issue of dynamics sensitive to initial conditions, such as the three-body problem in mechanics have not been discussed here. Let us just mention that in these cases, the impossibility of a perfect measurement of the initial conditions prevents empirical long term predictions and retrodictions because small differences have significant consequences.

Discrete Time

Dynamics defined by recurrence will now be discussed: $p(t+1)=g(p(t))$.

Mathematical Notion of Time Reversibility

The transitions t to $t+1$ and $t+1$ to t follow the same law, that is:

$$p(t+1) = g(p(t)) \rightarrow p(t) = g(p(t+1)).$$

Weak Time Reversibility

By analogy with the continuous case, changes in the values of the parameters of g in this weaker notion of time reversibility are allowed:

$$p(t+1) = g(p(t), \alpha_i, \ldots) \rightarrow p(t) = g(p(t+1), \alpha_i', \ldots).$$

Note also that the new parameters must have a physical or biological meaning.

Example: $p(t+1)=p(t)/2 \rightarrow p(t)=2p(t+1)$. In this case $\alpha=1/2$ and $\alpha'=2$. See the next case for a dynamic that does not meet this criterion.

Retrodictability

Retrodictability corresponds to the fact that the reverse dynamic is deterministic. A system is retrodictable when a function h exists that satisfies:

$$p(t+1) = g(p(t), \alpha, \ldots) \rightarrow p(t) = h(p(t+1), \alpha', \ldots)$$

Function h may be completely different from g (even though they are of course related).

Example: $p(t+1) = p(t)^2 \rightarrow p(t+1) = \sqrt{p(t)}$. This dynamic is retrodictable but not reversible, even in the weak sense.

Counter example: $p(t+1) = 10p(t) - \lfloor 10p(t) \rfloor$, where $\lfloor x \rfloor$ is the integer part of x (a digit is lost at every step). For example $P(0) = 0.97511\ldots$, $P(1) = 0.7511\ldots$, $P(2) = 0.511\ldots$ But with $P'(0) = 0.87511\ldots$, we get $P'(1) = 0.7511\ldots$, which is identical to P(1). This ambiguity prevents defining a deterministic reversed dynamic. This deterministic dynamic does not enable retrodiction.

Application to Models of Population Genetics

In this part, the reversibility of two classical models of selection in population genetics are discussed. In one case (asexual haploid population), the dynamic is reversible in the weak sense. In the second case (diploid, diallelic locus), the dynamic is retrodictable but not reversible.

Asexual Haploid Population: A Weakly Reversible Model

We start with a classical model of selection at a single locus in a haploid population. W_i is the relative fitness of the allele i and p_i is the proportion of allele i in the population. W(t) is the average fitness of the population at time t.

$$p_i(t+1) = p_i(t)\frac{W_i}{W(t)} \tag{13.16}$$

To assess the properties of this dynamic with respect to time reversal, let us write $p(t)$ as a function of $p(t+1)$, and let us compare the result with Eq. (13.16).

$$p_i(t) = p_i(t+1)\frac{W(t)}{W_i} \tag{13.17}$$

Equation (13.17) can also be written:

$$p_i(t) = p_i(t+1)\frac{1/W_i}{1/W(t)} \tag{13.18}$$

Here $1/W_i$ seems to play the same role as W_i in Eq. (13.16), therefore we can write $W_i^{'} = 1/W_i$. However, we have to verify whether 1/W(t) can be interpreted as the mean fitness in the reversed dynamic $W^{'}(t+1)$, where the new fitness coefficients are the $W_i^{'}$.

By definition, we have:

$$\frac{1}{\mathrm{W(t)}} = \frac{\mathrm{N(t)}}{\mathrm{N(t+1)}} = \mathrm{W}'(\mathrm{t}+1) \tag{13.19}$$

It is also possible to do this computation on the basis of the analytical expression of $\bar{W}(t)$ which leads to the same result. Then we can conclude that:

$$\mathrm{p_i(t)} = \mathrm{p_i(t+1)}\frac{\mathrm{W}'_\mathrm{i}}{\mathrm{W}'(\mathrm{t}+1)} \tag{13.20}$$

This equation has exactly the same form as Eq. (13.16). Therefore the dynamic is weakly reversible.

Sexual Diploid Population: A Retrodictable But Irreversible Model

A population with two alleles at a single locus is now considered. The relative fitnesses are $W_{11} \neq 0$ and $W_{22} \neq 0$ for homozygotes and W_{12} for heterozygotes, and $W_{11} > W_{12} > W_{22}$. W(t) is the average fitness of the population. p and q are the proportions of the first and second allele in the population. A classical model for this situation is then:

$$\mathrm{p(t+1)} = \frac{\mathrm{W_{11}p^2(t) + W_{12}p(t)q(t)}}{\mathrm{W(t)}} \tag{13.21}$$

We use $q(t) = 1 - p(t)$ and $\mathrm{W}(t) = W_{11}p^2(t) + 2W_{12}p(t)q(t) + W_{22}q^2(t)$.
This leads to a quadratic equation, where $p(t)$ is the variable:

$$\begin{aligned} 0 = {} & (\mathrm{W_{11} - W_{12} - p(t+1)(W_{11} - 2W_{12} + W_{22})})\mathrm{p^2(t)} \\ & + (\mathrm{W_{12} - p(t+1)2(W_{12} - W_{22})})\mathrm{p(t)} - \mathrm{p(t+1)W_{22}} \end{aligned} \tag{13.22}$$

The reduced discriminant determines the solutions of such an equation. If the discriminant is positive, there are two mathematical solutions and no solutions if it is negative.

$$\Delta' = \left(\mathrm{W_{22}W_{11} - W_{12}^2}\right)\left(\mathrm{p(t+1) - p^2(t+1)}\right) + \mathrm{W_{12}^2/4} \tag{13.23}$$

$$0 \leq p(t+1) \leq 1, \text{ thus } (t+1) - p^2(t+1) > 0.$$

If $\gamma = W_{22}W_{11} - W_{12}^2 > 0$, then clearly Δ' is positive. In the opposite case, it is simple to verify that the smallest possible value for Δ' is given by $p(t+1) = 1/2$, which leads to $\Delta' = W_{22}W_{11}/4 > 0$. Therefore, $\Delta' > 0$ always holds and the quadratic equation has two mathematical solutions. After computation we get:

$$p(t) = \frac{(W_{12} - W_{22})p(t+1) - W_{12}/2 \pm \sqrt{(W_{22}W_{11} - W_{12}^2)(p(t+1) - p^2(t+1)) + W_{12}^2/4}}{p(t+1)(W_{12} - W_{22}) + (W_{11} - W_{12})(1 - p(t+1))} \tag{13.24}$$

Because $W_{11} > W_{12} > W_{22}$, the denominator is always positive, and the relations between the roots and the coefficients show that the two solutions have different signs. The only solution which has biological meaning is therefore the positive solution:

$$p(t) = \frac{(W_{12} - W_{22})p(t+1) - W_{12}/2 + \sqrt{\gamma(p(t+1) - p^2(t+1)) + W_{12}^2/4}}{p(t+1)(W_{12} - W_{22}) + (W_{11} - W_{12})(1 - p(t+1))} \tag{13.25}$$

$$p(t) = h(p(t+1)) \tag{13.26}$$

Consider now time reversal. Equation (13.21) gives $p(t+1)$ as a function of $p(t)$. This corresponds to function g in the definition of reversibility above (Appendix A.1.2). $p(t+1) = g(p(t))$, is a rational function, that is to say a fraction of polynomials. When $\gamma \neq 0$, the square root in the definition of $p(t)$ as a function h of $p(t+1)$ implies that h is not a rational function. The difference between g and h is then more than just a change of coefficient. In this case, retrodiction is possible but time is not reversible in either the weak or the strong sense.

Interestingly, the case where the square root disappears, $\gamma = 0$, corresponds to $W_{12} = \sqrt{W_{22}W_{11}}$. Thus, in this case, the fitness of heterozygotes is the geometric average of the fitnesses of homozygotes, which can be interpreted as a form of linearity in the effects of the alleles. Reciprocally, when $\gamma \neq 0$, the change of the form of the dynamics by time reversal stems from the non-linearity in the effects of the alleles in heterozygotes.

References

Beatty, J. 1995. The evolutionary contingency thesis. In *Concepts, theories, and rationality in the biological sciences*, ed. G. Wolters and J. Lennox, 45–81. Pittsburgh: University of Pittsburgh Press.

Boltzmann, L. 1896. *Vorlesungen über Gastheorie*. Leipzig: Barth.

Brandon, R. 1997. Does biology have laws? The experimental evidence. *Philosophy of Science, 64, Supplement. Biennial Meetings of the Philosophy of Science Association*. Part II: Symposia Papers (Dec, 1997), S444–S457.

Crow, J., and M. Kimura. 1970. *An introduction to population genetics theory*. New York/ Evanston/ London: Harper & Row, Publishers.

Darwin, C. 1859. *On the origin of species*. London: Murray.

Demetrius, L. 1992. Growth rate, population entropy and evolutionary dynamics. *Theoretical Population Biology* 41: 208–223.

———. 2000. Thermodynamics and evolution. *Journal of Theoretical Biology* 206: 1–16.

Demetrius, L., S. Legendre, and P. Harremöes. 2009. Evolutionary entropy: A predictor of body size, metabolic rate and maximal life span. *Bulletin of Mathematical Biology* 71(4): 800–818.

Dollo, L. 1893. Les lois de l'évolution. *Bulletin de la Société Belge de géologie, paleontology, et hydrologie* 7:164–166. [English translation in Gould 1970, 210–212].

Edwards, A.W.F. 1977. *Foundations of mathematical genetics*. Cambridge: Cambridge University Press.

———. 2000. Sewall Wright's equation $\Delta q = (q(1 - q)\partial w/\partial q)2w$. *Theoretical Population Biology* 57: 67–70.

Ewens, W. 2012. James F. Crow and the stochastic theory of population genetics. *Genetics* 190 (2): 287–290.

Fisher, R.A. 1918. The correlation between relatives on the supposition of Mendelian inheritance. *Transactions of the Royal Society of Edinburgh* 52: 399–433.

———. 1941. Average excess and average effect of a gene substitution. *Annals of Eugenics* 11: 53–63.

Fisher, R. 1958 [1930]. *The genetical theory of natural selection*, 2nd ed. Oxford/New York: Clarendon/Dover.

Gayon, J. 1998. *Darwinism's struggle for survival: Heredity and the hypothesis of natural selection*. Cambridge: Cambridge University Press.

———. 2014. De la portée des théories biologiques. In *Problèmes théoriques et pratiques en biologie évolutionnaire*, ed. T. Martin, 13–52. Besançon: Presses Universitaires de Franche Comté.

Gould, S.J. 1970. Dollo on Dollo's law: Irreversibility and the status of evolutionary laws. *Journal of the History of Biology* 3 (2): 189–212.

Goux, J.M. 1979. Les modèles en génétique des populations. In *Élaboration et justification des modèles: Applications en biologie*, ed. P. Delattre and M. Thellier, vol. II, 565–571. Paris: Maloine.

Griffiths, P.E. 1996. The historical turn in the study of adaptation. *The British Journal for the Philosophy of Science* 47 (4): 511–532.

Haldane, J.B.S. 1954. The statics of evolution. In *Evolution as a process*, ed. J.S. Huxley, A.C. Hardy, and E.B. Ford, 109–121. London: Allen and Unwin.

Hardy, G.H. 1908. Mendelian proportions in a mixed population. *Science* 28: 49–50.

Hartl, D.L. 1980. *Principles of population genetics*. Sunderland: Sinauer Associates.

Huneman, P. 2013. Assessing statistical views of natural selection: Is there a room for non local causation? *Studies in History and Philosophy of Biological and Biomedical Sciences* 44: 604–612.

Jacquard, A. 1971. *Structures génétiques des populations*. Paris: Masson. [English translation: *The genetic structure of populations*. Springer, 1974].

Kelly, F.P. 2011. *Reversibility and stochastic networks*. Cambridge: Cambridge University Press.

Kingman, J.F.C. 2000. Origins of the coalescent: 1974–1982. *Genetics* 56: 1461–1463.

Lewontin, R.C. 1967. The principle of historicity in evolution. In *Mathematical challenges to the Neo-Darwinian interpretation of evolution*, ed. P.S. Moorehead and M. Kaplan, 81–94. Philadelphia: Wistar Institute Press.

Li, C.C. 1955. *Population genetics*. Chicago: University of Chicago Press.

Longo, G., and M. Montévil. 2014. *Perspectives on organisms: Biological time, symmetries and singularities*. Coll. 'Lecture notes in morphogenesis'. Springer.

Malécot, G. 1948. *Les mathématiques de l'hérédité*. Paris: Masson.

Matthen, M., and A. Ariew. 2009. Selection and causation. *Philosophy of Science* 76: 201–224.

Montévil, M., and M. Mossio. 2015. Biological organisation as closure of constraints. *Journal of Theoretical Biology* 372: 179–191.

Price, G.R. 1972. Fisher's 'fundamental theorem' made clear. *Annals of Human Genetics* 36: 129–140.

Roughgarden, J. 1979. *Theory of population genetics and evolutionary ecology: An introduction*. New York: MacMillan Publishing Co..

Simon, H. 1977. *Models of Discovery*. Dordrecht: Springer.

Simpson, G.G. 1964. *This view of life*. New York: Harcourt, Brace and World.

Sober, E. 1984. *The nature of selection*. Cambridge, MA: MIT Press.

———. 1997. Two outbreaks of lawlessness in recent philosophy of biology. *Philosophy of Science, 64, Supplement. Biennial Meetings of the Philosophy of Science Association. Part II: Symposia Papers* (Dec, 1997), S458–467.

Walsh, D., A. Ariew, and T. Lewens. 2002. The trials of life: Natural selection and random drift. *Philosophy of Science* 69: 452–473.

Watterson, G.A. 1976. Reversibility and the age of an allele. I. Moran's infinitely many neutral alleles model. *Theoretical Population Biology* 10: 239–253.

———. 1977. Reversibility and the Age of an allele. II. Two-allele models, with selection and mutation. *Theoretical Population Biology* 12: 179–196.

Weber, J.S., D. Depew, and J.D. Smith, eds. 1988. *Entropy, information and evolution*. Cambridge, MA: The MIT Press.

Weinberg, N. 1908. Ueber den Nachweis der Vererbung beim Menschen. *Jahreschriften des Vereins für Vaterländische Naturkunde in Württemburg* 64:368–382. Translated in Boyer, S.H. (Ed.), *Papers in human genetics*, 4–15. Englewood Cliffs: Prentice-Hall, 1963.

Williams, G.C. 1992. *Natural selection: Domains, levels and challenge*. New York: Oxford University Press.

Wright, S. 1921. Sytems of mating. *Genetics* 6: 111–178.

———. 1937. The distribution of gene frequencies in populations. *Proceedings of the National Academy of Science* 23: 307–320.

———. 1940. The statistical consequences of Mendelian heredity. In *The new systematics*, ed. J. Huxley, 161–183. Oxford: Clarendon.

Chapter 14
Macroevolution and Microevolution: Issues of Time Scale in Evolutionary Biology

Philippe Huneman

Abstract According to the Modern Synthesis (MS), population genetics, as the science of the dynamics of changing allele frequencies in a population, is the core of evolutionary biology since it explains the arising of adaptations by cumulative selection. Its scale is microevolution, namely, evolution of the population of one species within a timescale not too large, defined by a small window of variations and environmental changes. Microevolution constrats with macroevolution, that is, evolution above the level of speciation – such as the extinction or emergence of species and clades – which involves a longer timescale and therefore may assume large environmental changes. MS claimed that macroevolution is not different from macroevolution. This "extrapolationist" thesis formulated by Simpson has been challenged for three decades: by the "punctuated equilibrium" thesis, and recently by Evo-Devo. Here I question the reasons why the extrapolationist thesis is threatened by advances in paleobiology and evolutionary developmental theory. The paper essentially distinguishes between biological and mathematical reasons why there could be principled differences between microevolution and macroevolution. The former concern the nature of variation, which fuels natural selection: whether it's only made up by mutations and sexual recombination, or whether other developmental features should account for phenotypic variation; it ultimately relies on topological features of the genotype-phenotype maps. Mathematical reasons concern the modeling of chance events in microevolution: at larger timescales, models of chance (such as Gaussian distribution of fluctuations) may not be any more justified, and other models would be required, though at microevolutionary timescales all models would be in practice equivalent. This argument will be applied to recent evolutionary research on extinction time. It appeals to the distinction made by mathematician Mandelbrot between "wild randomness" and "mild randomness" as two distinct structures of randomness. I conclude by showing that the mathematical differences between micro and macroevolution are more general,

P. Huneman (✉)
Institut d'Histoire et de Philosophie des Sciences et des Techniques, CNRS/Université Paris 1 Panthéon-Sorbonne, Paris, France
e-mail: philippe.huneman@gmail.com

C. Bouton, P. Huneman (eds.), *Time of Nature and the Nature of Time*, Boston Studies in the Philosophy and History of Science 326,
DOI 10.1007/978-3-319-53725-2_14

and therefore may challenge the extrapolation thesis even if empirical facts do not support the biological differences.

14.1 Introduction

The issue of timescale lies at the heart of several key problems in evolutionary biology, regarding both its main ontological commitments and its fundamental epistemic rules. It is first raised by the question of the difference between what is called "microevolution", and "macroevolution". Microevolution concerns the transformation of traits in a population of a given species. Macroevolution concerns evolution above the species level: it includes the diversification of high level taxa, (mass) extinctions, origin and diversification of clades, etc. Speciation – the arising of new species, and the main processes likely to produce it – stands at the boundary between those two evolutions, which is sometimes called mesoevolution. Notably, when people think of the term "evolution" they typically think of macroevolution (think of Darwin's focus on "species", their origins and transformations). Micro-evolution, no less important, is the focus of population genetics and quantitative genetics, which in the 1930s, due to the combined research of Huxley, Haldane, Wright and Fisher, constituted a mathematical body of knowledge about the processes of evolution (especially by natural selection) that over many decades accounted for the very possibility of evolution.

The lingering open question concerning these two is this: whether macroevolution is to be understood in the same terms as microevolution. In other words, the question concerns whether they involve the same forces or causes. This is all the more important since *speciation* – the arising of new species – will be one of the major issues faced by evolutionary biology after the establishment of the so called Modern Synthesis, which constitutes the classical framework for addressing evolution, adaptation and diversity.[1] Does this require more than microevolution? And if no, are the beyond-species-level macroevolutionary patterns such as the distribution of extinctions, the changes in diversity values within and across clades, the gradual or discrete tempo of evolution as distinguished by Simpson (1944) likely to be accounted for by the principles and causes of microevolution? Simpson's answer was affirmative, and this is called the "extrapolation thesis": microevolution extrapolated to macroevolution. Yet this claim has been challenged for at least three decades on various grounds, and those challenges are themselves intertwined with general challenges that the Modern Synthesis currently faces.[2]

[1]For example Ernst Mayr is one of the founders of the Synthesis, and one of his major achievements is arguably the elaboration of a model of speciation focusing on the mechanisms of "reproductive isolation" (Mayr 1963).

[2]On those challenges see Gould (2002), Müller and Pigliucci (2011), Huneman and Walsh (2017). See also a two-sides paper published in *Nature* last year (Wray et al. 2014, Laland et al. 2014) under the title "Does evolutionary theory need a rethink?"

In this paper I consider those current major challenges for the extrapolation thesis. Some have famously been raised by Gould and Eldredge in the 1970s; they concern the overall pattern of macroevolution, and focus on gradualism. Others are about the issue of contingency in the general history of life, especially at geological timescales. I will start by detailing the extrapolation thesis and explaining its justifications. Then I will indicate some challenges raised by paleontologists and biologists in the last four decades. The second section of the paper considers those challenges, which have been extensively sketched by Stephen Jay Gould throughout his career. I'll introduce the famous thesis he elaborated with Eldredge, called "punctuated equilibria", and indicate how as a new way of construing phylogenetic patterns it may challenge the extrapolation thesis. Then I will discuss the plausible processes accounting for those patterns, and argue that there is a conceptual and formal argument to be made about the fact that those underlying processes should probably not be only microevolutionary. The two last sections consider macroevolution at the highest timescales (often called megaevolution) and focus on Gould's so called "contingency thesis", about a contingency of the history of life that contrasts with the directionality found in microevolutionary processes, mostly driven by natural selection. This thesis relies on the knowledge of extinction, especially mass extinctions, gathered by paleobiologists since the 1970s, which I will survey here. In the last section, after having sketched some models of extinction time research in microevolution, I introduce a novel formal argument about it, which recasts Gould's contingency claims in terms of the mathematics of randomness, and provides another argument against the extrapolation thesis.

14.2 The Extrapolation Thesis

Microevolution concerns populations of one species in a limited range of time and (genetic) variation. A key example of microevolution is what has been for a long time the most well studied instance of biological evolution, namely industrial melanism: moths in an English forest turning from light to black while, due to industrial pollution, the color of the trees on which they stand darkened (Kettlewell 1955).

In contrast, Eldredge says, "Macroevolution, however it is precisely defined, always connotes 'large-scale evolutionary change'" (1989, p. vii). In his widely used textbook, Mark Ridley writes accordingly:

> Macroevolution means evolution on the grand scale, and it is mainly studied in the fossil record. It is contrasted with microevolution, the study of evolution over short time periods, such as that of a human lifetime or less. Microevolution therefore refers to changes in gene frequency within a population (...). Macroevolutionary events are much more likely to take millions of years. Macroevolution refers to things like the trends in horse evolution (...) or the origin of major groups, or mass extinctions, or the Cambrian explosion (...) Speciation is the traditional dividing line between micro- and macroevolution. (Ridley 2004)

The Modern Synthesis (MS) is the framework that emerged from population genetics – which conciliated Darwinian selection and Mendelian genetics – and further gathered under the same concepts systematics, paleontology, some ecology (Ford 1975) and natural history (Dobzhansky 1951). Even though the actual scope and nature of the MS are still debated by biologists and historians alike, it is fair to say that the architects of the Modern Synthesis agreed on a couple of facts: the fact that evolution occurs in populations with a "mendelian constitution" (namely, the fact that variation is due to allelic mutation and recombination), and that natural selection, the only force creating adaptations, is in fact the "principal agent" (Huxley) of evolution.

In this framework, since they explain the process of evolution, population and quantitative genetics are the matrix of mechanistic models of evolution and they focus on microevolution. Population and quantitative geneticists study the changing frequencies of genotypes (or alleles) and of trait values in a population of a given species, due to what they call the distinct "forces" of selection, drift, migration and mutation. Models operate on a limited timescale and do not assume too wide a range of variation. The extrapolation thesis was formulated by George Gaylord Simpson and his work integrated paleontology into the Modern Synthesis (his own empirical research first concerned the phylogenies of *Equus*). Through extrapolation, paleontology could indeed be tied to a core of knowledge of evolution by natural selection that provided population genetics and quantitative genetics, and that focused on microevolution.

In those fields, a now classical graphical model has been introduced by Sewall Wright in 1932 under the name "fitness landscape" (or "adaptive landscapes"). According to two possible interpretations, these landscapes plot : (1) either various combinations of *alleles* in genotypes of *organisms*,[3] onto their fitness values (in ordonate), or (2) various *frequencies* of alleles in a population onto the resulting mean fitness *of the population* (in ordonate). In the former case (1) populations, as groups of organisms, are represented by clouds of points, while in the latter (2) populations are represented by points (Fig. 14.1), but in any case natural selection causes on the average those clouds or points to climb the so-called 'hills' on which they are located, since natural selection tends to increase fitness. The validity of this representation has been debated for some time – whether it's accurate or often misleading (e.g. Wilkins and Godfrey Smith 2009; Kaplan 2009) – in addition to the role played by Fisher's "fundamental theorem of natural selection". Proved in Fisher (1930), this theorem states a priori an intergenerational fitness increase of the part of the fitness change due to natural selection in a population (Frank and Slatkin 1992; Winther 2006). More recent concerns have been raised about the hyperdimensionality of those landscapes, and the conflations resulting from intuitions that come from 3-dimensional landscapes but do not apply to higher dimension spaces (Gavrilets 1999).

[3]An allele is the variable form of a given gene. A genotype is the whole or a given part of the genetic information of an individual.

Fig. 14.1 A Wrightian fitness landscape. The *x* and *y* axes are the frequencies of genes A and B; the vertical axis is the fitness of a population of those genes. The landscape shows the variation of the population fitness according to various values of the frequencies of genes A and B (Actual landscapes have many more dimensions)

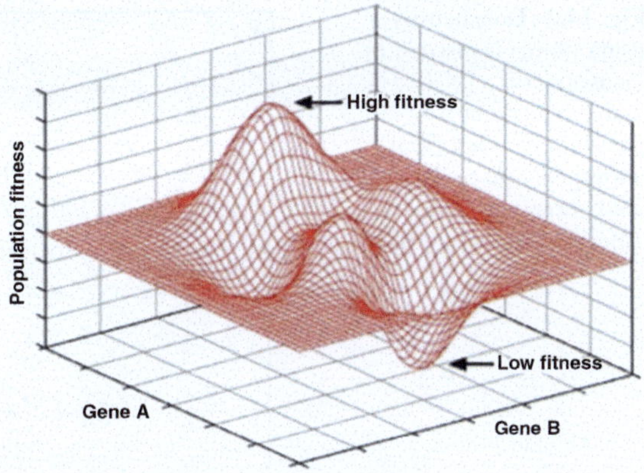

Fitness landscape, with global and local optima

However, fitness landscapes are a useful way to explicate the idea of microevolution/macroevolution extrapolation, since Simpson, after Wright, used them in a totally different context concerning macroevolution. Here, gene combinations are the coordinates of the landscapes, but adaptive peaks are where *species* (not populations) go. "Adaptive valleys" are deserted, since phenotypes[4] are too poorly adaptive; by contrast, peaks define specific niches, because they indicate genotypes that optimally fit their environments. The proximity between peaks, that is, between genotypes that are close and that are similarly optimal in their environment, means a possible evolutionary proximity.[5] Those fitness landscapes stand at the macroevolutionary scale (Fig. 14.2).

These adaptive landscapes are thereby defined at the scale of major families, or groups of clades. To this extent, such a landscape parallels what we could call a space of possible phenotypes, which would be very clustered since phenotypic forms tend to be clustered around what morphologists of the nineteenth century called types, as Darwin himself acknowledged (Darwin 1859, chap. 6). Yet the fitness landscape space, as compared to a space of possible phenotypes, is endowed with an additional dimension of fitness, and the blanks in the phenotypes space – that is, non-existence of phenotypes – are now reinterpreted as low-fitness regions. As Dobzhansky wrote, when he endorsed Simpson's interpretation of the fitness landscapes:

> Each living species may be thought of as occupying one of the available peaks in the field of gene combinations. The adaptive valleys are deserted and empty. Furthermore, the adaptive peaks and valleys are not interspersed at random. Adjacent adaptive peaks are arranged in groups, which may be likened to mountain ranges in which the separate pinnacles are divided by relatively shallow niches. Thus, the ecological niche occupied by the species

[4]The phenotype is the set of observable characters of an individual.

[5]Proximity in adaptive terms – technically, it means *analogies* rather than *homologies*. The quotation below by Dobzhansky provides examples.

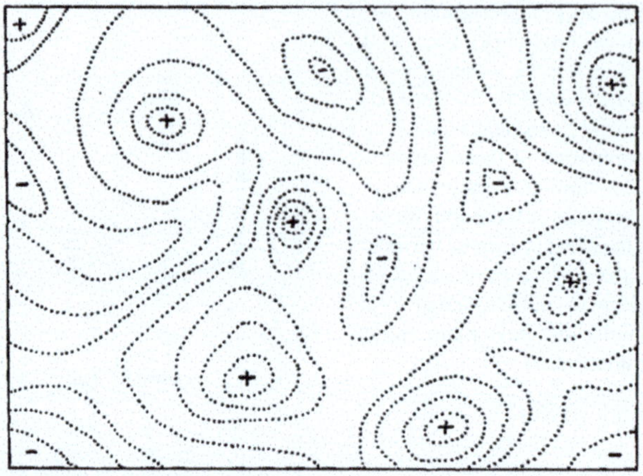

Fig. 14.2 Dobzhansky using fitness landscape as landscape of macroevolutionary possibilities (seen from above: "+" are peaks, "–" are bottom of the valleys). After Dobzhansky (1951)

"lion" is relatively much closer to those occupied by tiger, puma, and leopard than by to those occupied by wolf, coyote and jackal. The feline adaptive peaks form a group different from the group of canine peaks. But the feline, canine, ursine, musterine and other groups of peaks form together the adaptive range of carnivores, which is separated by deep adaptive valleys from rodents, bats, ungulates, primates and others (Dobzhansky 1951).

The way Simpson and Dobzhansky pull Wright's graphical model away from population genetics exemplifies the logics of the extrapolation thesis: what initially describes microevolution is still able to characterize the dynamics and processes taking place at a longer time scale. Expanding the dimensions and involving all species instead of a population of one species defines this scale-switch, but then everything, especially the hill-climbing processes, the role of the hills, etc., remains the same. Of course, one also assumes that the peaks are inhabited, which implies an overall greater role for natural selection, or, in other words, which is much more adaptationnist than are Wright and generally population genetics. One reason for this is that at those scales, selection may precisely have swamped the effects of drift – for instance through processes such as the "shifting balance theory" that Wright described (see Coyne et al. 1997 for an assessment.). But I just recalled this reinterpretation of the landscapes as an illustration of the way the extrapolation thesis proceeds – namely, by re-dimensioning modeling tools from microevolution.

To the contrary, pre-Modern Synthesis authors were often thinking that when high level characters appear, namely, the characters that distinguish species from one another, specific mechanisms were involved, distinct from those studied by population geneticists. In other words, according to those authors even though population geneticists could account for the change of colour of black moths, they could not account for the character that defines the species itself or the genera to which moths belong.[6] As Filipchenko (1927) wrote, "the origin of characters [that differentiate] the higher systematic categories requires some other factors than

[6]See comments in Amundson (2005).

the lower taxonomic units" (cited in Dobzhansky 1951). Filipchenko was an evolutionist, but he was working at a pre-MS period where Mendelians, who were seeing macromutations as the source of evolution, diverged from Darwinians, who saw selection as its prime engine (Gayon 1998; Beatty 2016). Filipchenko was himself an orthogenist – a view that Simpson later dismissed, according to which a trend within variation drives evolution; he was the teacher of Dobzhansky, and he also introduced the distinction between microevolution and macroevolution. Yet redimensioning landscapes in the conservative way Simpson and Dobzhansky do it rejects such a view and assumes the validity of the extrapolation thesis, which Dobzhansky put in the following terms: "we are compelled at the present level of knowledge to reluctantly put a sign of equality between the mechanisms of micro and macro evolution." ("Reluctantly" may indicate the regret of having to part company with his master Filipchenko (Burian 1994).)

Dobzhansky's argument in favor of extrapolation begins by stating "Evolution is a change in the genetic composition of populations. The study of mechanisms of evolution falls within the province of population genetics." Then he notices that evolutionary change occurs on various and very different scales: "Of course, changes observed in populations may be of different orders of magnitude ranging from those induced in a herd of domestic animals by the introduction of a new sire to phylogenetic changes leading to the origin of new classes of organisms. The former are obviously trifling in scale compared with the latter." (Dobzhansky 1951) But epistemically speaking, only the study of the former provides controlled access to the latter – either with observations, or with experiments, which he himself has done on great snail populations (Millstein 2009).[7]

Interestingly, Dobzhansky notices that the idea that microevolutionary changes are different in nature from macroevolutionary evolution ones are "popular among those who approach evolutionary problems on the basis of data of palaeontology and comparative anatomy". Some of those scientists indeed think that, "while the former can be understood in terms of the known genetic agents (mutation, selection, genetic drift), the latter involves forces that are experimentally unknown or only dimly discerned". Those agents could be "some directing forces either inherent in the organism itself or acting on it by some inscrutable means from the outside" – like orthogenesis – but escape experimental science, namely, a "precise definition which would make them subject to experimental test or to any kind of rigorous proof or disproof."

According to Dobzhansky (1951), we are epistemically constrained to approach macroevolution by inferring from microevolution, since: "it is obviously impossible to reproduce in the laboratory the evolution of, for example, the horse tribe, or for that matter of the genus *Drosophila*. All that is possible is to examine the

[7]Dobzhansky (1951) writes: "Experience shows, however, that there is no way toward understanding of the mechanisms of macroevolutionary changes, which require time on geological scales, other than through understanding of microevolutionary processes observable within the span of a human lifetime, often controlled by man's will, and sometimes reproducible in laboratory experiments."

evidence bearing on macroevolution which has been accumulated by palaeontologists and morphologists, and to attempt to decide whether it agrees with the hypothesis that all evolutionary changes are compounded of microevolutionary ones". The fact that such inferences are sufficiently explanatory will ultimately be an argument in favour of the extrapolation thesis. And precisely, adds Dobzhansky: "Simpson (1944) in palaeontology, and (...) Schmalhausen (1949) and Rensch (1947) in comparative anatomy and embryology", have done this task, and "found nothing in the known macroevolutionary phenomena that would require other than the known genetic principles for causal explanation". Hence, he concludes: "The words 'microevolution' and 'macroevolution' are relative terms, and have only descriptive meaning; *they imply no difference in the underlying causal agencies*."

This explanatory sufficiency is, for Dobzhansky, Simpson and their fellow advocates of the Modern Synthesis, empirically attested. So now an argument against the extrapolation thesis is that the current state of our knowledge does not so easily allow explaining macroevolutionary patterns and features from the intrinsically microevolutionary causes. And indeed, as now surveyed by Grantham (2007), "paleobiology has provided some challenging data, including evidence for mass extinction selection regimes that differ from background selection, species selection, passive diffusion as an explanation for evolutionary trends (McShea 1994), a tendency for higher taxa to preferentially originate in on-shore environments (Jablonski and Bottjer 1991), and developmental constraints. All of these findings *challenge the idea that we can smoothly extrapolate microevolutionary processes to explain macroevolutionary patterns*." (my emphasis)

It's easy to see how this occurs; for instance, Mc Shea in several papers (Mc Shea 1994, 2005) argued that no natural selection is needed to account for some trends detected in the fossil records – especially, an intra-clade and inter-clade trend towards increasing complexity. The patterns themselves are very compatible with a pure diffusion effect, far from the overwhelming role of natural selection, which is assumed by population genetics when it applies its models to microevolution and speciation events. Moreover, the concept of species selection, advocated by Gould (2002) or Jablonski (2008a) among others, means that there is selection *of* species (against other species) in virtue of properties *proper to species themselves* – such as variability, sexuality, polymorphism, spatial range – (and not to the species' individuals), and this process accounts for properties of clades and families (Gould and Lloyd 1999). For instance, a sexually reproductive species may outcompete an asexual one in a changing environment because of its being sexual (which entails more variability, hence a better chance to find the variants better adapted to a drastic environmental change), and being sexual is not a property of the individuals but of the species itself. In contrast, microevolution, taking place in pools of organisms of genes of a species, doesn't have room for this *species* selection.

The paleobiological challenge to the extrapolation thesis appears crucial in the context of current controversies over the status of the Modern Synthesis (Pigliucci and Muller 2011; Wray et al. 2014 vs. Laland et al. 2014; Huneman and Walsh

2017) to the extent that the centrality of population and quantitative genetics as a science of processes of evolution entailed that microevolutionary processes were mostly driving evolution overall (Bateson 2017; Depew 2017). Therefore, one is entitled to reflect on general types of challenges to the extrapolation thesis as a typology of key challenges to the Modern Synthesis.

As one of the major biologists whose work pursued and then assessed for decades the prospects for the Modern Synthesis and the need for a renewal, Stephen Jay Gould provides us with an articulated set of such critical arguments. There are indeed three main reasons for which macroevolution should be something other than "successive rounds of microevolution" (Erwin 2000). The first two concern either the *patterns*, or the *processes* of evolution – using here a distinction that is classically made among evolutionists.

- First, the issue of *gradualism*. Microevolution is gradual. This is due to the fact that, as Fisher already pointed out, the larger the mutations, the higher the chances they'll hugely affect several traits, and therefore disrupt the integrity of the organism. Hence most of the mutations likely to be retained by selection will be small mutations. Therefore microevolution will be gradual. But, regarding patterns of macroevolution, some have claimed that they are not gradual.
- Second, the *process* issue. As indicated, microevolution accepts as basic processes mutation, migration, (organismic or genic) selection and drift. Allowing for high-level selection in the form of species selection or clade selection (Williams 1992) means accepting a novel explanatory process, which therefore exceeds the process framework of microevolution.

Those two issues have been strongly raised by the paleobiological thesis first formulated by Gould and Eldredge (1977) and called the *punctuated equilibria*.

The third issue concerns macroevolution at the highest timescale – for instance the history of life across geological periods. It requires that one distinguishes two senses of macroevolution – the higher scale one being called here *megaevolution* (as Gould sometimes does). It is mostly affected by the famous challenge stated by Gould in the form of the puzzle: "replaying the Tape of Life" (Gould 1989). According to him, no repetition of the history of life would yield the same outcome as life on Earth, and this contrasts with microevolution where many events are somehow predictable, to the extent that they are selection-driven, and therefore would be recurrent in any repetition, since selection is a directional force oriented through fitness increase (Gillespie 2004; Huneman 2014b; Gayon and Montevil, Chap. 13, this volume).

In the next section I'll reflect on the two first challenges to the extrapolation thesis. The upshot is the following: even if punctuated equilibria are indeed at least sometimes a correct view of macroevolutionary patterns, this challenges gradualism but may not require different processes than microevolutionary ones. Yet I'll offer a formal argument to say that very probably on large timescales the extrapolation thesis fails.

In the last section, I'll consider the contingency challenge. I'll try to make sense of the contingency argument by considering the evolutionary research on extinctions, since mass extinction is at the heart of Gould's argument about megaevolution. I'll suggest a formal argument to say that, for mathematical reasons, the timescale of megaevolution may somehow require different modeling practices than the timescale of microevolution.

14.3 Biological Challenges to Extrapolation Thesis: Gradualism and Particular Processes

14.3.1 Phylogenetic Patterns, Gradualism and Punctuations

Gould and Eldredge (1977) proposed a new reading of the fossil record. Darwinian evolutionism holds a gradualist view of evolutionary change. According to Darwin, incremental changes on a very long time scale lead, through the accumulation of selected variations, to all the adaptations one witnesses in the living world. The Modern Synthesis shared this view; moreover, it justified it by the notion that variations in general are mutations or recombination, and are small, as I explained above. The supporters of the non-gradual view were the so-called "saltationists", e.g. Mendelians like De Vries, and they were precisely anti-Darwinian, in the sense that, for them, macro mutations were driving evolution, and not natural selection (Beatty 2016). So gradualism seems well entrenched within Darwinism.

In his *Origins* book, Darwin spends one chapter explaining why the fossil records presents discontinuities – while in principle evolution should be gradual (Darwin 1859, chap. 8). This includes stratigraphical considerations, biological arguments (especially the fact that intermediary forms have fewer chances to survive a long time since both competitors at the two extremes will drive them extinct, and therefore they will not be represented in the record), and geological reasoning. Gould and Eldredge's radical thesis is that the fossil record is not incomplete – so in fact, the phylogeny is actually discontinuous. Eldredge and Gould (1972) offer powerful arguments both from the latest empirical paleontology and from philosophy of science (including parsimony, and an appeal to the kuhnian notion of paradigm). According to punctuated equilibria, evolutionary history is made of very long periods of "stasis", in which the lineages mostly diversify by fine-tuned adaptation to various circumstances (hundreds of millions of years), and rapid stages of evolutionary change (a few million years) (Fig. 14.3a). Major evolutionary change, for instance the arising of new body plans (the emergence of the chordates, or even of deuterostomes from protostomes), extinction of many major branches, multiplication of new clades occur during rapid phases. This concerns also the multiplicity of phenotypic traits called "evolutionary innovations", which are key events for the following evolutionary success and adaptive radiation (e.g. the wings of birds, the gills of fish). In the punctuated equilibria view, therefore, one should not explain why in the distinct stratigraphical strata, some

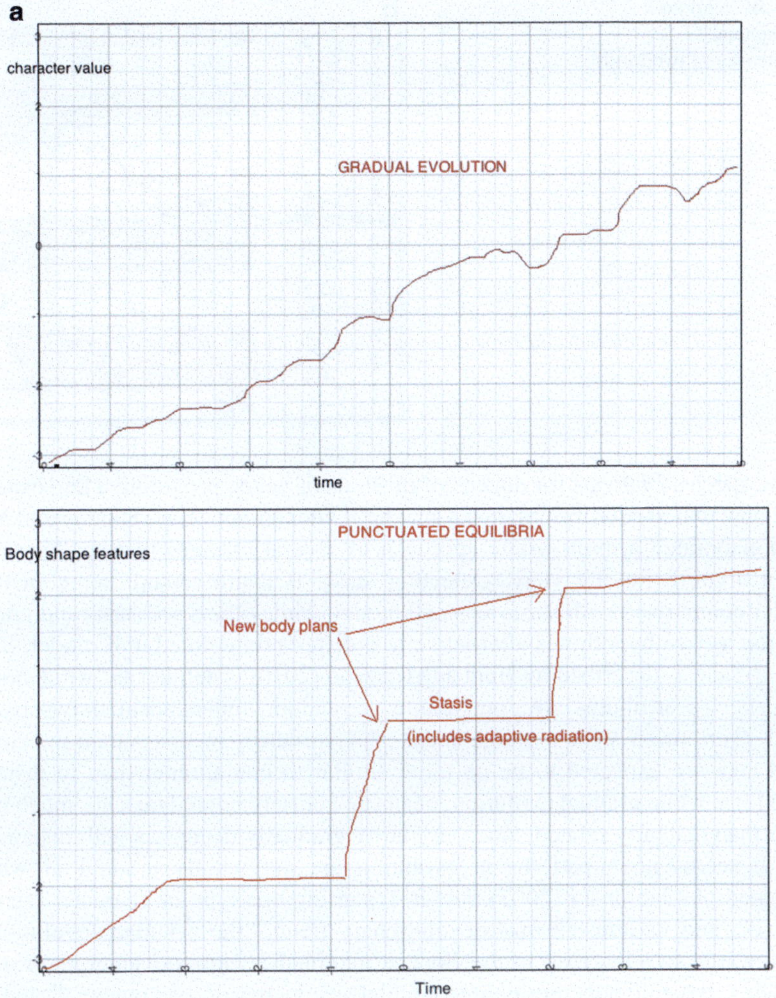

Fig. 14.3a Punctuated equilibria vs. gradualism

intermediary forms are missing (Fig. 14.3b); rather, one has to understand why evolution is itself discontinuous.[8]

There are several ways to measure stasis, which can be predicated either at the level of lineages (e.g. Cheetham 1986), or at the higher level of clades (Pagel et al. 2006). In the former case, Eldredge (1971) measured the lack of variation in

[8]Eldredge (1971) anticipated some of the radical claims made in Eldredge and Gould (1972), especially by studying an allopatric-speciation based model accounting for phylogenetic patterns that would yield non-gradual evolution. I thank Alexandre Peluffo for having brought this paper to my attention.

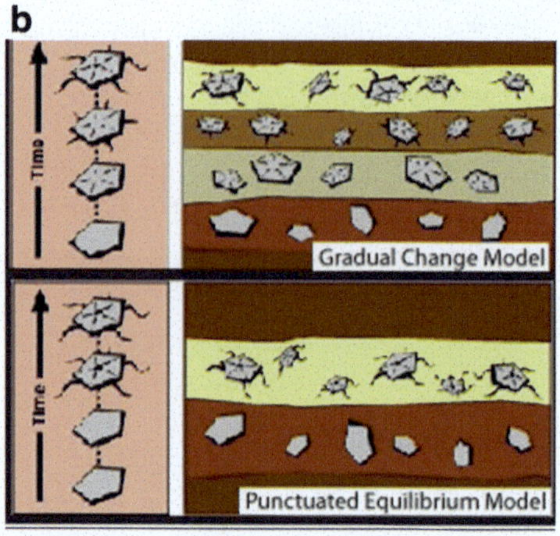

Fig. 14.3b Various stratigraphical consequences of two kinds of change

trilobite species *Phacops* for dozens of millions of years, providing a key example for the punctuated equilibrium theory that he'll later co-conceive. Stanley and Yang (1987) elaborated a multivariate approach to several lineages that has been used later in arguing for the view. Detecting stasis is controversial; some methods consider comparison with stasis as a null model, while others see stasis and gradual evolution as two hypotheses alternative to a differently defined null model. Stasis within a clade as described by Hunt and Carrano (2010) is defined as "white noise," where species attributes are comparable to random, independent draws from a normal distribution with a stable mean and variance. In this situation, closely related taxa are expected to be no more similar to one another than to distantly related taxa, while gradual evolution means that closely related taxa are much more similar than distantly related, since similarity gradually decreases with distance.

As it seems now, patterns in evolution are not wholly gradual or wholly punctuated. As Jablonski (2007) puts it, regarding one type of character (size) in one clade, and its attested macroevolutionary trend, "the net macroevolutionary trend towards size increase in the Eocene mammal *Hyopsodus* emerges from an underlying dynamic containing three gradual size increases, one punctuational size increase, one period of size stasis and three gradual size decreases, i.e. gradualistic change was random with respect to the macroevolutionary outcome."

Though thought provoking, and followed by a huge controversy in evolutionary biology, the punctuated-equilibria view of evolutionary patterns was not absolutely new.[9] Gould and Eldredge were touching on matters familiar to biologists studying speciation, which has been the bulk of evolutionary biology since the rise of the

[9]Notice also that discontinuous change was already acknowledged by Simpson (1944) who named it "quantum evolution". The point was that according to him, even if in principle plausible, quantum evolution was not the rule in nature and only concerned rare and rapid diversification and abrupt transition to different "adaptive zones". Simpson reacted very harshly to the punctuated equilibria thesis, as it is documented by his private correspondence with Gould (Cain 2009), since he objected strongly to the claim of novelty from the authors.

Modern Synthesis. Biologists indeed used to contrast two modes of speciation: sympatric speciation, where a population of a species progressively allows for a set of variants within itself to become a new species, and allopatric speciation, which means that some individuals from a species get into a new environment and evolve in a slightly different way due to the fact of different selective pressures. Then, they progressively reach a point where the genotypic systems of the daughter population and the parent population are not compatible any more, and therefore inter-population reproduction becomes impossible. Such impossibility of generating fecund offspring indeed is the hallmark of a species difference according to Mayr's widely shared "biological species concept". Mayr himself was the champion of allopatric speciation, claiming that most speciation in nature is of this type (Mayr 1963). But in this manner, as he argued, speciation events can be involved in many various phylogenetic pathways, and may generate patterns compatible with punctuated equilibria. Eldredge and Gould (1972) themselves relied on allopatric speciation, taking place in peripheric populations; this would indeed contrast with the "norm for a species during the heyday of its existence as a large population" (p. 117) which is stasis, because allopatric speciation in isolated populations cannot happen. Hence Mayr was not convinced by the theoretical novelty of punctuated equilibria, even though he would not object to the content of the theory.[10] This may mean that a tension existed between the current Mayr-inspired theories of speciation and the extrapolation thesis, and punctuated equilibria theory brought this tension to the foreground.

14.3.2 Punctuated Patterns and Accounting Processes

Yet, a phylogenetic pattern by itself being punctuated or gradual does not entail a particular generating process. Even if likely candidates exist, a question for punctuated equilibria biologists is about finding the key processes responsible for these patterns. It might be that they require other processes than the ones posited by microevolutionary theory – which tends to generate gradual change – but this is not obvious. Granted, specific combinations of those processes may yield those particular patterns. About the fact that stasis seems not derivable from the constant presence of selective forces in microevolutionary settings (see also Hunt et al. 2015), Jablonski (2007) notes:

> Empirical extrapolation [from patterns to processes] appears to break down, for example, in the mismatch between the demonstrated potential of most populations for rapid net change and the prevalence of net morphological stasis in many lineages over long time-scales. This mismatch across scales *need not require novel forces* to limit phenotypic change over most

[10]As he wrote in retrospect: "punctuated equilibria, which at first sight, seem to support saltationism and discontinuity, are in fact strictly populational phenomena, and therefore gradual (Mayr 1963). They are in no respect whatsoever in conflict with the conclusions of the evolutionary synthesis" (Mayr 2001, p. 298).

> of the duration of a species, but it shows that short-term, localized observations on the evolutionary responsiveness of living populations are poor predictors of species-level behavior over millions of years[11] (my emphasis).

Indeed, several processes are likely to produce a pattern made up of punctuations. Some of them will be understandable along the Modern Synthesis in microevolutionary terms: one can think of a series of adaptations in some species, followed by sets of allopatric speciations, rounds of coevolution and then positive feedbacks which increase the rate of adaptations, yielding the potential of evolutionary bursts of novelty. Geological events may make allopatric speciation more vivid with the diversification and multiplication of possible founder's effects. And finally the pattern of macroevolution will display a punctuation stage, which will contrast with a more ordinary regime in which those microevolutionary regular processes do not find themselves in such a condition. Actually, that is close to what Gould and Eldredge themselves hypothesized, by relying on Mayr's theory of allopatric speciation – which is the reason why Mayr himself did not see this theory as revolutionary (as mentioned above). Gould and Eldredge (1977) indeed saw allopatric speciation, added to the newly discovered role of regulatory genes in microevolutionary change, as likely to explain punctuations in a way compatible with population genetics,[12] and therefore reject the idea that their theory is an avatar of saltationism.

Yet some punctuated equilibria episodes, according to Erwin (2000) are actually not explainable by appealing to such kind of processes:

> Microevolution provides no satisfactory explanation for the extraordinary burst of novelty during the late Neoproteozoic- Cambrian radiation (...), nor the rapid production of novel plant architectures associated with the origin of land plants during late Devonian (...) followed by the origination of most major insect groups (...).

But some argued that stasis is even more problematic for the extrapolation thesis. How could microevolutionary processes, namely selection and drift with continuous mutation and migration, not yield constant change, and instead allow for extremely long periods of near constancy? Stanley and Yang (1987) used comparison to actual geographic variants in order to show that at larger time scales stasis occurs, and that it contrasts with the microevolutionary diversification processes intensively studied, which account for geographic distributions. Two of the most cited cases of intra-lineage stasis, the bryozoan *Metrabdotus* first studied by Cheetham, and the fossil freshwater mollusks from the Turkana basin (Williamson 1981), could hardly be expected under the mutation rates and selective pressures that underpin microevolutionary processes. More strikingly, the shapes of *Drosophila* wings did not change under 50 million years, while the range of genetic

---

[11] My emphasis; notice how this contradicts what Dobzhansky says about the only way to address macroevolution.

[12] "Regulatory changes in the timing of complex ontogenetic programs seem far more promising and potentially rapid, in conformity with our punctuational predilections. (. . .) We are pleased that some recent molecular evidence, based on regulatory rather than structural gene changes, supports our model" (Gould and Eldredge 1977, p. 138)

variants underlying wing shape has been huge; and mammalian body temperature has kept constant between 37 °C and 38 °C, over dozens of million years (Hansen and Houlé 2004): how could this be accounted for given the huge diversity of genetic variants available on large timescales, and the diversity of environments across which, diachronically and synchronically, mammals have been adapted to?

Notice also that stasis can be seen at the level of genes: some of them have been conserved across geological periods, like *Pax 6*, which is involved in the development of eyes and brains, and has been conserved across the bilaterian phyla.[13] This contrasts with the variation of the genome along huge periods of time, which is due not only to selection but to ordinary mutations. As the neutral theory of evolution by Kimura (1983) indeed has shown, the nucleotidic-level constitution of the genome changes constantly, even with no selection, at a constant pace that allows geneticists to talk of "molecular clock". So it is all the more challenging for population geneticists to consider those genes that are so deeply conserved and are often crucial.

For all those reasons people often speak of the "paradox of stasis". Many explanations for stasis have been developed (Estes and Arnold 2007). The most compatible with population genetics usual explanatory tools is stabilizing selection. It might be for instance that *Pax 6* is so adaptively important that it has been conserved against the differentiation pressure exerted by population genetic processes. Stabilizing selection is in microevolution one of the major forms of selection (together with directional selection, which is the one likely to transform traits and then organisms), and it is arguably the most frequent. Yet even though one assumes stabilizing selection, the problem is that this selection only ensures the fit between the population's mean phenotype values and the environmental demands: it maintains populations on adaptive peaks in the landscape. However, it cannot as such lead to stasis, except if the adaptive optimum, yielded by environmental demands, remains almost the same, which should not be taken for granted, as Hansen and Houlé (2004) emphasize.[14] Hence one would need an additional mechanism to account for the constancy of environmental optima across large periods of time (that is, the stability of adaptive peaks), given that environments are unanimously acknowledged as varying over the macroevolutionary timescales. Yet, as Kaplan (2008) argues, many mechanisms can yield such results; thus, ascribing "stabilizing selection" as the cause of stasis will not answer this question and decide among those putative mechanisms.

At the other extreme, in the wake of Gould's and Lewontin's famous 1979 paper on constraints that govern variation and prevent selection to reach all best possible traits, some biologists argued that stasis suggests the existence of underlying constraints. Yet constraints are in general *relative* to a timescale (Maynard Smith et al. 1985) and therefore, no argument can be given for constraints that would be

[13]The class of bilaterian animals includes all animals showing some symmetry, which encompasses both deuterostomes and protostomes.

[14]Even though some biologists designed powerful models that still ascribe stabilizing selection a major role in this, e.g. Eldredge et al. (2005).

absolute.[15] In a nuanced manner, according to Hansen and Houlé (2004), the constraints that underlie stasis arise from genetic covariance among characters under selection operating in different directions, which eventually decreases the amount of available phenotypic variation: "epistatic interactions tend to restrict variation under selection". Selection fixing a series of genes would actually negatively impinge on other genes in a way that makes the overall phenotypic organismal variation very small. In this sense, the massive genetic variation assumed by population geneticists working on microevolution becomes a very restricted variation, once one considers phenotypic variation of whole organisms at the level of macroevolution. In turn, these constraining effects of epistasis appear only at timescales of macroevolution, since selection on several loci acts on extant variation at microevolutionary scale and indeed changes gene frequencies, but its negative effect on overall variation emerges only at higher timescales. Such suggested mechanism for stasis therefore is mostly proper to macroevolution.

Hence, to sum up, according to current paleobiology some punctuated patterns would require wholly novel processes: it might be processes at *a higher level than organisms population*, such as the species selection (favored by Gould 2002) that I mentioned before, or it might be processes *emphasizing the mechanisms of variation at the developmental or molecular level and the constraints stemming from them.*

The latter are generally bracketed in microevolutionary theories since they only consider genotypes and phenotypes but not development (Huneman 2010a; Walsh 2015). Classical MS explanations indeed focus on the effect of genic or organismic selection; as it has been often argued, the crux of the debates between early Darwinians and Mutationists about evolution was the respective explanatory roles of selection and variation (e.g. mutation). Darwinians emphasized selection, and assumed that variation is so abundant that it does not play any explanatory role (Beatty 2016). The issue of the role such variation-producing processes play by themselves at macroevolutionary timescales therefore involves the assessment we should make of the Modern Synthesis today.

The mechanisms of variation, which includes any way of producing new genes, and especially new genomes, organizing genomic architecture, etc., may *pace* the MS be actually irreducible to single allelic mutation (Kirschner and Gerhart 2005), and yield very sparsely distributed variations (Huneman 2017). In this case, they may play a crucial explanatory role for evolution, and, since variation will not be homogenously and isotropically distributed, non-gradual phylogenetic changes, discontinuities, and finally punctuations and stasis can be expected in macroevolution (the latter possibly according to the mechanism hypothesized by Hansen and Houlé (2004)).

For such reasons, Jablonski argued that

> the fossil record should be used more extensively to test hypotheses on the macroevolutionary consequences of the *architecture of developmental systems*. Whenever phylogenetic analysis can be combined with developmental data to characterize major

[15]Except purely physical constraints like gravity, but here we talk of genetic constraints, or developmental constraints bearing on genetic systems (e.g. Wake 1991).

> developmental differences among clades (usually by bracketing deep phylogenetic nodes with extant species), paleontologists can assess the macroevolutionary role of those differences. We would like to know, for example, *whether the tempo and mode of large-scale phenotypic evolution varies with such developmental factors* as: genome organization dominated by multiple, slightly divergent copies of genes versus single-copy genes with large batteries of regulatory binding sites versus genes generating many isoforms via alternative splicing or translation initiation (all of these being ways to expand the effective genome size; not mutually exclusive, but apparently varying in importance among clades). (Jablonski 2009, my emphasis[16])

Gould himself, in the 1980s and 1990s, was also a supporter of the hypothesis of novel processes to account for those macroevolutionary new patterns – besides species selection, he investigated the consequences of developmental mechanisms as possible promoters of major evolutionary change, above and beyond what can provide allelic mutation and recombination as fuel for selection. In *Ontogeny and phylogeny* (1977) he focused on heterochrony, which is the temporal rearrangement of developmental sequences (shortening, adding, deleting sequences) and showed that it can indeed yield major evolutionary changes (see also Nicoglou, this volume, on developmental time). But our recent knowledge of molecular mechanisms of development, as well as, more generally, the architecture of the genome and the way it looks like a complex adaptive system rather than like a set of instructions (Walsh 2015), provides us with a whole class of developmental mechanisms likely to generate variation. Briefly said, the current scientific understanding and investigation of the genomic system, which includes the way it requires complex genomic networks to regulate the expression of all genes in accordance with intra-organismic and extra-organismic environmental demands (Davidson 1986) – not to speak of the epigenetic factors, which in the short term adapt gene expressions in cells to those demands (Jablonka and Raz 2009) – brings up a new battery of explanatory processes for macroevolutionary patterns. As Valentine and Jablonski (2003) pointed out: "periods of *relatively rapid genomic reorganization* in response to whatever selective factors were in play create *new architectural norms*." This architectural creation instantiates, at the level of the genomes, the body plans shift that were seen as the main objects of rapid evolution in punctuation periods, in the early years of punctuated equilibria theory (Davidson and Erwin 2006).

Hence, the validity of the extrapolation thesis now seems rather to hang upon the biological challenge to traditional microevolutionary processes – namely, whether or not novel ones should be postulated in order to make sense of punctuation and stasis. This is of course a plainly empirical question, and part of it revolves around the issue of the role of variation-producing-mechanisms in evolution, to which the next section is devoted.

[16]*Tempo* and *modes* are the concepts Simpson (1944) introduced to address evolutionary change. The tempo is the rate of evolution of something, for instance the amount of change per million years in a given character. The mode is, more generally, the way evolution occurs in changing populations, and it's not only quantitative (unlike tempo): for instance, "quantum evolution" and "gradual evolution" are modes of evolution.

14.3.3 A Formal Argument About Development, Selection and Macroevolution

Even though one will legitimately consider that this is an empirical issue, there is a way to address this question in very general terms – in order to parse the possible empirical situations likely to be found into two classes. Based on that view, suggested in Huneman (2010a), I now propose a first theoretical argument against the extrapolation thesis.

The view suggested, in a quick way, is the following. Think of genotypes as points in a genotype space G, and think of phenotypes as points in a phenotype space P (both are discrete spaces). Now, think of possible developmental pathways as points in an abstract "developmental space" D. Huneman (2010a) considers the set of possible applications from G to D, and then from D to P. Intuitively, the realized subsets in those sets, meaning the extant applications between genotypes, phenotypes and developments can be of various sorts. They may preserve lots of the topological, metric and other features of the initial set of points (for example, they would map close genotypes onto close developmental paths, etc.) (Fig. 14.4a). Or they may disrupt those features.

When the applications are such that all those features are preserved, when applying G subsets to D subsets and then D subsets to P subsets (Fig. 14.4a), it means that the developmental space as such is not very relevant for our explanations: all relevant explanations are only concerned by the phenotype and the genotype spaces. Therefore development is not so explanatorily relevant for evolution – for instance, we can consider that variation is evenly, homogeneously distributed, no singular features of variation should be addressed, or, genetic effects are mostly additive effects (Huneman 2010a). In this case, the Modern Synthesis assumptions are correct, and from that we may infer either gradualism or (at least) the fact that no other processes than the ones proper to microevolution should be incorporated into our explanations of macroevolution (provided we assume that high level processes such as species selection, considered above, are left aside). The extrapolation claim then has good prospects.

However if the G→D→P applications are less conservative, what happens in the Developmental space will be crucial for our explanations (Fig. 14.4b), and it means that the bracketing of development (which is an assumption of the Modern Synthesis) is not so correct. In this case, given that other plausible processes relative to development and developmental variation should be considered in order to explain evolution,[17] the extrapolation thesis is threatened.

I call "type a" applications the conservative applications, and "type b" the other ones. Once again, it is an empirical issue to decide in which world we live – i.e., is a or b the dominant type of G→D→P application? However, based on this I'd like to

[17]As in the above Valentine and Jablonski (2003) quote about developmental processes impinging on macroevolution.

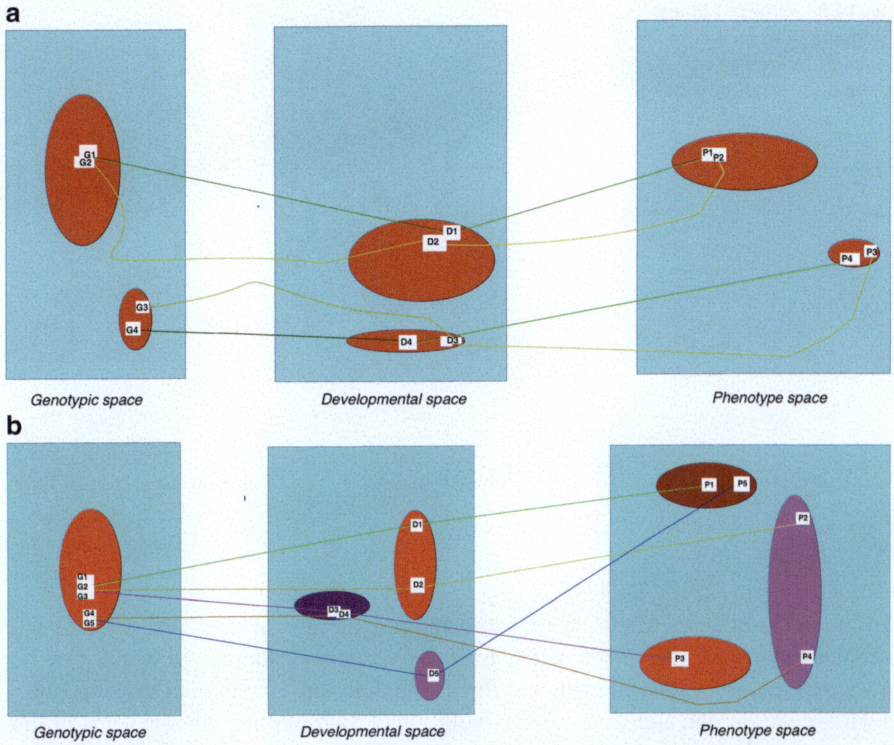

Fig. 14.4 (**a**) Type a applications: topological or metric properties of the initial space are conserved. (**b**) Type b applications. Those properties are disrupted by the G→D and/or the D→P applications

suggest one conceptual argument to undermine the extrapolation thesis. It's a simple mathematical consideration, which is the following.

Consider regions of the genotype space more extended than the ones considered in usual population genetics models, for instance when one looks at traits conserved in many clades, which means very wide regions of the phenotype space, and then of the other spaces. Then, I claim that *the larger a region, the higher the chances that it will not be "well behaved", namely that the inter-spaces applications will be of type b*. Why should one think this way? Because the relative frequency of *type a* applications among all possible applications decreases when the size of spaces increases, since the amount of *possible* relations between elements of the spaces is increasing when the size of the space considered is increasing.[18]

This is a purely a priori argument, of course: it says that when macroevolution is considered the chances that randomly taken G→P→D applications are of *type a*,

[18]Remember, the genotypes are dots in the space; the whole reasoning assumes that we deal with *discrete* sets.

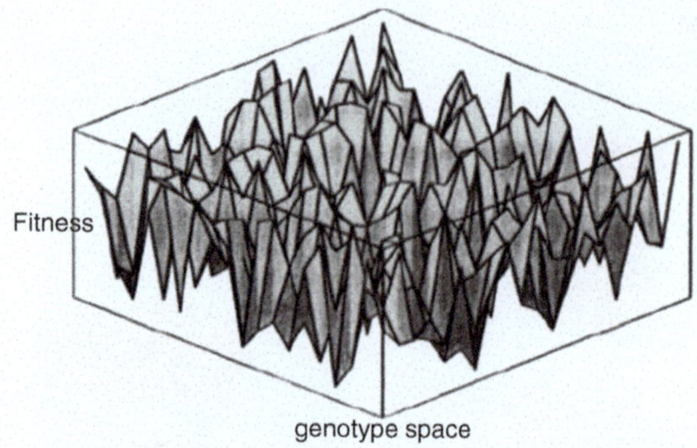

Fig. 14.5 A rugged fitness landscape. The selection-driven hill-climbing of populations is not possible

are much lower than when we focus on microevolution; it does not say that the actual structure of this triple space isn't such that developmental space can be bracketed, but only that, in principle this is much less probable than the opposite.

Hence, when one addresses macroevolution, which implies that the sizes of the phenotype and genotype spaces considered are much larger than in microevolution, one is much less justified in making the simplifications proper to microevolutionary theory, which is centered on population genetics that assumes bracketing of development. It entails that there are many chances that a major causal role is played by other processes, mostly relevant to the developmental variation-producing mechanisms.

A parallel argument to this one could be elaborated in relation with fitness landscapes. We saw (Sect. 14.2) that Simpson and Dobzhansky extrapolated Wrightian fitness landscapes proper to microevolution into multispecies macroevolutionary fitness landscapes. This meets a problem of the same kind as the principled argument I just sketched, for the following reason.

We now know that fitness landscapes should be relatively smooth if adaptive evolution is likely to occur on them. If they are too rugged (Fig. 14.5), as Kauffman in the 1990s famously showed, then natural selection is not capable of making populations climb to global fitness peaks (Kauffman 1993), even through various mechanisms such as the ones Wright was envisaging (e.g., the "shifting balance theory" (Wright 1932; Coyne et al. 1997)). Therefore a condition for natural selection to be the main driver of evolution, and finally to yield adaptive evolution, is that fitness landscapes should not be too rugged.

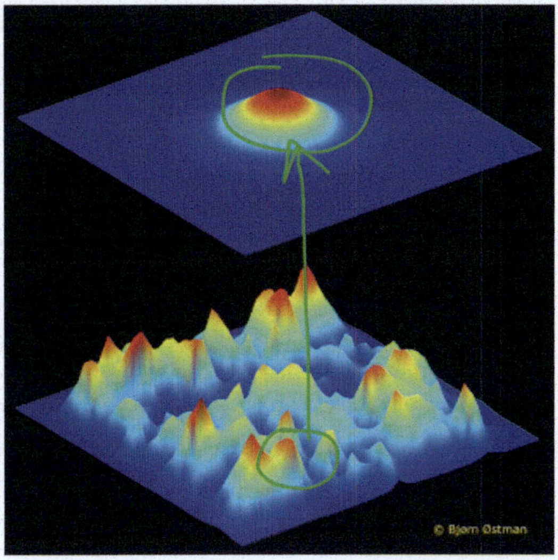

Fig. 14.6 A simple fitness landscape, which, when zooming out, proves to be part of a rugged fitness landscape

Now, here is the problem with Simpson's extrapolation: it is perfectly possible that *locally* the fitness landscape is smooth, but that it's part of a larger landscape, which is *globally* rugged. Zooming out from the microevolutionary smooth fitness landscapes gets you into a very rugged global fitness landscape (Fig. 14.6).[19] If this is the case, then, the conclusions that population genetics as a science of microevolution is able to draw regarding adaptive evolution are not likely to hold when it comes to macroevolution. Moreover, given that in principle rugged fitness landscapes are more frequent in the set of possible landscapes than smooth landscapes, this type of zooming out is perfectly plausible, and could even be expected in the absence of other, independent, empirical evidence.

Thus, we have here another instance of the logics-based worry that extrapolation claims lose some of the formal properties supporting the validity of the teachings based on population and quantitative genetics regarding evolution. Of course, the issues about what fitness landscapes *actually* exist, as well as the issue of the kinds of applications that hold between the wide G/D/P spaces, are both empirical issues. They cannot be solved by considering the question of what is in principle the most probable, even though formal arguments are helpful in showing why the extrapolation thesis cannot be taken for granted.

Nevertheless, those are logical arguments *about features of the biological reality*; they are not purely mathematical arguments, since they concern features of development, genotypes, landscapes, etc. I will now turn to another set of worries concerning the extrapolation thesis, which are not about gradualism and its accounting processes, but about the directionality and predictability of evolution at the largest scales (namely, far above the scale of speciation).

[19]See also Wilkins and Godfrey-Smith (2009) on zooming in and out landscapes.

14.4 Megaevolution, Contingency and Directionality

14.4.1 General Issues About Megaevolutionary Patterns

Paleobiology in the 1970s and the 1980s emerged as the project of building a theory of the history of life by capturing the patterns of phylogenies at various scales, which means, extracting the regularities in distributions – and by accounting for those patterns in terms of the relevant processes. This project was very clear for its promoters, Steven Jay Gould, Thomas Schopf, David Simberloff and David Raup, who met several times in the 1970s for long working sessions in Schopf's country house. In the letters they kept exchanging for years during the elaboration of their program, and especially the so-called MBL null model I'll consider below, they state their objectives and comment on them. In a 1972 letter, Schopf writes:

> I think that in our meetings care should be given to the question of the initial problems to be explained. Prejudicing the issue as little as I can, these general topics appear to me at the moment: 1. Organismal diversity through times; 2. Morphological themes through time; 3. Chemical themes through time; 5. Phylogeny through time. If these are part of what we want to understand, then we want to ask what are the processes underlying these patterns and what are their long term equilibrium consequences; the processes include: speciation theory, including population genetics and the species equilibrium. 2; the constraints imposed by size, shape and habitat or organized protoplasm. 3. The unity (or disunity) of biochemical pathways, including modes of reproduction. 4. Is there an equilibrium model of phylogenetic development?

So the paleobiological program as it emerged in the 1970s[20] defined a large set of issues that went well beyond the sole issue of gradualism vs. discontinuity: considering macroevolution at various scales, especially well above the scale of speciation, it included new questions such as:

Whether phylogeny displayed trends of some sorts (regarding body size, mass, complexity, etc.) (e.g. McShea 1994, 2005, see discussion in Turner 2015);
Whether evolution at large scale is predictable (Gould 1989; Conway-Morris 2010; Beatty 1995, 2006);
Whether clades and their distributions, their origination and their extinction display specific patterns, and how constantly (e.g. Jablonski and Sepkoski 1996; Valentine et al. 1999; Foote 2003; Jablonski 2007, 2008b).

Interestingly, Schopf, Gould and Raup saw the whole project as "heretical" for the paleontology of the times because it was a "search for a kind of timeless generality" in a "science so deeply committed to historicity", as Schopf writes in this letter.

The latter issue I mentioned covers several important questions, which requires differentiating among meanings of "diversity". Evolutionary biologists distinguish "diversity" *stricto sensu*, namely the amount of different species or clades, that

[20]On this program see Ruse and Sepkoski (2011), Huss (2004).

some call "richness", and "disparity", which is diversity mitigated by taxonomic distance: a clade is more disparate than another if its subclades are more distant, even though the latter has more subclades than the former (Gould 2002; Sterelny 2007). The question thus concerns the possible patterns of disparity/richness ratio evolution within clades. Some paleobiologists claim that in general clades are initially more diverse, and later on, when they are prone to extinction, they show much less diversity, but one could refine this idea by stating that the ratio disparity/diversity gets inverted during the existence of the clade, starting with much more disparity than diversity. An important issue in paleobiology hence concerns the plausible processes accounting for this pattern. Another general set of patterns concerns extinctions (Jablonski 2000): how do they affect the diversity and disparity within clades? Are there regularities about that? How do clades recover disparity and richness after quantitatively important extinctions?

Patterns may also concentrate upon the distribution of what paleobiologists and some evo-devo biologists call "novelties", namely, novel qualitative traits (with respect to formerly existing characters) (Mayr 1963) that in some cases, labeled "key innovations/novelties", are likely to trigger (or at least are correlated to) phylogenetic increases in diversity and/or adaptive radiation. The question consists in uncovering plausible patterns of the emergence, distribution and diffusion of key novelties. Some challengers of the Modern Synthesis argued that in the classical MS framework of evolutionary biology, novelty is by principle not accountable, and that the emergence of novelty constitutes an explanandum distinct from adaptive evolution (that is, the proper explanandum of the microevolution-centered population genetics) (Müller and Newman 2005).

Considering those issues have thus shown that punctuated equilibria is not the only paleobiological pattern that raises issues for an account of the history of life in classical terms of microevolutionary processes. As Jablonski asserted, "stasis is not necessary for large-scale trends to be shaped by more than just selection and other processes at the organismic level" (Jablonski 2009).

Once the patterns are established, as we saw, paleobiologists investigate their regularity in clade distributions. I'll just focus on two issues here, which are particularly significant regarding extrapolation claims and megaevolution.

14.4.2 Challenges of Megaevolution Patterns

14.4.2.1 Accounting for Regularities in the Shapes of Clades

Regarding the shapes of clades and a possible regular disparity/diversity pattern that would characterize them, one major advance has been made by the so called MBL (Marine Biology Laboratory) model, elaborated by Gould, Raup, Schopf and Simberloff in 1972. This model has been hugely influential, because it was one of the first achievements intended to set paleontology as a major evolutionary discipline (Huss 2009) along the lines of the project Schopf, Raup and their friends have

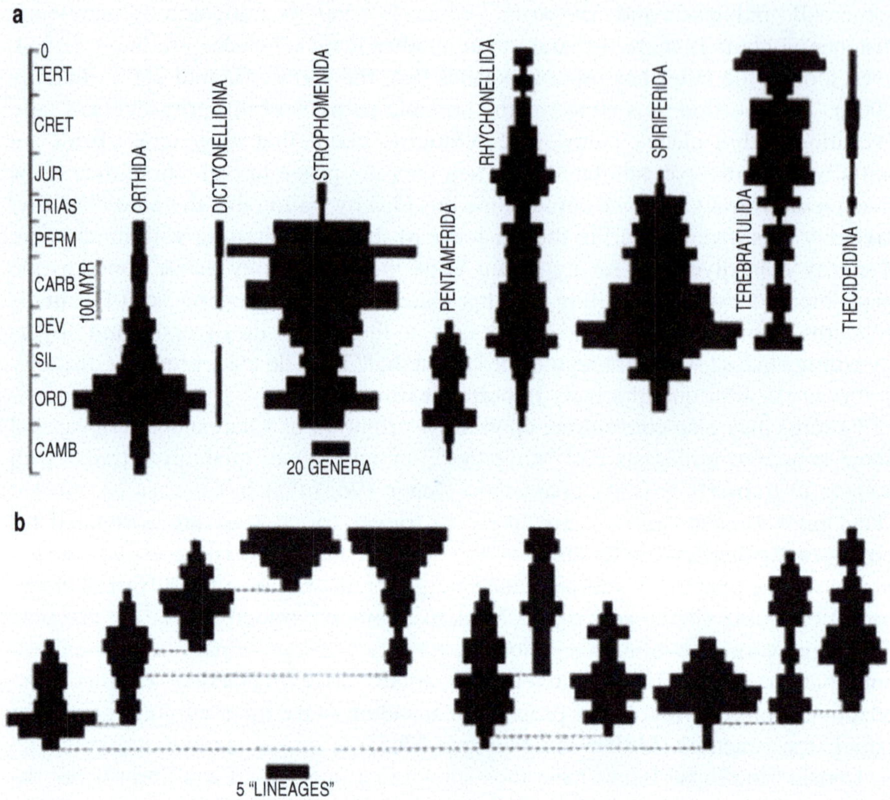

Fig. 14.7 Comparison between patterns of clade size evolution (**a**), and distributions produced by stochastic MBL model (**b**) (Raup et al. 1973)

undertaken – and not just as a part of geology, as it used to be. Another reason for its influence is its being an important "neutral" model, at a time where neutral models were highly debated in evolutionary genetics (Nitecki and Hoffman 1987). "Neutral" here means "with no selection" – exactly as in the "neutral evolutionary theory", advanced by Kimura and debated at this period, in which one considered the evolution of a system made up of alleles with no fitness differences (Kimura 1984). In the MBL model, species are evolving in discrete time, but have equal chances of surviving and speciating at each time point. Hence, survival, speciation and clade diversification at each time step are stochastic. The intention of the model was precisely to assess what would be an evolution at the level of clades – and the phylogenetic patterns – if selection were not playing a significant role at this level. The results were impressive and at the same time ambiguous. The shapes of clades that are produced in stochastic simulations are not so different than what is attested to in the fossil record (Fig. 14.7). On the other hand, they are not exactly the same.

Later on, ecologist Stephen Hubbell, in his groundbreaking book (Hubbell 2001) and in a later paper (Hubbell 2009) reflected upon this model. Hubbell elaborated on an ecological theory in which biodiversity patterns, in ecology, are not due to natural selection but mostly to stochastic effects. In his view the MBL model and his neutral ecology are close parents; but he argues that an MBL model in which fitness equality is predicated at the level of *individuals* and not at the level of species would produce clade patterns much closer to the data than the extant MBL model. This is exactly like what he has himself done in his neutral ecology, since the switch in predicating fitness equivalence is definitional of his model, as he argues in Hubbell (2001, chap.1). Another issue with the MBL model is that it does not generate the five attested mass decimations – but I'll expand later on this.

In any case the MBL model is one possible account of some phylogenetic patterns; it can be used as a null hypothesis for estimating the role of natural selection in shaping those regularities (Huss 2004). It is rivaled by an account due to prominent paleontologist James Valentine in the 1980s. According to him, the emergence of clades required the invasion of empty ecospaces; large ones were required for the origin of the highest taxa, medium- sized ones for taxa of intermediate rank, and so on. The ecospace mosaic, composed of tessera (representing niches), was invaded by species-level lineages, but as it filled up, the opportunity to produce novelties was progressively reduced, hence the drop in disparity. Thus diversity among higher taxa was regulated ecologically. Selection drives the whole process.

Another hypothesis would emphasize the role of species level or clade level selection as Gould himself asserted. In this case, even though classical gene-level or organism-level selection drives microevolution, and plays a huge role in speciation, these patterns proper to megaevolution – even if they are not due to random processes – rely on processes that don't have effects at microevolutionary levels: for instance, clades that encompass more variety will remain longer because variety buffers against environmental changes, which should be large at the largest timescales.

The history of life at very large scales is thus an object of controversy, and in any case, the fact that natural selection seems to be overwhelming at the microevolutionary level does not *ipso facto* entail that the patterns characterizing such history will be wholly driven by natural selection.

14.4.2.2 Contingency of Evolution and the Mass Extinctions

A second very general issue regarding mega-evolution and mass extinctions has been famously formulated by Steve Gould in *Wonderful life*. As he says, what would happen if we "replay the tape of life"? Would we meet the same species and families? Or "almost the same", meaning that we would get intelligent creatures like us, marine creatures (like fish), animals and plants and then carnivores and herbivores etc.? In other words, is evolution at the largest scale wholly contingent, meaning that its outcomes could be otherwise, and could be non-existent? Or is it somehow a necessary process, whose major steps, patterns and orientations (if any) are robust across all possible contingent starting initial conditions we could think of?

Clearly, if selection were wholly driving the "evolutionary play" (as Hutchinson called it), such questions would not have to be formulated. In each replay, the species reach the same adaptations, since they are defined by the environmental demands, which are initially the same in general: gathering resources, exploiting sources of energy, etc. It might be that the outcomes are not exactly identical – for instance, we'll get creatures which have different kinds of eyes, and a nervous system differently wired and elaborated, or a hereditary material not made of DNA – but life would still evolve seeing and planning creatures (Huneman 2010b). Criticizing Gould, Dennett (1995), following Dawkins (1982), called such general types of adaptation the "good tricks", and claimed that whatever the variations in initial conditions selection was always likely to discover "good tricks", instantiated in any particular matter and shape.

But Gould's argument is that, whatever those good tricks are, evolution is massively contingent since some *mass extinctions* happen for no reason connected to the adaptive capacities of species. For this reason, extinctions play a major role in those questions about megaevolution. There could be a catastrophe, planetary or astronomical, and then what saves a species from becoming extinct is just luck (no species at all have evolved adaptation for dealing with astronomic catastrophes, anyway, so evolved capacities due to natural selection don't make any difference). The paradigm of such contingent extinctions is the extinction of dinosaurs at the end of the Cretacean (or K-T extinction), caused probably by the aftermath of an asteroid colliding with Earth. The mass extinctions randomly decimate clades and therefore reduce the disparity of clades (Jablonski 2001, 2005). The subsequent history of life is therefore wholly contingent upon who randomly survived.

Some authors contested Gould's view, for instance, Conway-Morris – ironically, one of the investigators of the Burgess Shale, whose rediscovery is the main argument in Gould's book. Where Gould claimed that megaevolution is essentially contingent, Conway-Morris (1998) insists that evolution is predictable. He pinpoints convergences (independent clades evolving the same trait) as an argument in this sense. But, commenting on Lenski's bacterial evolution experiments, which have been going on for 3 decades (Lenski and Travisano 1994), philosopher John Beatty emphasized that the order of mutations, which is contingent, is important in determining what evolution by natural selection can reach, and therefore, this contingency plays an irreducible role in mega-evolution (Beatty 2006).

Notice that the contingency Gould is dealing with is not exactly what population geneticists are well acquainted with, namely, "random genetic drift". Even though this concept has been widely discussed and is still philosophically opaque or controversial (Plutynski 2007; Matthen 2009), drift clearly relates to what statisticians call "sampling error", since it is strictly related to the size of the population: the smaller the population, the higher the chances that the outcome of evolution as predicted by fitness values will not be reached (see Beatty 1994;

Millstein 2002; Huneman 2015). But what goes on with mass extinction is not proportional to population size. In case of a major volcanic episode, for instance, large populations will be as affected as the small ones, no matter their adaptedness. Hence, what underpins "mass extinctions" does not seemingly pertain to drift.

"Contingency" has several meanings, as Beatty (2006) usefully highlights. For now I just indicate that it means both a kind of unpredictability, and the fact that what's contingent is "contingent upon" some events. Of course, everything is plausibly contingent upon something: Kant, Claude Bernard and Cournot concurred in saying that this is not an empirical truth but a necessary assumption for empirical science. But the contingency we deal with here means that some major evolutionary facts, such as the K-T extinction, are contingent *upon other facts* that are themselves *not relevant for microevolution* – such as a cosmic collision, which is never represented in classical genetic models, since it's not an environmental parameter. In this sense, contingent facts such as mass extinction are not predictable since our evolutionary models don't represent the facts they are contingent upon. Of course they may be predictable for astrophysicists, but this is not the biology viewpoint. And, even if evolution would be predictable after the mass extinction, precisely because of our models that design predictions based on fitness values, the overall Tape of Life is not predictable, due to events that are not given in the beginning, and that therefore would not be recurrent across several hypothetical iterations of such Tape. To this extent, it's not unreasonable to say that mega-evolution is contingent in itself, and that its outcomes are not predictable.

Granted, microevolution itself is contingent, because of drift: when a population is small, one cannot predict the outcomes of an evolutionary dynamics, and it will never be the same if we replay the tape many times. But these effects of drift almost disappear when the population is large enough – in this case, the dynamics becomes predictable given the fitness values.

What interests us here is that, while microevolution will often be driven by natural selection and display patterns of directionality due to the overwhelming role of natural selection, which is by definition oriented towards an increase in fitness,[21] if Gould is right macroevolution does not inherit those features. By contrast, it is affected by a major dimension of contingency, which

[21]There is a large controversy over whether natural selection maximizes population fitness, initiated by Fisher (1930) and his "fundamental theorem of natural selection", and still going on now, but this is not the place to develop it. See Grafen (2007), Huneman (2014a) for a contemporary defense of the view, and Lehmann and Rousset (2014) for a critique. In the current context it's enough to indicate a link between selection and directionality: if selection drives microevolution, then in many cases we can expect a maximal (inclusive) fitness phenotype, and in even more cases we can predict the outcome, even if it's not a fitness maximizing outcome (for instance because the genetic structure prevents this maximization, even if the genotypic frequencies under selection are predictable).

precludes extrapolating the lessons of microevolution to megaevolutionary patterns.

The contingency thesis is harshly contested (Dennett 1995; Dawkins 1982; Conway-Morris 2010) for various and sometimes opposite reasons (Huneman 2010b); in the following section, I'll delve into more detail about the question of mass extinctions in paleobiology, in order to elaborate in the last subsection a formal argument likely to support the anti-extrapolation claim included within the contingency thesis.

14.5 Extinctions and Their Causes (Gould Meets Mandelbrot)

14.5.1 Mass Extinctions; Paleobiological Hypotheses

As I said, a major topic of nascent paleontology was extinctions: their distribution, their causes, and their relations. Especially *mass extinctions* came to the fore, because they don't easily relate to selective disadvantages (most species are well adapted: how could they so suddenly go extinct together?). Granted, given the nature of the evolutionary process, extinctions occur all the time; each species, and then each clade, is deemed to extinction on the long run. However, during the history of life, at least five episodes have been detected in the fossil record, in which for a very short amount of time a huge proportion of the extant species (up to 70%

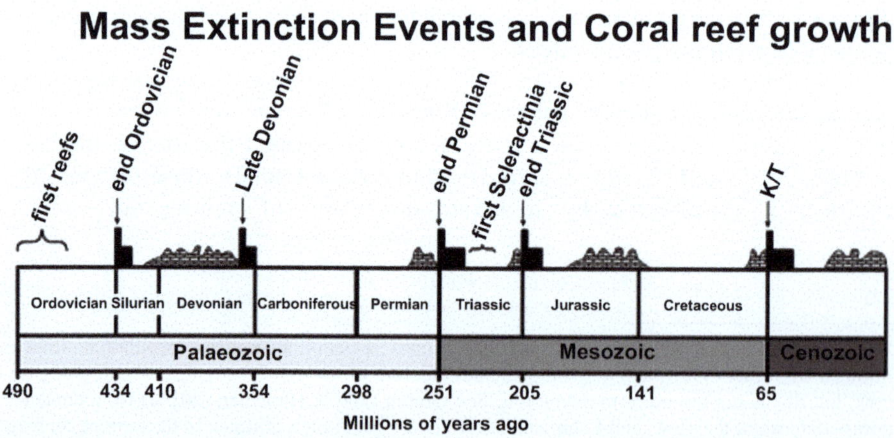

Fig. 14.8 Distribution of mass extinctions

sometimes, and even 95% of marine life during the Permian-Trias extinction) goes extinct (Fig. 14.8).

Extinctions, their width, regularities and distributions emerged as a central problem from the very beginnings of paleobiology. In a 1978 letter to his coworkers David Raup wrote: "I'm becoming more and more convinced that the key gap in our thinking for the last 125 years is the nature of extinction." And he gets very critical about the Modern Synthesis, then

> if we take neo-Darwinian theory at face value, the fossil records makes no sense. That is, if we have adaptation through natural selection or species selection and extinction through competitive replacement or displacement, then we ought to see a variety of features in the fossil record that we do not seen such as: (a) clear evidence of progress, (b) decrease in evolutionary rates (both morphologic and taxonomic), (c) possible a decrease of diversity (at least within an adaptive zone.) Now, we do not see these things because we are too dumb, or the record is lousy, or there are features of the evolutionary mechanisms that prevent the approach to a steady state. The last of these is the conventional explanation (...). In the conventional wisdom, evolutionary change is always adaptive and extinction is always related to a fitness problem (either with regard to the physical or biological environment) and we do not see a slow down in evolution or evidence of an optimization of the whole system because the system is so damned complicated. My trouble is that I don't believe the conventional scenario. My candidate explanation is that extinction is random with respect to fitness. (Raup to Schopf, 28 Juanuary 1979)

Among the paleobiologists I mentioned, Raup was perhaps the most committed to randomness in mass extinction processes, as the last sentence quoted illustrates. Nonetheless, a widely admitted feature of those extinctions, which Gould later emphasized, is that, being quick at geological timescales, they also affect species that seem to have been correctly adapted, species that were "ecologically tolerant and occurring in great numbers in all parts of the world" (Raup 1994). If microevolutionary processes were at work here, those extinctions should have been gradual and slow – accumulation of disadvantages in the face of changing environment, so to say. However given that it's not the case, as it has been documented especially for the late Cretacean extinction (which included the extinction of the dinosaurian clades), the accounting processes may well be novel processes.[22]

Importantly, natural selection is defined by selective pressures, which are by definition recurrent and regular environmental parameters (Huneman 2015). Populations may adapt slowly and gradually to these parameters' values, via the selection of small variations. "Most species have evolved ways of surviving anything that their environment can throw at them, as long as the stress occurs frequently enough for natural selection to operate." (Raup 1994). Thus, mass

[22]This extinction was indeed massively studied, especially because of the controversy about the causes of the disparition of the dinosaurs, which was revivified by the hypothesis of an asteroid impact, elaborated by Luis Alvarez in the 80s – an asteroid often later identified with the meteor responsible of the Chicxulub crater in Yucatan. We know that in no sense were the dinosaurs groups' diversity declining just before the extinction.

extinctions seem, by contrast, to require "rare physical events" to which populations cannot adapt. Raup emphasized that those "stresses" triggering mass extinctions should be "experienced on time scales short enough for natural selection to act." But if selection – hence adaptive values, or fitness – does not predict the results of mass extinctions, those appear as random from the viewpoint of subsistent or extinguished species. Considering large time mega-evolution here confers randomness an increasing role in the explanation of evolutionary outcome. Such a relation between large time scales and the random character of the surviving species connects the anti-extrapolation thesis with the contingency thesis. My own suggestion, later on in the next section, will develop what's in this relation.

Once mass extinctions have been identified, the question becomes: what should account for them? The debates especially concentrated on the opposition between internal causes, such as development- and ecology-based factors – and external factors, such as non-biological forcing. The collision with an asteroid, which has been held as responsible for the meteorological disturbances triggering the extinction of the dinosaurs, appears as the paradigm of such a rare external physical event, as a forcing likely to involve a contingent mass extinction.

Several interesting theories have thereby been proposed as a framework for our understanding of large extinctions – a feature characteristic of megaevolution. Raup and Sepkoski (1984) noticed that there was some regularity in the interval of large (not mass only) extinctions, and the burst of biological disparity that often follows then. The mean period between large extinctions is 26 Million years. Such regularity is puzzling – no life-based rhythm is likely to be so slow. Only astronomic events and cycles are like this, hence the idea that large extinctions could be coupled to a cosmic cycle, such as the regular rotation of asteroids or comets that would regularly visit the neighborhood of the Earth and trigger a cascade of geological events (the so-called "Nemesis hypothesis"). This is quite speculative, and as Jablonski noted, "the jury is still out on whether these pulses of evolutionary inventiveness and, just as important, the cessation of these pulses derive mainly from developmental or ecological factors, although environmental triggers and ecological feedbacks are currently in favor." (Jablonski 2007 – but the Nemesis hypothesis sounds now abandoned)

Some focused on the notion of "ecological feedback" included in these views. In a work using dynamical systems theory, Solé (2002) advanced an "ecological perspective" on mass extinctions. He sees patterns of mass extinctions as ecological patterns of chaotic response. This conception starts with the remark that not all meteor craters on Earth have been followed by mass extinctions. So it could be that cosmic cycles affect the biosphere, not each time, but only once a threshold of environmental perturbation has been reached by the effects of the successive impacts of cosmic objects on Earth. Even if celestial bodies were regularly affecting the Earth, a genuine drop of diversity would occur only after 5 or 6 impacts, when a kind of threshold is reached (Fig. 14.9).

Whatever future science decides on those theories, it seems that the distributions of extinctions in megaevolution, and especially the contingency of evolution that derives from those extinctions, is not likely to be explained through a

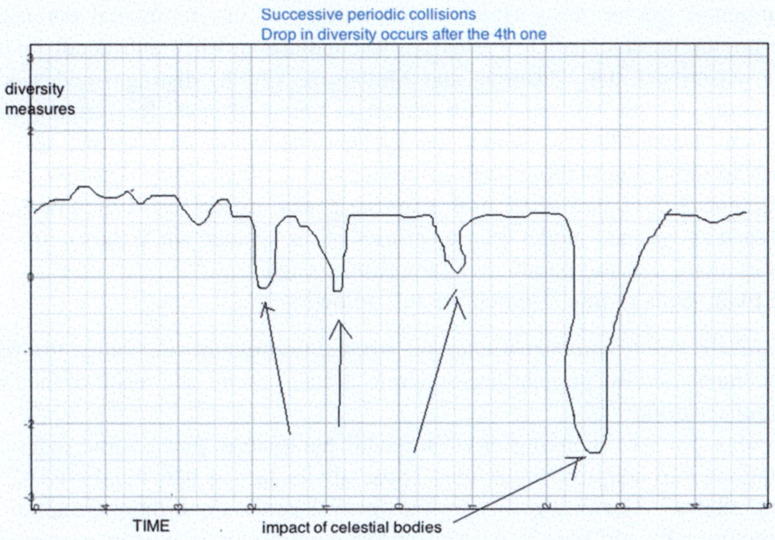

Fig. 14.9 Threshold model of ecological feedbacks. The mass extinction is "prepared" by iterated cosmic cyclic events, but occurs only when a threshold is reached

microevolution-based framework. But the contingency thesis itself, understood as the unpredictability of evolution, is still controversial; therefore the extrapolation thesis could be safe if indeed the objectors of the thesis are proved right in the end.

In the last section, I'll elaborate an interpretation of the contingency thesis that is intrinsically connected to the issue of timescales, and that will therefore count as a mathematical argument against the extrapolation thesis; this argument, unlike the former one about developmental spaces (Sect. 14.3), is exclusively mathematical. In order to explain it I'll describe in more detail evolutionary research on extinction in general, and what is known as "evolutionary rescue".

14.5.2 Extinction Time Research

Since the 1980s, there have been a number of studies devoted to extinction time. The main question, is how population of a given species responds to fluctuations, and when and why it should go extinct? Since the 80s, population and quantitative geneticists have contributed massively to this investigation (Lande and Orzack 1988; Lynch and Lande 1993; Lande 1988; Burger and Lynch 1995; Bells and Collins 2008; Chevin and Lande 2010; Chevin et al. 2010). This proves to be all the more important because of global climate change, whose severity have been increasingly measured since the first of those studies, and which means a great global shift in the environment. This program employs two major concepts, environmental stochastictity (random fluctuations of environmental parameters) and demographic stochastictity (random fluctuations in birth and death rates for various reasons). Later, Hastings and Melbourne (2008) distinguished kinds of stochastictity, which will be often modeled as binomial laws. Researchers model the dynamics according to which the species change in an adaptive way as a response to those fluctuations, and sometimes go extinct. They consider that the environment defines one or several adaptive peaks, in the terms of fitness landscapes seen above, and model the trajectories of the species while the environment fluctuates along varying parameters (amplitude, rate etc.).

The main ideas in this research are the following:

- *Sensitivity to environmental change*: when environment fluctuates, the fitness peaks move and the population somehow "tracks" them, as a result of the fact of natural selection;
- *Rescuing alleles:* changing environmental parameters make some deleterious mutations into adaptive mutations. Those alleles turned beneficial are the "rescuing alleles". The probability to find them depends upon the population size. Predicting why and how a species can avoid extinction is thereby a question of finding the conditions for a rescue effect (Bell and Collins 2008)
- *Lag behind optimum*: the populations can't follow instantaneously the optimal phenotype; hence there is a time lag between the initial state of adaption to its environment, and the final state of a species adapted to a changed environment. During this time lag, the population is sub-adaptive; and the time lag is partly

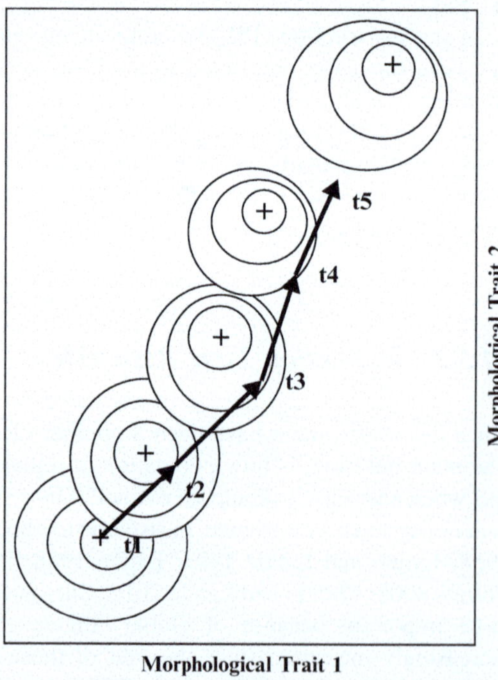

Fig. 14.10 A simple case of a species tracking the fluctuation-driven adaptive optimum in a fitness landscape. At t5, the time lag implies that the species won't be able to reach the 5th optimum, and hence goes extinct

determined by its mean growth rate. For instance, during a local warming correlated to global climate change, a mountain butterfly species will climb up, so that it can live in a chiller place (warmth is inversely proportional to height) (Devictor et al. 2012). Yet in the same time, the temperature of this new location warms up too, so when it has settled, it has to leave and climb higher to adapt, and so on and so on. But given that the warming is quicker than the evolution of the butterfly species, at some point the population will be stuck at an altitude where it's too warm for the butterflies, but they can't have time to adapt, and the species gets extinct.

From this, it results that the mean long run growth rate of a population is often the best predictor of adaptation of the population (Burger and Lynch 1995): it itself depends on generation time. But of course, if the time lag is too long, so that the species cannot keep track of the moving adaptive optimum, then it is deemed to go extinct (Fig. 14.10).

A key feature of those studies is the way researchers model stochastictity. Even though demographic and environmental stochastictity are different, for instance they may have different timescales, and also may include different heterogeneous sources of stochasticity (as Gillespie 2004 pointed out), it is frequent that their randomness is modeled by varieties of the binomial law (with parameters n, t, B (n,t)). Those random variables following a binomial law add up to produce a general stochastic dynamics, which can be modeled as a mix of diffusion processes followed by the optimal peak (Lande 1988).

In turn, a plausible approximation of the distribution of fluctuations is the normal distribution (Box et al. 1978) in cases where the parameter n is high, which would be generally the case here. It means that stochasticity sources in those researches are represented generally as fluctuations around a mean, which yields a drift of the adaptive optimum; thus the general question is the conditions under which a population is likely to follow this peak for a given time. For instance, Bell and Collins (2008) write "Provided that the environment changes rather smoothly, with little stochastic variation around its expected value, this critical value [characterizing the variables that determine the time lag in optimum tracking] is given approximately by $k_{crit} = R\ (2\ V_\lambda \ln \lambda^*)^{0.5}$" (where k is the rate of change of the optimum, λ^* is the "maximal rate of increase attained when the population mean phenotype matches the optimal value", and $V_\lambda = \sigma^2{}_p + \omega^2(\sigma^2{}_p$ being the variance of the phenotypic character governing the viability, and ω^2 being the width of the fitness function ruling stabilizing selection that acts on this character.))

I will use this framework to introduce a suggestion regarding the treatment of randomness at large timescales, and to argue that Gould's contingency thesis about the unpredictable effects of mass extinctions could be formulated in terms of a challenge to modeling habits regarding stochastictity. I'll appeal to a distinction made by Benoit Mandelbrot a while ago, in the context of financial mathematics, and will apply it to extinction research. As a result, I'll argue that this provides a mathematical reason for resisting extrapolation when one switches from small to very large timescales.

14.5.3 Mild Randomness, Wild Randomness, the Contingency Thesis and the Extrapolation Thesis

Mandelbrot used to distinguish what he called "wild" randomness from "mild randomness". The latter is the most usual kind of randomness, and concerns anything likely to be viewed as fluctuations around a mean. For instance, randomly picking someone and measuring her size will be correctly approximated by a normal distribution centered on the mean human size. Clearly, a huge part of the randomness cases in daily life are like this; they allow us to dismiss most extreme deviations from the mean, since by definition those are very extremely improbable. Mandelbrot's argument was that some other cases of randomness are very different, and that modeling them as "mild randomness" can lead to a severe misrepresentation of what actually takes place – especially in the case of economics, where the consequences of those mistakes are serious, since this is about unpredictable financial crashes (Mandelbrot 1997)

To get an idea of "wild randomness", think of non-Gaussian distributions such as the distribution of wealth. It's more likely to be a scale-free distribution: few very, very rich; some very rich; about 10 times more "just rich" people; then 10 times more "average wealthy" people, etc. We end up with the famous picture of the 0.1% on one end of the ladder, and 99.9% at the other hand.[23] But even if someone extremely rich is very rare, this rarity is not exactly like the rarity of someone who is 2.25 m high – since the former may have a very important impact on the economic system, while the latter won't have any relevance for the sizes of the others. Roughly said, wild randomness is a kind of randomness where extremely rare events cannot be easily dismissed. For instance, in financial economics, those rare events, which may be connected to stock-market crashes, should not be left out from the description of the system at first approximation, or one risks a mischaracterization of the logics of financial crises. In natural sciences, the mean temperatures pertain to mild randomness, whereas if you randomly pick an earthquake, regarding their magnitude earthquakes are distributed in a way proper to wild randomness (Fig. 14.11).

An important feature of this distinction is that *locally* it is not easy to distinguish on empirical grounds between wild and mild randomness, i.e. between distributions that are close to a normal one, and distributions that exhibit the features of wild randomness (important extreme events etc.) (Fig. 14.12). In other words, the former can be a good approximation of the latter, as long as we remain in a small range of time and parameter values.

Let's return to evolutionary studies. Most of the studies about extinction time are microevolutionary approaches. They model stochastictity as Gaussian approximations, therefore in our current terms they only handle mild randomness. Suppose now that we want to expand the timescale. Granted, on small timescales the proper

[23]In practice, things are not always like this, and the effects of taxes, in particular, may hugely impinge on this distribution (happily).

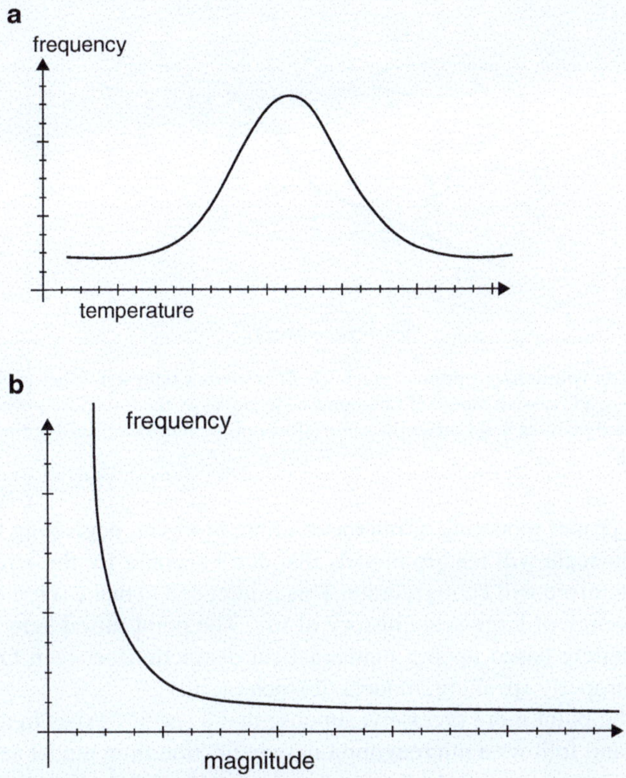

Fig. 14.11 Mild randomness and wild randomness: temperature (**a**) vs. earthquakes (**b**)

Fig. 14.12 Mild randomness, locally (on the *right*), correctly often approximates wild randomness (on the *left*): chunks from the above graphs on temperature and earthquake magnitudes

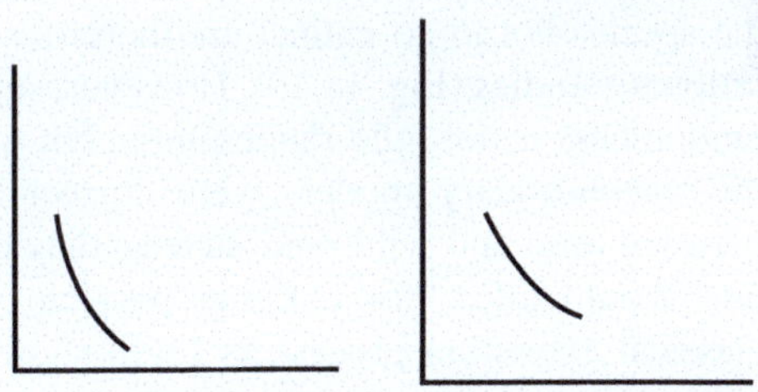

modeling of randomness, i.e., the choice between wild or mild randomness, may not be consequential since one approximates the other, so that both options will yield comparable predictions. Yet, if one shifts to a much higher timescale, then the two kinds of randomness are very different and the models using them will yield very different predictions. Incorrectly modeling wild randomness as mild randomness will then obfuscate the occurrence and consequences of extreme events.

Intuitively, that is what happens with the extrapolation to macroevolution or megaevolution, as criticized by Gould's and others' theories of mass extinction. Events such as those "rare physical events" correlated to mass extinctions may not

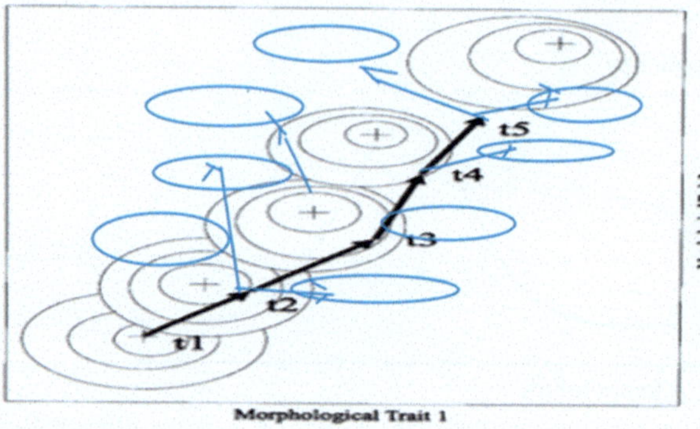

Fig. 14.13 Microevolutionary optimum tracking. *Blue arrows* represent other probable fluctuations, according to a Gaussian model of randomness. In any case, the trajectory towards extinction can differ but the dynamics will remain the same (slow optimum tracking until optimum gradually gets out of reach)

be part of the proper modeling of microevolution, however, neglecting them on the megaevolution scale will lead to models that don't account for the effects of such events, and therefore will be inaccurate. This is intended to make sense of the claim of the contingency of large-scale history of life. The point raised here is that such claim is ultimately based upon a mathematical distinction between kinds of randomness or, properly speaking, models of stochastictity.

To make the point more precisely, suppose that a species exists in a fluctuating environment and follows a microevolutionary extinction time model as it has been sketched above; then expand the timescale, and consider a very long evolution. The stochastictity remains classically modeled in a somehow Gaussian way, and the optimum tracking process will be the same as in the usual case; the conditions for the species not to go extinct are therefore the same as discussed in the microevolutionary studies (Fig. 14.10). However, suppose that the nature of actual biological randomness is not mild randomness, but wild randomness – then, even though the microevolutionary models were correct because of the approximation relation between mild and wild randomness, they can't hold at the longer timescale. Here, one should allow for extreme possible fluctuations, and this may disrupt the classical extinction process as I will show below. Basically, introducing the wild randomness changes the process of tracking optimum; at long timescales, one has to choose between wild and mild randomness in the foundations of the model.

In the classical microevolutionary case, a rough description of the findings is the following (Fig. 14.13). Populations slowly move in order to follow an optimum that environmental and demographic stochasticity have displaced. (Actually *demographic* stochasticity tends to move the population away from the optimum, while *environmental* stochasticity moves the optimum away from the population, but this difference does not matter here.) The time lag between optimum switching and the species reaching it may increase if, for instance, the population moves slightly slower than the optimum: at each optimum move, the population will reach

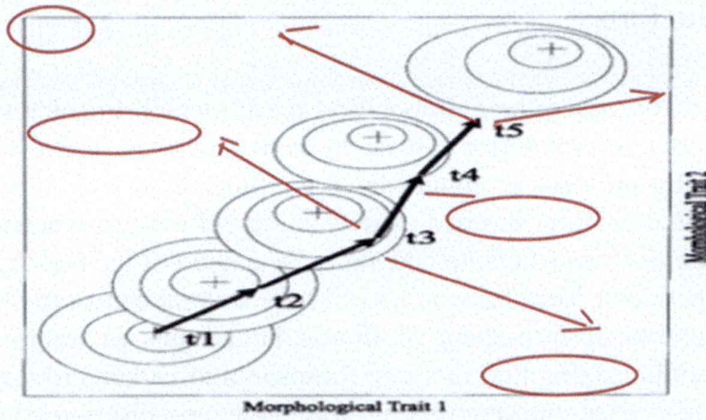

Fig. 14.14 Macroevolutionary model of extinction with wild randomness. Extreme jumps have to be considered in the model at each time step (in *red* instead of the black circles). Taking at each time step the mean between fluctuations in order to define new fitness peak is misleading. Optimum can be suddenly lost by the species, leading to fast extinction

some slightly suboptimal phenotype, which will become more and more suboptimal. Finally, the optimum moves out of the reach of the population, and this means extinction. There is a gradual move towards extinction, which can be more or less fast depending upon the parameters that define the species' time lag.

Suppose now that we introduce wild randomness. This process of optimum tracking takes place in macroevolutionary time. Here, it's not possible to consider the post environmental change new optimum as a mean between smooth small fluctuations like in the case of wild randomness, since extremely large fluctuations should be envisaged at any timestep. At some point, a huge fluctuation may occur, which gets immediately out of reach of the progressive gradual process of optimum tracking. So here there is no gradual extinction following a progressive loss of optimality, but a sudden extinction because that optimum gets "instantaneously" out of reach. The dynamics are much different (Fig. 14.14) Hence, at those scales, conflating both types of randomness leads to misrepresenting a dynamics. And the second kind of dynamics yields indeed patterns much closer to the empirical view of the phylogenetic patterns gathered by paleobiology.

Hence I claim that Gould's contingency thesis can be construed as a vindication for wild randomness in megaevolutionary or macroevolutionary time. Given that extinction studies at the microevolutionary scale are based on mild randomness, this is an argument against the extrapolation thesis. Microevolutionary modeling is more likely to go wrong when we switch up timescales; and inversely, macroevolution will appear as contingent when compared to the dynamics of selection, mutation and drift modeled by microevolutionary population genetics, which is predictable to the extent that selection dominates the dynamics.

This is a purely mathematical argument about the nature of randomness in the short term and the long term. However, it is not devoid of empirical content; actually, the justification for arguing that wild randomness is indeed closer to the genuine nature of randomness in macroevolutionary time is the fact that mass extinctions exist and that they are not predicted by microevolutionary models.

14.6 Conclusions

In this chapter, I discussed the extrapolation thesis, which in some sense reduces macroevolution to microevolution. This is in large part an empirical thesis, and one could wonder why philosophy should have anything to say on it all besides the interesting fact that shifting timescales may let novel processes into play, or not, that is, in the case the thesis fails, to take note once again of the fact that our science is timescale-dependent. Yet I have proposed here several conceptual arguments.

I used the graphical formalism of fitness landscapes to frame some of the developments, emphasizing that the way Simpson and Dobzhansky recast macroevolution in a microevolution-compatible frame have precisely used it. After having surveyed some of the arguments related to the patterns of macroevolution advanced by post-1980s paleobiologists to claim that as Erwin (2000) puts it, macroevolution is more than successive rounds of microevolution, I insisted on the fact that even if patterns such as punctuated equilibria are evidenced, it does not entail the need for acknowledging novel processes.

But I provided two formal reasons for thinking that the extrapolation thesis is very likely to be flawed. The first one concerns the debates over the role of development and constraints in microevolution. If one assumes that population genetics is entitled to bracket development – as it has been doing with various justifications since Fisher – and therefore, to put in the background developmental constraints, then an aspect of the extrapolation thesis consists in saying that this is also valid for macroevolution. My formal argument concerns what happens to this biological assumption when one shifts timescales. Because of the properties of the phenotype-development-genotypes maps, I claim that the validity of the development-bracketing is less likely to hold when one switches to macroevolutionary scales. It's not an empirical argument, it's a formal argument about the plausibility of the claim – but a formal argument that concerns biological features.

The second argument concerns Gould's contingency thesis, a property of history of life on the very long term (here called megaevolution). The suggestion is that the legitimacy in using mild randomness models of stochasticity when dealing with microevolution is threatened when we turn to evolution at much larger timescales. The proper stochastictity modeling here is plausibly wild randomness. To that extent, the evolutionary dynamics modeled by microevolution models is not likely to hold when we turn to megaevolution. According to this argument, Gould would be right for very formal reasons, which are much less about biology than about the consequences of timescale shifting upon mathematical modeling of randomness.

The latter argument against extrapolation therefore does not have the same nature as the former. It is much less bounded to biological facts. Therefore it is certainly stronger than the other. In other words it could be that our biological world is such that features a priori weakly probable of the GDP maps are realized, and we could know this empirically – but the link between wild randomness and time seems harder to be defeated. In this case, the extrapolation thesis would have a mixed validity: correct for macroevolution defined as an evolution that includes speciation and clade diversification, it would become incorrect when one turns to megaevolution, as the grand history of life.

Acknowledgements The author warmly thanks Jean Gayon, Annick Lesne, Scott Lidgard and Mael Montevil, for helpful discussions, as well as audiences at the ISHPSSB 2015 meeting in Montréal. He is also grateful to Andrew Mc Farland for a thorough language check of the manuscript, and to Sébastien Dutreuil and Christophe Bouton for criticisms and suggestions on a first draft. This work has been done with the support of the grant ANR 13 BSH3 0007 "Explabio".

References

Amundson, R. 2005. *The changing role of the embryo in evolutionary thought*. Cambridge: Cambridge University Press.

Bateson, P. 2017. Evolutionary theory evolving. In *Challenging the modern synthesis: Development, adaptation and inheritance*, ed. P. Huneman and D. Walsh. New-York: Oxford University Press.

Beatty, J. 1994. Chance and natural selection. *Philosophy of Science* 51 (2): 183–211.

———. 1995. The evolutionary contingency thesis. In *Concepts, theories, and rationality in the biological sciences*, ed. G. Wolters and J.G. Lennox. Pittsburgh: University of Pittsburgh Press.

———. 2006. Replaying life's tape. *Journal of Philosophy* 103: 336–362.

———. 2016. The creativity of natural selection? Part I: Darwin, Darwinism, and the mutationists. *Journal of the History of Biology*. doi:10.1007/s10739-016-9456-5.

Bell, G., and S. Collins. 2008. Adaptation, extinction and global change. *Evolutionary Applications* 1: 3–16.

Box, G.E.P., W.G. Hunter, and J.S. Hunter. 1978. *Statistics for experimenters*. New York: Wiley.

Burger, R., and M. Lynch. 1995. Evolution and extinction in a changing environment: A quantitative-genetic analysis. *Evolution* 49: 151–163.

Burian, R.M. 1994. Dobzhansky on evolutionary dynamics: Some questions about his Russian background. In *The evolution of Theodosius Dobzhansky*, ed. M.B. Adams. Princeton University Press: Princeton.

Cain, J. 2009. Ritual Patricide: Why Stephen Jay Gould assassinated George Gaylord Simpson. In *The paleobiological revolution*, ed. M. Ruse and D. Sepkoski, 345–363. Chicago: University of Chicago Press.

Cheetham, A.H. 1986. Tempo of evolution in a Neogene bryozoan: Rates of morphologic change within and across species boundaries. *Paleobiology* 12 (2): 190–202.

Chevin, L.M., and R. Lande. 2010. When do adaptive plasticity and genetic evolution prevent extinction of a density-regulated population? *Evolution* 64: 1143–1150.

Chevin, L.-M., R. Lande, and G.M. Mace. 2010. Adaptation, plasticity, and extinction in a changing environment: Towards a predictive theory. *PLoS Biol* 8 (4): e1000357.

Conway-Morris, S. 1998. *The crucible of creation: The Burgess shale and the rise of animals*. Oxford: Oxford University Press.

———. 2010. Evolution: Like other science it is predictable. *Philosophical Transactions of the Royal Society, B: Biological Sciences* 365 (1537): 133–145.

Coyne, R., N.H. Barton, and M. Turelli. 1997. Perspective: A critique of Sewall Wright's shifting balance theory of evolution. *Evolution* 51: 643–671.

Darwin, C. 1859. *The origin of species*. London: John Murray.

Davidson, E.H. 1986. *Gene activity in early development*. Orlando: Academic Press.

Davidson, E., and D. Erwin. 2006. Gene regulatory networks and the evolution of animal body plans. *Science* 311 (5762): 796–800.

Dawkins, R. 1982. *The extended phenotype*. Oxford: Oxford University Press.
Dennett, D. 1995. *Darwin's dangerous idea*. New York: Simons & Shuster.
Depew, D. 2017. Natural selection, adaptation, and the recovery of development. In *Challenging the modern synthesis: Development, adaptation and inheritance*, ed. P. Huneman and D. Walsh. New York: Oxford University Press.
Devictor, V., C. van Swaay, T. Brereton, L. Brotons, D. Chamberlain, J. Heliölä, S. Herrando, R. Julliard, M. Kuussaari, Å. Lindström, J. Reif, D.B. Roy, O. Schweiger, J. Settele, C. Stefanescu, A. Van Strien, C. Van Turnhout, Z. Vermouzek, M. WallisDeVries, I. Wynhoff, and F. Jiguet. 2012. Differences in the climatic debts of birds and butterflies at a continental scale. *Nature Climate Change* 2: 121–124.
Dobzhansky, T. 1951. *Genetics and the origin of species*, Columbia University Biological Series. 3rd revised ed. New York: Columbia University Press.
Eldredge, N. 1971. The allopatric model and phylogeny in Paleozoic invertebrates. *Evolution* 25 (1): 156–167.
———. 1989. *Macroevolutionary dynamics: Species, niches and adaptive peaks*. New York: McGraw-Hill.
Eldredge, N., and S.J. Gould. 1972. Punctuated equilibria: An alternative to phyletic gradualism. In *Models of paleobiology*, ed. T.J. Schopf. San Francisco: Freeman Cooper.
Eldredge, N., J. Thompson, P. Brakefield, S. Gavrilets, D. Jablonski, J. Jackson, R. Lenski, B. Lieberman, M. McPeek, and W. Miller. 2005. The dynamics of evolutionary stasis. *Paleobiology* 31 (S2): 133–145.
Erwin, D. 2000. Macroevolution is more than repeated rounds of microevolution. *Evolution & Development* 2 (2): 78–84.
Estes, S., and S. Arnold. 2007. Resolving the paradox of stasis: Models with stabilizing selection explain evolutionary divergence on all timescales. *American Naturalist* 169: 227–244.
Fisher, R. 1930. *The genetical theory of natural selection*. London: Oxford University Press.
Foote, M. 2003. Origination and extinction through the Phanerozoic: A new approach. *Journal of Geology* 111: 125–148.
Ford, E.B. 1975. *Ecological genetics*. London: Chapman and Hall.
Frank, S.A., and M. Slatkin. 1992. Fisher's fundamental theorem of natural selection. *Trends in Ecology and Evolution* 7:92–95.
Gavrilets, S. 1999. A dynamical theory of speciation on holey adaptive landscapes. *American Naturalist* 154: 1–22.
Gayon, J. 1998. *Darwinism's struggle for survival: Heredity and the hypothesis of natural selection*. Trans. M. Cobb. Cambridge, MA: Cambridge University Press.
Gillespie, J. 2004. *Population genetics*. New York: Oxford University Press.
Gould, S.J. 1977. *Ontogeny and phylogeny*. Cambridge, MA: Harvard University Press.
———. 1989. *Wonderful life. The Burgess shale and the nature of history*. New York: Norton.
———. 2002. *The structure of evolutionary theory*. Chicago: University of Chicago Press.
Gould, S.J., and N. Eldredge. 1977. Punctuated equilibria: The tempo and mode of evolution reconsidered. *Paleobiology* 3: 115–151.
Gould, S.J., and E. Lloyd. 1999. Individuality and adaptation across levels of selection: How shall we name and generalize the unit of Darwinism? *PNAS, Proceedings of the National Academy of Sciences* 96: 11904–11909.
Grafen, A. 2007. The formal Darwinism project: A mid-term report. *Journal of Evolutionary Biology* 20: 1243–1254.
Grantham, T. 2007. Is macroevolution more than successive rounds of microevolution ? *Palaeontology* 50 (1): 75–85.
Hansen, T., and D. Houlé. 2004. Evolvability, stabilizing selection, and the problem of stasis. In *Phenotypic integration: Studying the ecology and evolution of complex phenotypes*, ed. M. Pigliucci and K.A. Preston, 130–150. Oxford: Oxford University Press.
Hastings, A., and J. Melbourne. 2008. Extinction risk depends strongly on factors contributing to stochasticity. *Nature* 454: 100–103.

Hubbell, S.P. 2001. *The unified neutral theory of biodiversity and biogeography*. Princeton University Press: Princeton.

———. 2009. The neutral theory of biodiversity and biogeography and Stephen Jay Gould. *Paleobiology* 31 (2): 122–132.

Huneman, P. 2010a. Assessing the prospects for a return of organisms in evolutionary biology. *History and Philosophy of the Life Sciences* 32 (2/3): 341–372.

———. 2010b. Topological explanations and robustness in biological sciences. *Synthese* 177: 213–245.

———. 2014a. Formal Darwinism and organisms in evolutionary biology: Answering some challenges. *Biology and Philosophy* 29: 271–279.

———. 2014b. Selection. In *Handbook of evolutionary thinking in the sciences*, ed. T. Heams, P. Huneman, G. Lecointre, and M. Silberstein. Dordrecht: Springer.

———. 2015. Inscrutability and the opacity of selection and drift: Distinguishing epistemic and metaphysical aspects. *Erkenntnis* 80: 491–518.

———. 2017. Why would we call for a new evolutionary synthesis? The variation issue and the explanatory alternatives. In *Challenging the modern synthesis: Development, adaptation and inheritance*, ed. P. Huneman and D. Walsh. New York: Oxford University Press.

Huneman, P., and D. Walsh. 2017. *Challenging the modern synthesis: Development, adaptation and inheritance*. New York: Oxford University Press.

Hunt, G., and M. Carrano. 2010. Models and methods for analyzing phenotypic evolution in lineages and clades. *Paleontological Society Papers* 16: 245–269.

Hunt, G., M. Hopkins, and S. Lidgard. 2015. Simple versus complex models of trait evolution and stasis as a response to environmental change. *Proceedings of the National Academy of Sciences of the United States of America* 112: 4885–4890.

Huss, J. 2004. *Experimental reasoning in non-experimental science*. PhD Dissertaton, The University of Chicago.

———. 2009. The shape of evolution: The MBL model and clade shape. In *The paleobiological revolution*, ed. M. Ruse and D. Sepkoski, 326–345. Chicago: University of Chicago Press.

Jablonka, E., and G. Raz. 2009. Transgenerational epigenetic inheritance: Prevalence, mechanisms, and implications for the study of heredity and evolution. *Quarterly Review of Biology* 84: 131–176.

Jablonski, D. 2000. Micro- and macroevolution: scale and hierarchy in evolutionary biology and paleobiology. *Paleobiology* 26(Suppl. to No. 4): 15–52.

———. 2001. Lessons from the past: Evolutionary impacts of mass extinctions. *Proceedings of the National Academy of Sciences of the United States of America* 98: 5393–5398.

———. 2004. Extinction: Past and present. *Nature* 427: 589.

———. 2005. Mass extinctions and macroevolution. *Paleobiology* 31(Suppl. to No. 2): 192–210.

———. 2007. Scale and hierarchy in macroevolution. *Palaeontology* 50: 87–109.

———. 2008a. Species selection: Theory and data. *Annual Review of Ecology, Evolution, and Systematics* 39: 501–524.

———. 2008b. Biotic interactions and macroevolution: Extensions and mismatches across scales and levels. *Evolution* 62: 715–739.

———. 2009. Paleontology in the twenty-first century. In *The paleobiological revolution*, ed. M. Ruse and D. Sepkoski, 471–517. Chicago: University of Chicago Press.

Jablonski, D., and J.J. Sepkoski. 1996. Paleobiology, community ecology, and scales of ecological pattern. *Ecology* 77: 1367–1378.

Kaplan, J. 2008. The end of the adaptive landscape metaphor? *Biology and Philosophy* 23: 625–638.

———. 2009 The paradox of stasis and the nature of explanations in evolutionary biology. *Philosophy of Science* 76 (5): 797–808.

Kauffman, S. 1993. *The origins of order: Self-organization and selection in evolution*. New York: Oxford University Press.

Kettlewell, H.D.B. 1955. Selection experiments on industrial melanism in the *Lepidoptera*. *Heredity* 9: 323–342.
Kimura, M. 1983. *The neutral theory of molecular evolution*. Cambridge: Cambridge University Press.
Kirschner, M., and J. Gerhart. 2005. *The plausibility of life: Resolving Darwin's Dilemma*. New Haven: Yale University Press.
Laland, K., T. Uller, M. Feldman, L. Sterelny, G.B. Müller, A. Moczek, E. Jablonka, and J. - Odling-Smee. 2014. Does evolutionary theory need a rethink? Yes: Urgently. *Nature* 514: 161–164.
Lande, R. 1988. Genetics and demography in biological conservation. *Science* 241: 1455–1460.
Lande, R., and S.H. Orzack. 1988. Extinction dynamics of age-structured populations in a fluctuating environment. *PNAS* 85: 7418–7421.
Lehmann, L., and F. Rousset. 2014. Fitness, inclusive fitness and optimization. *Biology and Philosophy* 29: 181–195.
Lenski, R., and M. Travisano. 1994. Dynamics of adaptation and diversification: A 10,000-generation experiment with bacterial populations. *PNAS* 91: 6808–6814.
Lynch, M., and R. Lande. 1993. Evolution and extinction in response to environmental change. In *Biotic interactions and global change*, ed. P. Kareiva, J. Kingsolver, and R. Huey, 234–250. Sunderland: Sinauer.
Mandelbrot, B. 1997. *Fractals and scaling in finance: Discontinuity, concentration, risk*. New York: Springer.
Matthen, M. 2009. Drift and "Statistically abstractive explanations". *Philosophy of Science* 76: 464–487.
Maynard Smith, J., R. Burian, S. Kauffman, P. Alberch, J. Campbell, B. Goodwin, R. Lande, D. Raup, and L. Wolpert. 1985. Developmental constraints and evolution. *Quarterly Review of Biology* 60: 265–287.
Mayr, E. 1963. *Animal, species and evolution*. Cambridge, MA: The Belknap Press at Harvard University Press.
———. 2001. *What evolution is*. New-York: Basic Books.
Mc Shea, D. 1994. Mechanisms of large-scale evolutionary trends. *Evolution* 48 (6): 1747–1763.
———. 2005. The evolution of complexity without natural selection: A possible large-scale trend of the fourth kind. *Paleobiology* 31 (2): 146–156.
Millstein, R.L. 2002. Are random drift and natural selection conceptually distinct? *Biology and Philosophy* 17 (1): 33–53.
———. 2009. Concepts of drift and selection in 'The Great Snail Debate' of the 1950s and early 1960s. In *Descended from Darwin: Insights into the history of evolutionary studies, 1900–1970*, ed. J. Cain and M. Ruse, 271–298. Philadelphia: American Philosophical Society.
Müller, G., and S. Newman 2005. The innovation triad: An evo-devo agenda. *Journal of Experimental Zoology* 304(B6): 487–503
Nitecki, M., and A. Hoffman. 1987. *Neutral models in biology*. Oxford: Oxford University Press.
Pagel, M., C. Venditti, and A. Meade. 2006. Large punctuational contribution of speciation to evolutionary divergence at the molecular level. *Science* 314: 119–121.
Pigliucci, M., and G. Müller. 2011. *Evolution: The extended synthesis*. Cambridge, MA: MIT Press.
Plutynski, A. 2007. Drift: A historical and conceptual overview. *Biological Theory* 2 (2): 156–167.
Raup, D. 1994. The role of extinction in evolution. *Proceedings of the National Academy of Sciences of the United States of America* 91 (15): 6758–6763.
Raup, D., and J. Sepkoski. 1984. Periodicity of extinctions in the geologic past. *Proceedings of the National Academy of Sciences of the United States of America* 81 (3): 801–805.
Raup, D.M., S.J. Gould, T.J.M. Schopf, and D. Simberloff. 1973. Stochastic models of phylogeny and the evolution of diversity. *Journal of Geology* 81: 525–542.
Ridley, M. 2004. *Evolution*. Cambridge: Blackwell.

Simpson, G.G. 1944. *Tempo and mode in evolution. Columbia classics in evolution*. Reprint ed. New York: Columbia University Press.
Solé, R. 2002. Modelling macroevolutionary patterns: An ecological perspective. In *Lecture notes in physics 585*, ed. M. Lassig and A. Valleriani, 312–337. Berlin: Springer.
Stanley, S.M., and X. Yang. 1987. Approximate evolutionary stasis for bivalve morphology over millions of years: A multivariate, multilineage study. *Paleobiology* 13 (2): 113–139.
Sterelny, K. 2007. *Dawkins Vs Gould: Survival of the fittest*. Cambridge: Icon Books.
Turner, D. 2015. Historical contingency and the explanation of evolutionary trends. In *Biological explanation: An enquiry into the diversity of explanatory patterns in the life sciences*, ed. C. Malaterre and P.A. Braillard, 73–90. Dordrecht: Springer.
Valentine, J.W., and D. Jablonski. 2003. Morphological and developmental macroevolution: A paleontological perspective. *International Journal of Developmental Biology* 47: 517–522.
Valentine, J.W., D. Jablonski, and D.H. Erwin. 1999. Fossils, molecules and embryos: New perspectives on the Cambrian explosion. *Development* 126: 851–859.
Wake, D.B. 1991. Homoplasy: The result of natural selection, or evidence of design limitations? *The American Naturalist* 138: 543–567.
Walsh, D.M. 2015. *Organisms, agency, and evolution*. Cambridge: Cambridge University Press.
Wilkins, J., and P. Godfrey-Smith. 2009. Adaptationism and the adaptive landscape. *Biology and Philosophy* 24: 199–214.
Williams, G.C. 1992. *Natural selection: Domains, levels and challenges*. Oxford: Oxford University Press.
Williamson, P.G. 1981. Palaeontological documentation of speciation in Cenozoic molluscs from Turkana Basin. *Nature* 293: 437–443.
Winther, R.G. 2006. Fisherian and Wrightian perspectives in evolutionary genetics and model-mediated imposition of theoretical assumptions. *Journal of Theoretical Biology* 240: 218–232.
Wray, G.A., H.E. Hoekstra, D.J. Futuyma, R.E. Lenski, T.F.C. Mackay, D. Schluter, and J.E. Strassmann. 2014. Does evolutionary theory need a rethink? No, all is well. *Nature* 514: 161–164.
Wright, S. 1932. The roles of mutation, inbreeding, crossbreeding and selection in evolution. *Proceedings of the sixth annual congress of genetics* 1: 356–366.

Chapter 15
The Timing of Development

Antonine Nicoglou

Abstract The question of whether it is possible to fit together the developmental and evolutionary explanations raises a number of difficulties. In a sense, it is possible to consider that the problems concerning the development of the individual have nothing to do with those related to the evolution of organisms over time (Wallace B, Can embryologists contribute to an understanding of evolutionary mechanisms? In: Bechtel W (ed) Integrating scientific disciplines: case studies from the life sciences. Springer. pp 149–163. Retrieved from http://link.springer.com/content/pdf/10.1007/978-94-010-9435-1_9.pdf, 1986). If one describes the development as the temporal trajectory of an individual from the zygote to adult, then the timescale of the individual development appears to be radically different from the evolutionary time scale. This chapter aims to show that the time dimension is an essential element to explain the proximal mechanism of development, and that it remains unspecified if not still largely ignored by biologists. I suggest that by focusing on the characters rather than on the "developmental stages", developmental biology, while approaching evolution, nonetheless and paradoxically lost sight of the actual temporal dimension its process (Beer, G.R. (de) 1930. *Embryology and evolution*. Gloucestershire: Clarendon Press; Hamburger V, Hamilton HL, J Morphol 88(1):49–92, 1951). Therefore, consideration and characterization of the timing of development remain to be done: it requires to analyze its peculiarities and the way they have been, or may be apprehended. This focus on the developmental time will allow us to emphasize the importance of time for the explanation of the developmental process.

15.1 Introduction

The issue of whether and how a gathering of developmental and evolutionary explanations should be achieved raises difficulties. In 1986, biologist Bruce Wallace argued, in a chapter entitled "Can embryologists contribute to an understanding of evolutionary mechanism?" in the book *Integrating Scientific Disciplines*,

A. Nicoglou (✉)
CRPMS, Institut Jacques Monod & IHPST, Paris, France
e-mail: antoninenico@gmail.com

C. Bouton, P. Huneman (eds.), *Time of Nature and the Nature of Time*, Boston Studies in the Philosophy and History of Science 326,
DOI 10.1007/978-3-319-53725-2_15

edited by William Bechtel, that "problems concerned with the orderly development of the individual are unrelated to those of the evolution of organisms through time" (Wallace 1986, p. 149). Wallace opposed two kinds of scientists: those who deal with individuals, their appearance or their morphologies – that he called the "developmentalists" – and those who deal with genetic programs, and their frequencies – that he called the "geneticists." The two camps were wrestling for control of the word "evolution" in biology, and this led, basically, to two opposite and incommensurable views on evolutionary biology: the first one emphasizing that "evolution was a matter of orchids, butterflies, warm bloodedness, mating systems, and so on" and the second one emphasizing "genetic programs whose somatic products either interact sufficiently adaptively with the prevailing environment or fail in death or sterility" (Wallace 1986, p. 162).

If Wallace's distinction sounds today, with the rise of evolutionary developmental biology's studies (or Evo-Devo to its practitioners, whose aim was to gather data from molecular developmental biology and evolutionary biology), rather old-fashioned or even wrong to many biologists and philosophers (e.g., Bateson et al. 2011; Laland et al. 2008; West-Eberhard 2003), his position was somewhat reminiscent of Mayr's own distinction between the "functional biologist" – who was "vitally concerned with the operation and interaction of structural elements, from molecules up to organs and whole individuals" (equivalent to Wallace's "developmentalist") – and the "evolutionary biologist" – who was considering that "every organism, whether an individual or a species, is the product of a long history, a history that dates back more than 2000 million years" (Mayr 1961, p. 360) (equivalent to Wallace's "geneticist," since, for Wallace, the geneticist was the evolutionary biologist interested in the long term evolutionary story).

To these two approaches, Mayr associated two types of causes involved in the evolutionary explanation in each case: respectively, the *proximate* and the *ultimate* causes. He suggested that each biological phenomenon could be explained through sets of causes. However, none of the two types of causes – proximate or ultimate – subsumed the other. Wallace took Mayr's idea a step further by suggesting that there was a kind of *incommensurability*[1] between developmentalist and geneticist explanations of biological phenomena, because the two kinds of explanations were "equally valid" although "incompatible" (Wallace 1986, p. 162). Such a view led him to the idea that developmentalist and geneticists pursued different goals. Moreover, one substantial outcome of Mayr's thesis was that evolutionary biology

[1]Wallace did not explicitly referred to the term "incommensurability" in the article, but to the term "incompatibility." However, the picture he uses to illustrate his argument (those of a well-known analogy of an optical illusion, an example used by psychologists: a sketch, where you might see in one moment an old woman, in the next a beautiful young lady, but at no time both images simultaneously) implicitly features incommensurability (see Wallace 1986, p. 163).

[2]See Ricqlès', Chap. 11 in this volume.

alone was genuinely "biological" and thus lied with biological time (the remainder – including developmental biology – being only physics!).[2]

Therefore, concerning ultimate causes, Mayr indicated: "These are causes that have a history [. . .]" (Mayr 1961, p. 363), suggesting at the same time that it was not the case of proximate causes. If we take the hypothesis of incommensurability between developmental and evolutionist[3] explanations of biological phenomena seriously, and if we see time as a decisive factor only in the characterization of ultimate causes (at the basis of evolutionist explanations), then, the characterization of proximate causes (at the basis of developmental explanations) may appear independent of time.

In fact, after Mayr's distinction, proximate causes have often been detached from any historical, or even, any temporal context in biology. Proximate causes have often been compared to any other reproducible physico-chemical mechanism, since the rules governing these phenomena are not dependent on time. In the same vein, Block von Wülfingen et al. have recently argued in a special issue on "temporalities of reproduction" (2015) that Evelyn Fox Keller, in her book *Refiguring Life: Metaphors of Twentieth-Century Biology* (1996), showed that the temporal regime of molecular biology (a disciplinary field devoted to proximate causes analysis) was that of eternity. Therefore, since the mid-1980s, the understanding of molecular mechanisms has highly changed the way to apprehend (or rather to not apprehend) temporality in developmental biology.

In the present chapter, I will challenge the assumption concerning the idea that, even when development would be depicted as a process in time, time remains less significant from an epistemic point of view in the study of development, compared to the study of evolution. I will argue that, in development, time might play a greater role than only be a variable in physic or chemistry equations. I will show that taking into account the *timing of development* (i.e., the variations in form/manifestation that time can take during development) might be as much significant as taking into account proximal causes (e.g., genetic or molecular mechanisms) when trying to explain developmental process. It also means that, in such a view, time would not necessarily be limited to a proximate cause (as it is sometimes depicted), and could nevertheless be seen as a determinant factor in explaining development. Furthermore, I will show that time disappeared from developmental explanations approximately when the discipline of Evolutionary Biology distanced itself from those of Developmental Biology and that reassessing the significance of time in explaining development could actually help better understand evolutionary biology processes with development.

In the foreword of a book published in 1993 and entitled *Developmental Time and Timing*, Gilbert Gottlieb argued:

---

[3]We distinguish "evolutionist explanations" with "evolutionary explanations". "Evolutionist explanations" are linked to the geneticists' view on biological explanations, whereas "evolutionary explanations" include both geneticists and developmentalists' view on these same biological explanations.

> It remains for the future to determine if timing can be seen as truly causal [. . .]. Whether the explicit use of cause itself creates more difficulties than benefits for developmental thinking is also a problem for consideration, especially when we need to avoid a sources-of-variation framework for empirical as well as conceptual analysis. Traditional causal thinking can lead the unwary to perhaps overrate the influence of a single variable rather than to recall the coaction of multiple influences that is the hallmark of developing systems. On the other hand, causal-analytic thinking does render manageable and sensible what might otherwise seem unmanageable and chaotic (and thus immune to analysis and understanding) [a criticism often addressed to the developmental systems biology way of thinking]. (Turkewitz and Devenny (eds.) [1993], p. x)

The present chapter provides an attempt to show in which way the analysis of the timing of development may be a specific factor for a renewed explanatory framework of developing systems without necessarily referring either to the sole traditional causal thinking ("I explain because I found the causes") or to the often "unmanageable and chaotic" view of the developmental systems biology.[4] In this regard, I discuss the basis of Mayr's distinction (between proximate and ultimate causes), while not necessarily denying Wallace's challenge (the actual distinction between developmental and evolutionary explanations).

If doubts remain as to the meaning of a distinction between – lets say – "evolutionist explanations" and "developmental explanations," a certain heuristic value of such a distinction is granted – at least to start defining and understanding explanatory patterns in biology (e.g., see Beatty 1994, p. 333; Amundsen 2001; Calcott, thesis for arguments 2006). For instance, it can be argued that evolutionary and developmental timescales can be decoupled and described on their own. Indeed, since development can be depicted as the trajectory of an individual from the zygote stage to the adult stage, in a process in time, its timescale, at least from this standpoint, appears clearly decoupled from the evolutionary timescale which requires several (at least two) generations (e.g., Reiss 2003; Hall 1999). Furthermore, the developmental process, itself, may include various processes/mechanisms at distinct time- (e.g., developmental stages, developmental events, etc.) and space- scales (e.g., molecular, cellular, etc.), which can be further analyzed on their own. Therefore, the "timing of development" reflects both the specificity of the developmental time compared to the evolutionary time, and the manifestations (i.e. duration and shape) time may take during development.

Once it is assumed that developmental timescale might be decoupled from evolutionary timescale, the heuristic value of analyzing time in a developmental/

[4]Developmental systems biology uses computer simulation of multicellular development as a research methodology to understand the function of the very complex processes involved in the development of organisms. This includes simulation of cell signaling, multicellular interactions and regulatory genomic networks in development of multicellular structures and processes (e.g., French flag model by L. Wolpert 1969).

proximate perspective can be claimed. If space scales have being highly analyzed, at least in a mechanistic perspective by biologists (e.g., Davidson 1991) and philosophers (mainly through the description of the different biological levels of organization, e.g., Craver and Darden 2013), the analysis of timescales, by themselves, have been overlooked.

In order to identify and analyze timescales in development, I first focus on the paradigmatic notion in biology of "developmental stage", (each of the periods of cell division and fetal development or, more broadly, each of the periods of the development cycle of an organism) which has been widely used by "developmentalists" since the beginning of descriptive embryology in the early nineteenth century and remains quite significant in practice to characterize model organisms. The description of each stage is linked to a more or less precise time (i.e. a duration) and such time relies a lot on external conditions (i.e. temperature). Therefore a close attention to the notion of "developmental stage" in the literature will help me to understand, first, why depth analyses of "the timing of development" have slowly disappeared from developmental studies, and, second, to see how a renewed definition of "developmental stage" based on our past and current knowledge could help to offer a new account of time in development.

In Sect. 15.2, I analyze when and why the attention has shifted from the notion of "developmental stage" to those of "character"[5] in developmental biology. The shift – that emerged in the renewed work of Gavin de Beer on heterochrony in the 1940s – led, in the 1970s–1980s, both to the gradual disappearance of time in the explanation of development and to the rise of correlated studies in developmental and evolutionary biology (Gould 1977, 1988; Alberch et al. 1979). In this section, my argument is mainly that since the first attempts for a synthesis between data in embryology, genetics and evolution the explanation of developmental stages, through the notion of heterochrony, has relied on underlying causal mechanisms; time being only seen as a general framework without being further detailed.

Section 15.3 looks at how time is currently analyzed in developmental explanations. With this purpose, I propose a renewed characterization of time in development by showing how one can identify, and distinguish different timings (duration and shape aspects of time) within developmental processes. I argue that the renewed characterization of time in development suggests the necessity to refer to a concept of time, proper to development. I, then show how this concept of time paves the way for an explanatory framework different from those of traditional causal thinking or those of developmental systems biology.

Finally, Sect. 15.4 summarizes the arguments of the chapter and draws some conclusions about the explanatory role of the timing of development for evolutionary biology.

---

[5]Notion used in biology to characterize the traits or set of traits that distinguish one species, one family, one plant/animal from another. The set of observable characters from an organism is its phenotype, which is both due to hereditary factors (genotype) and changes in the environment.

15.2 The Mismeasure of Time in Development

15.2.1 From Developmental Stages to Characters

In order to understand the reasons why, since the mid-twentieth century, the analysis of time have been overlooked in comparison to the analysis of characters (and their causes) in developmental biology (e.g., Hamburger and Hamilton 1951), one has to look at the genealogy of the discipline of developmental biology and especially at its relationships with other disciplines such as evolutionary biology and genetics. Developmental biology is probably one of the oldest disciplines of biology, rooted in embryology, but it is also the discipline that has received more recently (from the 1980s) further major strides. In some ways it came into existence as a science in the eighteenth century with the works of Caspar Friedrich Wolff[6] (1774) in descriptive embryology. In the beginning of the twentieth century it was revolutionized by the rise of genetics (1900) and even more, from the 1940s, by molecular biology. The turning point in the study of time in development – which resulted in a shift from the notion of "developmental stage" to the notion of "character" – has to be understood in this specific historical framework.

In 1930, Gavin de Beer, a British zoologist who first compiled data from embryology, genetics and evolutionary biology thought that "character's inheritance" – and not "developmental stages comparison among species" – was the essential factor to draw a link between ontogeny and phylogeny[7] (De Beer 1940, p. 6–7). At the same period he also provided the first critic against Ernst Haeckel's theory of recapitulation, which had been important in the discussions for almost a century. In his theory, Haeckel stipulated "ontogeny recapitulates phylogeny": he claimed that an individual organism's biological development (its ontogeny), parallels and summarizes its species' evolutionary development (its phylogeny) (Haeckel 1866). In other words, he considered that during its development, the individual organism goes through different stages, which correspond to the adult stages of all his ancestors. For instance, gill slits that are transitory in human embryos would be an adult character in a certain fish, which was itself an ancestor of hominids (Schmitt 2004, p. 43).

The notion of stage has a long history in developmental biology that almost started with the beginnings of embryology. Since Haeckel, the establishment of stages has been associated to the possibility of easy comparisons across taxa. Haeckel was the first one who described gastrulation – one of the most famous embryonic stages in animal development – when the single-layered embryo (the blastula) becomes reorganized into a tree-layered embryo (the gastrula at the origin of the different cell and tissue types). This idea has to be associated with the notion

[6]Caspar F. Wolff is known as the "father of embryology" (see Needham 1959).

[7]"Ontogeny" and "phylogeny" were the terms used (mainly before Darwin's theory of evolution by the means of natural selection) to characterize and distinguish the developmental process from the evolution of species process.

of "layers" (the different primary tissues of the embryo) that Pander first identified and described in 1818. The theory of recapitulation dates back almost from the beginnings of embryology since it was already formulated by C. F. Wolff's students, Johann F. Meckel and Étienne Serres in the 1810s and 1820s in the form of a law (it is known under the name of "Meckel-Serres law"[8]). It was still quite distant from the theory of evolution as Darwin formulated it in 1859.

De Beer offered one of the first formal critics of Haeckel's theory by rejecting some of its principles. One of his critics was on the notion of "stages" itself. De Beer argued that instead of focusing on the "stages," or on any arbitrary sequence of time in each of the processes (ontogeny or phylogeny), one should rather focus on the "characters" of the developing organisms independently of their timing of appearance. De Beer's critic was based on one specific aspect of Haeckel's theory: Haeckel's attempt to introduce the temporal dimension in order to unify the problem of the hierarchy of repeated parts – the fact that more complex parts arose after more simple ones – and those of recapitulation – the fact that same morphological stages appeared in different species (see Schmitt 2004, p. 296). The temporal dimension was essential for Haeckel because he could then show that not only the organism recapitulates the whole organic universe during its development (what was the tacit acceptation of the term "recapitulation" and what de Beer was actually rejecting), but also that the adult organism recapitulates itself in each of its different parts (the idea that Meckel-Serres's law had introduced). Haeckel's view then led both to a temporal and a spatial recapitulation (Schmitt 2004, p. 84).

De Beer expressed objections against Haeckel's attempt, arguing:

> In the first place it was pointed out (by Agassiz, Keibel, and Mehnert among others) that the order in which characters appeared in phylogeny is not always faithfully reproduced in ontogeny. For instance, teeth were evolved before tongues, but in mammals now tongues develop before teeth. To this alteration and reversal of sequence of stages the term *heterochrony* is applied [...]. It [...] suggests that it is not legitimate to speak of a 'stage' being shifted back to a later or on to an earlier period in the life-history. It is not the 'stage' which is shifted *en bloc*, but certain characters which may be peculiar to the stage. It must also be remembered that what an embryologist calls a 'stage' is merely an arbitrarily cut section through the time-axis of the life of an organism. A 'stage' is thus really an abstraction of the four dimensional space-time phenomenon which a living organism is. (De Beer 1940, p. 6–7)

By focusing on the characters instead of the stages, de Beer called into question an "arbitrarily" periodization of development to be related to an "arbitrarily" periodization of evolutionary "stages" and preferred to focus on characters' appearance independently of their developmental stages. For de Beer, the stage was a

[8]Meckel-Serres law attempted to provide a link between comparative embryology and a "pattern of unification" in the organic world. They argued that fetal deformities result when development prematurely stops. These arrests characterized lower life forms, through which higher order organisms progress during normal development. The Meckel-Serres law stipulates that the embryos of higher order organisms progress through successive stages in which they resemble lower level forms. (http://embryo.asu.edu/pages/meckel-serres-conception-recapitulation, L. O'Connell 2013)

"section" arbitrarily cut along the lifespan time axis of an organism. Furthermore, he thought that if it was possible to have a mosaic or a modular view concerning organisms, it was not possible concerning the embryo. Therefore, he chooses to focus on organisms' characters and not on embryos' stages because it allowed him to draw a link between genetics, embryology and evolutionary biology thanks to a decoupling between the "characters' appearance" and the "characters' becoming". If embryology alone was dealing with "characters' becoming", the three disciplines – embryology, genetics and evolutionary biology – had to do with "characters' appearance".

In the same vein, Müller-Wille and Rheinberger (2012) have argued (formulated more recently by Bock von Wülfingen et al. 2015) that in the early twentieth century, "the emerging field of genetics advocated a concept of temporality that explicitly disregarded issues of 'history' (in the sense of past genealogies of traits) and was much more concerned with the future of a population. [...]" The authors interpreted this as a shift from a diachronic to a synchronic view, as a "shift from a focus on the vertical (genealogical or evolutionary) line of descendants towards a focus on the horizontal (genetic) relationship within population" (Bock von Wülfingen et al. 2015, p. 6–7). In such a view, it seems that the "character" became a *reference standard* to be used across all fields of biology, while the notion of "stage" fell in disuse. Such a shift, from the notion of stage to the notion of character, has also to be understood in a context where Darwin had already given a certain attention to characters with his theory of evolutionary biology through natural selection. But the emphasis on characters would mostly be reinforced, after Darwin, with the rise of genetics and the importance given to phenotypes (Johannsen 1911).

From the 1940s, embryology in combination with molecular genetics became "developmental biology", particularly with Salome Gluecksohn-Waelsch (1938, see Fig. 15.1) and Conrad Hal Waddington's (1940) works, in which they attempted to precise the origin of the mutations affecting some characters during the first development stages (Hall and Olson 2003).

From their different observations of modified developmental processes, they managed to induce the role of genes involved in the characters' expression (e.g., Fig.15.1: the expression of the tail of the mouse). Finally, with the rise of "geneto-developmental biology",[9] the study of characters' development through genetic studies became one of the most important fields of research in the discipline, whereas in the past, and from von Baer (1827), the study of developmental stages had been the research topic of first importance in embryology (see Abzhanov 2013 for a detailed review on this topic).

Indeed, in the early nineteenth century, von Baer had shown that general structural relations were likewise formed before the most specific appeared. This

[9]I refer to this notion of "geneto-developmental biology" instead of those of developmental biology to highlight the fact that developmental biology, from the beginning, has mainly focused on genetics and molecular biology and not so much on non-genetic processes of development.

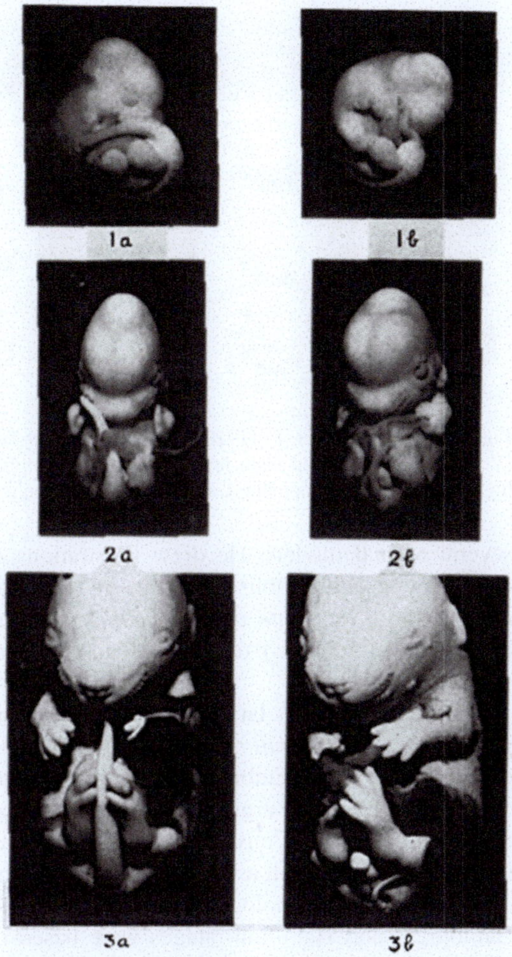

Fig. 15.1 From S. Gluecksohn-Waelsch, 1938, "The Development of Two Tailless Mutants in the House Mouse," the six pictures show either "normal embryo" (*a*), either "tailless embryo" (*b*) at respectively age 11 days in a lateral view (*1*), age 13 days in a ventral view (*2*) and age 15 days in a ventral view (*3*). In this paper, she identified the genes involved in forming the tail of the mouse embryo

became the main principle of what was then called "von Baer's laws": i.e. the specification from homogeneous to heterogeneous. The law explained that embryonic forms of highly developed species would repeat embryonic forms of less developed species. The principle was applied at the level of the whole organism and embryologists looked at which point every more complex structure appeared, calling it a stage. Finally, in the 1920s, the first videos of organism's entire embryonic development, and filmed in real time, exemplified the pinnacle of descriptive embryology and the focus on developmental stages.[10]

Before that, in 1895, Franz Keibel (1861–1929) completed the first important description of different morphological stages, based on analyses and comparison of

[10]For example, the following video shows the development of a salamander) https://en.wikipedia.org/wiki/File:Embryonic_development_of_a_salamander,_filmed_in_the_1920s.ogv.

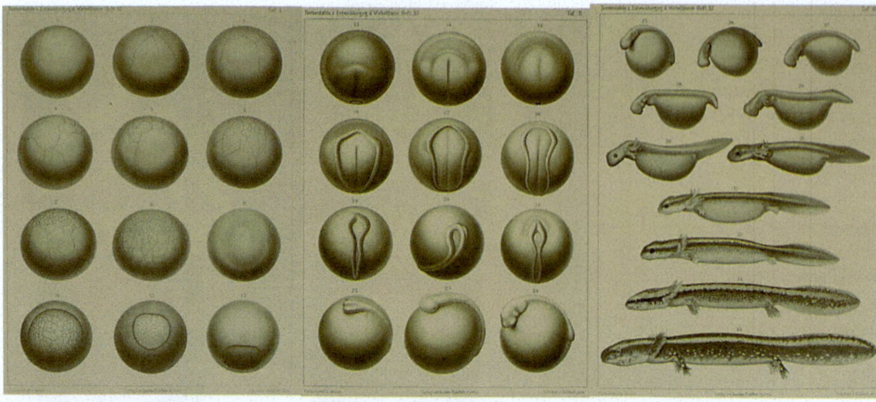

Fig. 15.2 From Keibel, F. (1910), *Normentafeln zur Entwicklungsgeschichte der Wirbelthiere*, (Vol. Hft.11). Retrieved from http://www.biodiversitylibrary.org/item/35263.Illustrations showing the different developmental stages of the common mudpuppy (*Necturus maculosus*)

several animal models. He drew illustrations of embryos at different stages of development and published them in the 16-volume text *Normentafeln zur Entwicklungsgeschichte der Wirbelthiere* (Normal Plates of the Developmental history of Vertebrates) (e.g., see Fig. 15.2 for the complete illustration of the common mudpuppy).

More than 50 years later, in 1951, Viktor Hamburger and Howard L. Hamilton published in the *Journal of Morphology* "A Series of Normal Stages in the Development of the Chick Embryo" claiming that an update of Keibel's illustrations was necessary. They were arguing that Keibel's illustrations remained incomplete and difficult to interpret. Conversely, in their illustrations, some "visible features"[11] of the chick embryo (such as the growth of different limbs: legs and/or wings) helped to assign the "morphological stages". Indeed, in contrast with Keibel' stages, Hamburger and Hamilton' stages were based on some specific growth and size changes of the characters (see Fig. 15.3), and not on an arbitrary periodization of the development of the whole organism, defined in hours or days (see Fig. 15.2).

Therefore, Hamburger and Hamilton initiated a tradition where developmental stages' definition relied on *visible* changes in the *growth or size* of precise characters instead of on a periodization of the development of the whole organism over time.

First, the conceptual and epistemological shift from "developmental stages" to "characters" helps to explain how developmental biology moved closer to evolutionary biology (Amundson 2005, p. 249). When de Beer published his book *Embryos and Ancestors*, in 1930 appeared also the idea that evolutionary change

[11] Adam R. Navis (2007) has recently shown that: "the main requirements of [Hamburger & Hamilton's] project were that the stage had to be easily identifiable by visible features, and they required the smallest possible differences between the features to avoid confusion" (Concerning this question see more at: http://embryo.asu.edu/pages/stages-chick-development#sthash.leTBR20X.dpuf).

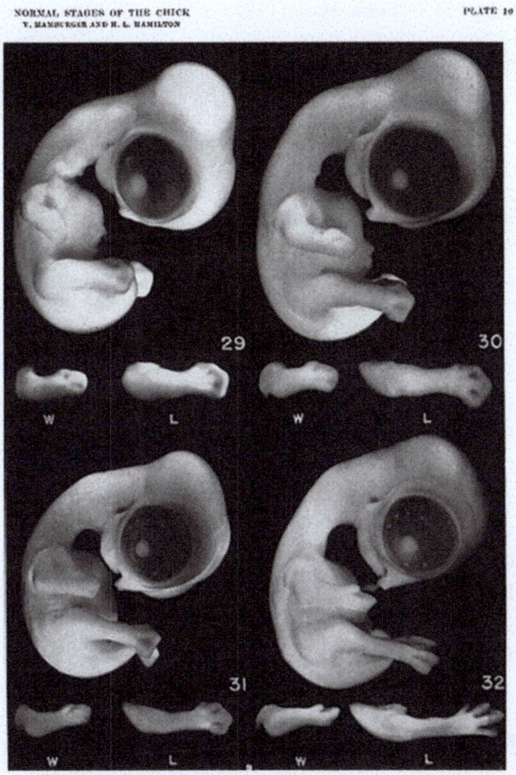

Fig. 15.3 From Hamburger and Hamilton (1951), "A Series of Normal Stages in the Development of the Chick Embryo." Plate showing chick embryos at stages 29 to 32, and, more specifically, the progressive development of their wings and legs (the characters used to analyze the progressive development of the embryos)

could take place in some body parts (could affect some characters) without simultaneously inducing changes in other parts (or characters). This principle, called "mosaic evolution", put a certain emphasis on the character-centered view of evolution – also linked thereafter to a gene-centered view. Therefore, by having a character-centered view in development, followed then by a gene-centered view, developmental biology came closer to evolutionary biology.

Second, this shift (from "developmental stages" to "characters") helps also to explain why the analysis of development timing has turned to the study of growth characters. It appeared more convenient, and more precise, to compare the size and the growth of morphological characters rather than the temporal sequences of developmental processes within/between species, but also within/between closed families (see Table 15.1).

However, the reasons of the epistemological shift – from stages to characters – do not entirely explain why developmental biology has lost track of its temporal dimension. To understand this aspect, I will now focus on the modifications of the notion of heterochrony (which has to do with the "change in the sequence of developmental events over time") from its first use by Haeckel in the latter half of the nineteenth century until its most recent use in biology.

Table 15.1 From Hall (1999, p. 371). Table comparing relative duration of various early developmental processes as measured by t_o, the "biological time" itself determined by the duration in hours of one cell cycle based on the synchronized early phase of cleavage divisions in fishes and amphibians. It illustrates how difficult it might be to have to deal with temporal sequences when trying to compare different species (what evolutionary biology generally seeks to achieve when comparing developmental processes)

Period	Fishes	Amphibians
Fertilization to first cleavage furrow	3	3
Fertilization to fall in mitotic index	12–14	
First cleavage furrow to fall in mitotic index	14–20	10–26
Fall in mitotic index to gastrulation	6–16	12–27
Fertilization to onset of RNA synthesis	10–18	
Fertilization to onset of gastrulation	18–29	
Fertilization to 10 pairs of somites	50–61	
Fertilization to end of epiboly	35–66	
Gastrulation to slit-like blastopore	18–19	12–25
Slit-like blastopore to fusion of neural folds	7[a]	12–21
Gastrulation to ten pairs of somites	28–34	
Gastrulation to end of epiboly	17–37	
Fertilization to fusion of neural folds	45[a]	47–67

Based on data in Dettlaff et al. (1987) for 7 species of fishes and 8 species of amphibians.
[a]Based on single species, *Acipenser güldenstädtii*

15.2.2 *Heterochrony Among the Ages*

Looking back at the notion of heterochrony will help us to understand what scientists meant to express with this notion, and what has change or has been progressively lost with the raise of developmental biology. In other words, I will track, thanks to heterochrony, what happened to the issue of time in development.

In 1866, Heackel referred to the notion of heterochrony to call attention to the fact that "displacement in time in the germ-history" (i.e. one event occurring before another, or one event occurring later than it was supposed to occur…) was "a gradual disarrangement in the original phylogenetic sequence, caused by embryonic adaptation" (Haeckel 1866). Through the notion of heterochrony ("*Heterochronie*") Haeckel introduced a temporal dimension in the study of morphogenesis.

Already before him, in 1858, Rudolph L. Virchow[12] (1821–1902) had referred to the notion of "*heterochronia*" in a book on cellular pathology where he analyzed the different abnormalities caused by the disarrangement of tissues either in space or time. He called it his "doctrine of the heterology". Virchow used the term "*heterochronia*" to describe the situation when a tissue is seen as pathological

[12]Virchow was, in the mid-nineteenth century, at the origin of the cellular theory.

because it appears at a different time of when it is usually observed – "*an aberration temporis*" (Virchow 1860, p. 63–64). For Virchow, the term "heterochronia" was used to characterize a modification in the usual timing of tissue ontogeny. The comparison was between pathological and normal tissues development timing, and not between ontogenetic and phylogenetic timings. Virchow was Haeckel's professor of histology and was the one who had introduced him to the notion of heterochrony.

On his side, Haeckel's morphology was based on a will, even an obsession – current at that time[13]: to establish the "unity of life," and thus, to discover the relationships and parallels behind the apparent diversity of organic forms (Schmitt 2004, p. 42). Haeckel used the notion of heterochrony ("*heterochronie*") slightly differently from Virchow because he had different purposes. He embedded the term of heterochrony – usually associated with ontogeny and more precisely with the displacement of morphogenic stages – "within an explicitly phylogenetic paradigm" (Hanken 2015, p. 98).

By the early twentieth century, Haeckel's view became untenable to many leading embryologists because of the recapitulation doctrine. However, some of his critics, like de Beer in 1930 and Stephen Jay Gould in the late 1970s, reformulated the notion of heterochrony and gave it a new impetus with their respective classifications of evolutionary patterns based on different types of heterochronic phenomena.[14] Both of their analyses have been extremely influential for the current understanding biologists have of the notion of heterochrony (concerning this claim, see Smith 2001; Hanken 2015).

In 1930, de Beer published *Embryos and Ancestors*, a book on embryology and evolutionary biology – that we previously mentioned – which received attention from his contemporaries. The book was reissued in 1940 with few modifications. It is in this book that the conceptual shift from "stages" to "characters" I analyzed above appears. But what I didn't mention yet is that the shift had to do with another change: from a Haeckelian heterochrony to a "deBeerian heterochrony" (e.g., Raff and Wray 1989), a change that I will now describe in further detail.

In de Beer's view, heterochrony was not simply strictly related to a change in the timing of morphogenic stages, but it was, more broadly, related to any change in any temporal sequence be it developmental or along evolutionary process. Moreover, in the early beginnings of genetics, de Beer attempted not only to link ontogeny with phylogeny (as Haeckel tried before him),[15] but he also wanted to connect embryogenesis and evolutionary biology to genetics. Around the same period, in 1932, evolutionary biologist Julian Huxley explored in his *Problems of Relative Growth* the implications of changes in developmental timing in a context

[13]In the *Naturphilosophie*, a philosophical trend initiated by German philosophers (Goethe, Kant, Schelling, Hegel. . .) in the second half of the eighteenth century which influenced Haeckel's theorization (see Schmitt 2004).

[14]See Huneman's Chap. 14 in this volume.

[15]See Tassy's Chap. 12 in this volume.

where genetics and evolutionary biology were associated. Huxley saw "rate-genes" (this is the name given to the "controlling genes" that determine the expression of a character) as possible regulators of differential growth and, hence, of morphological differentiation over time. Therefore he "used the concept [of rate genes] in an explanation of how the appearance of recapitulation could be caused by ordinary genetic factors, if these included genes that determine rates of physiological processes" (Amundson 2005, p. 192). De Beer, in turn, paid attention, not only to the specific timing of expression of the genes during embryogenesis, but also to their rate of expression. However, de Beer had a different concept of rate genes as "genes that changes their rates". Going from tissue- or structure-scale to molecular-scale, he tried to draw a link between the timing of expression of a structure and the rate of action of the genes. De Beer argued:

> [...] by acting at different rates, the genes can alter the time at which certain structure appear [during ontogeny]. This conclusion is of considerable interest, for it enables us to see how heterochrony can take place. It may be imagined that a character A was evolved in phylogeny before a character B, but there is no reason why in ontogeny character B should not arise before, simultaneously with, or after, character A, according to the relative speeds of the genes at work [for the expression of characters during ontogeny], and the environmental conditions. A character which appeared in the adult stage of the ancestor might develop early in the ontogeny of the descendant, and vice versa. (de Beer 1940, p. 18)

Therefore, with de Beer, heterochrony was not anymore related to the change in the timing of expression of morphogenic stages – as it had been the case with Haeckel –, but to the effects of the change in the speed of expression of the genes during ontogeny, analyzed through character's expression changes. These character's expression changes were at the basis of the establishment of categories of evolutionary patterns. De Beer's offered "eight possibilities of variation in the way in which structure [or character] may appear in an ontogeny as compared with the ontogeny of the ancestor". In Fig. 15.4, I show de Beer eight representations and I confront them with other representations of the same events but this time focusing on character's change over the time of the individual development (and not on the relationship of phylogeny with ontogeny) in order to show how time remains a secondary factor in de Beer's analysis which focuses on character's change.

It seems that Gould (1977), thereafter, will not discuss the fact that de Beer's renewed concept of heterochrony was analyzed at the "character level" (and not at the stage level). Therefore, after de Beer, heterochrony will continue to be analyzed through character's change. Moreover, because for de Beer heterochrony relied on the rate of expression of genes rather that on the displacement of temporal sequence, it implied that:

> Now, the principle of heterochrony will make it possible for any structure to appear later or earlier *as well as at the same corresponding time*, when compared with a previous ontogeny. [...] Heterochrony, therefore, provides eight possibilities of variation in the way in which structure may appear in an ontogeny as compared with the ontogeny of the ancestor. (de Beer 2007[1940], p. 28, my emphasis, which explains why de Beer refers to "deviation" as an heterochronic pattern, whereas Gould will exclude it).

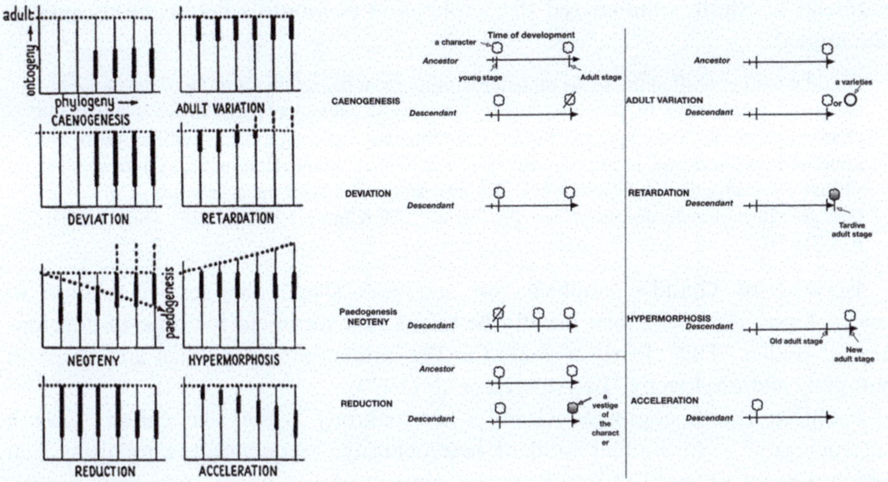

A B

Fig. 15.4 (**a**) The eight possibilities of variation in the way in which structure (set of characters) may appear in an ontogeny as compared with the ontogeny of the ancestor, by de Beer and that (**b**) I reinterpret on a straight line basis, which focuses on the appearance of a character along the timing of the individual's development. The linear model underlines the implications of de Beer's new concept of heterochrony, which focuses on a comparison of a character's expression ⬡ during ontogeny between ancestor and descendant (which itself reflects the rate-gene expression), instead of comparing the timings of expression of morphogenic stages in different species (what was Haeckel's conception of heterochrony)

De Beer's conception of heterochrony led to a neglect of the timing of ontogeny and an emphasis on the genes/characters' expression during development and its possible consequences for evolutionary biology. Gould followed de Beer's conception of heterochrony since he also completely ignored differentiation (the change of the characters during developmental process, independently of any temporal perturbation) in his model. However, Gould will try to simplify de Beer's model by suppressing some categories (see Gould 1977, p. 221–228).

In addition, Gould put a specific focus on allometric growth.[16] He saw it as a fundamental process, what allowed him to formulate a "compellingly simple clock model for heterochrony" (Gould 1977, p. 246–262). Gould choose to call it a "clock model", trying in this way to put a certain emphasis on temporality, which had clearly disappeared from de Beer's representation. However, despite the shape of Gould's model (a half sphere similar to a half clock with its hands), the multiple dimensions involved in the model (i.e. age, size, shape) make it difficult to interpret, and all the more so for its temporal dimension. In a paper of 2001, biologist

[16]Allometry studies the relationship of body size to shape and anatomy. It was popularized by D'Arcy Thompson's book, *On Growth and Form*, in 1917 and then further developed, through statistical analysis (J. Huxley 1932).

Kathleen K. Smith summarized the implication of Gould's heterochrony model. She argued:

> Gould therefore shifted the focus on heterochrony from the relative timing of developmental events to changes in size and shape. The almost exclusive focus on size and shape changes as the important heterochronic phenomenon was a significant redirection of the concept by Gould and is almost universally accepted today. Heterochrony has become virtually synonymous with allometry. 'Heterochrony was born because patterns of covariant growth permeate ontogeny and phylogeny' (McKinney 1999, p. 149). (Smith 2001, p. 170)

Because of Gould's emphasis on size and shape changes in relation to heterochrony, size will then usually be taken as a surrogate for time in developmental studies. This clearly appeared in the influential paper "Size and shape in ontogeny and phylogeny" by Alberch et al. (1979).

Kathleen Smith compared Gould's heterochrony – that she called "growth heterochrony" – to another kind of heterochrony: "sequence heterochrony," in which a developmental trajectory is conceptualized as a series of discrete events. Smith claimed: "growth heterochrony have limited the focus of studies of heterochrony to global (whole body) events and relatively late processes" (even if recent studies appear to be counterexamples to this account, e.g. Moss 2007 on heterochronic genes; Poethig 2009 on small RNAs and developmental timing or Frasch 2008 on microRNA and temporal identities).

And Smith to argue that,

> Most critical events in development occur early when size and shape may not be the appropriate reference points. These early events include such things as the initial differentiation and patterning of the major elements of the body, appearance of segmental and regional identity, patterns of regulatory gene expression, induction and signaling cascades, cell and tissue specification and differentiation, and the differentiation of skeletal elements and organ systems. Because these events do not rely on size and shape parameters they are by definition excluded from the kind of analyses presented by Gould (1977) [...] (Smith 2001, p. 172)

In conclusion, it appears that through their objections to Haeckel, de Beer and Gould provided important propositions about the relationships between developmental and evolutionary processes (by combining ontogeny with phylogeny on scientific basis where Haeckel had, at best, a successful aphorism); nevertheless, it is also clear that they were both robbing heterochrony from its explanatory power (Smith 2001). More precisely, since they actually claimed that they were giving an explanatory power to heterochrony in both development and evolutionary biology, it seems that their reformulation of the notion of heterochrony (in order to distance themselves from Haeckel's recapitulationist view) led to a mismeasure of time in development. Therefore, in order to give back to time all its importance in development, it seems more than relevant to offer a renewed characterization of development timing. This requires now a precise analysis of development timing's specificities and of the way they can be explored.

15.3 Reconsidering Time in Development

15.3.1 Current Descriptions of Time in Development

John Reiss, in *Keywords & Concepts in Evolutionary Developmental Biology* (2003), offered a definition of "time" for development. Despite the fact that the term appeared in a book devoted to evolutionary developmental biology (Evo-Devo), he first gave a characterization of time quite specific to developmental process. Somehow, like Rudolph Virschow who compared pathological and normal tissues and unlike Haeckel who compared ancestors and descendants, Reiss compared mutant phenotypes and wild phenotypes. Under the term "developmental time," Reiss distinguished within a single species four distinct aspects of developmental timing through comparisons with possible mutant phenotypes. Two involved aspects of *sequence* and two of *duration* (Reiss 2003, p. 359; see Figs. 15.7 and 15.8). In a previous paper, in 1989, Reiss had already summarized the basis of his distinction between sequence and duration. He argued that "time considered as sequence deals only with the relations "before," "after," and "simultaneous with," whereas time considered as duration also requires some measure of the *interval* between successive events" (Reiss 1989, p. 171).

Concerning time of development considered as a sequence, Reiss borrowed the definition from Alberch et al. who first defined it in 1979 and then Alberch alone defined it again in 1985. While Gould had essentially proposed in 1977, through his clock model, a static model of heterochrony (following de Beer's conception), Alberch argued that the individual development could be qualified as a "sequence of stages arranged in temporal order" and for which the description of the transformation was either (1) "continuous" (i.e. studies of allometric shape change), or (2) "discrete" (i.e. in the case of the transformation of complex character states, e.g. from symmetry to asymmetry), or (3) in sequences of differentiation of new elements (i.e. what he called "the sequence of developmental events", e.g. often represented in life cycles of model organisms) (Alberch 1985, p. 47; Alberch et al. 1979).

From Alberch's definition, it necessary follows that a "developmental sequence" can only describe a "sequence of stages". And then, in what he called d_1, $d_{2\ldots}$ d_i – "a sequence of differentiation processes" – each d_i refers to a stage. However, Alberch finally argued:

> For convenience we will let each d_i refer to both the process itself and to a discrete stage in each process for the structure or organism under consideration: the appearance of the first differentiated cell in a series of structures, the first indication of mineralization in each of a series of skeletal elements, of the first observable stage in a series of developmental processes. Each d_i is thus a developmental event. (Alberch et al. 1979, p. 297)

Consequently, as Kathleen Smith has brilliantly highlighted, for Alberch, the sequence of developmental events was (and it often remains so in current biology) commonly confused with the sequence of developmental stages (Smith 2001). While the event can be defined as any noticeable phenomenon during the

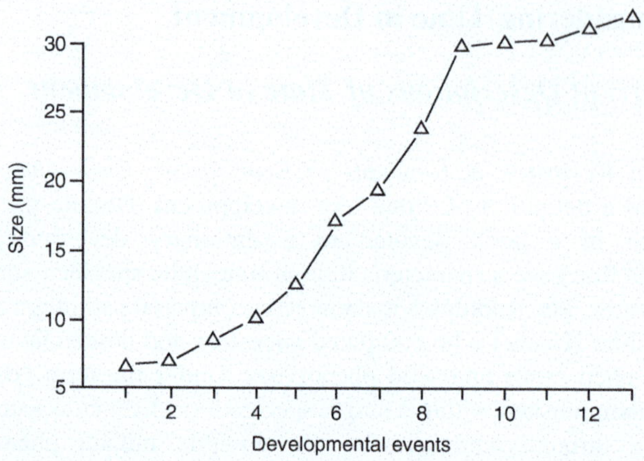

Fig. 15.5 Graph by Alberch et al. 1979, showing developmental events in the urodele *Triturus vulgaris* as a function of size. The first appearance of a bone is recorded as a developmental event (Each Δ), as follows: 1. splenial, vomer, palatine; 2. dentary; 3. pterygoid, squamosal, premaxillary; 4. prearticular; 5. frontal; 6. parietal; 7. opisthotic; 8. first basibranchial, orbitosphenoid; 9. prootic; 10. quadrate; 11. nasal; 12. maxilla, parasphenoid; 13. prefrontal

development of an individual of a given species, the stage remains a phenomenon, which can be observed during developments of individuals from different species.

From Hamburger and Hamilton, developmental stages became associated to growth and size of definite characters of a species (e.g., size of the primitive streak, or somite's number) and time became secondary in its account. Alberch et al. followed Hamburger and Hamilton's tradition, linking developmental stages to the appearance of specific characters (see Fig. 15.5), with the further purpose[17] to map developmental stages onto potential phylogenetic consequences (see Fig. 15.6).

On her side, Smith argued that developmental stages should clearly be contrasted with developmental events. She thought that while developmental stages permitted, in principle, comparisons of animals with different rates of development, it remained quite difficult to determine stages equivalents across taxa (against Haeckel's view). In this vein, Richardson (1995) had shown that embryonic stages are subjected to shifts in developmental timing during evolutionary process so that different organs develop at different times in different species. Because of this observation, it appeared difficult to identify any single conserved stage even for all species with the same body plan (Richardson et al. 1997, p. 93). Therefore, from this moment, one could consider that a sequence of developmental stages was not anymore the appropriate way to characterize the timing of development (if what

[17]Hamburger and Hamilton (1951) argued, "the preparation of a series of normal stages of the chick embryo [...] are proving to be increasingly valuable in medical research, as in work on viruses and cancer." Their main purpose was clearly not to establishing phylogenies between species but rather to establish a standardized view of the development of some animal models.

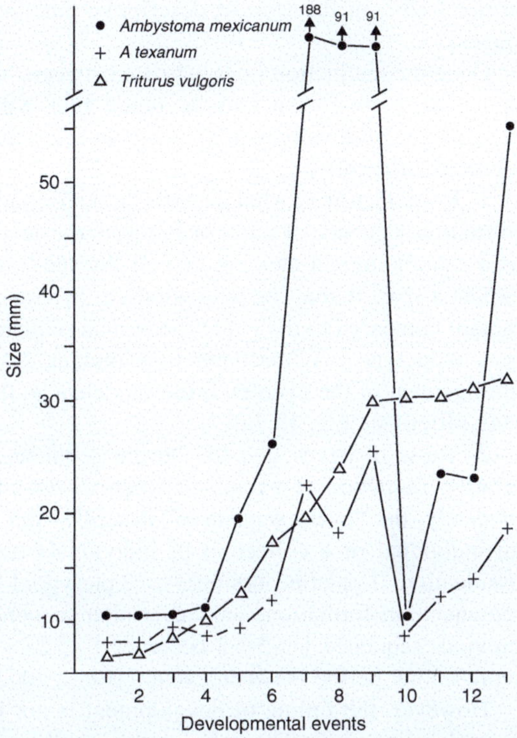

Fig. 15.6 Graph by Alberch et al. 1979, showing developmental events as a function of size in three species of urodeles. The reference is *Triturus vulgaris* (Δ), and the other species are compared to it

one had in mind was the potential evolutionary consequences of the timing of development), whereas a sequence of developmental events was (example of an event: first heartbeat). A precise characterization of the development timing potentially influent for evolutionary explanations seems to require a clear distinction between event and stage.

Smith defines "developmental sequence" as a series of *individual events* and considered that "developmental sequence analysis [allowed to] compare individual events and [made] no a priori assumption about the linkage of events in conserved stages" (Smith 2001, p. 180). In her conception, any event occurring during development – and, thus, not necessarily implying size and shape (i.e., quantitative events), which could be compared between species (i.e., through allometric studies) –, such as heart beating or hatching (i.e., qualitative events), can be taken care of in the analysis. Recently, researchers at Plymouth University have found that the timing of key developmental events – such as the first heartbeat, the formation of the eyes or the egg hatch – differs markedly between individuals in a species of aquatic snail, but also that the timing appear to be heritable (it is passed from mother to offspring) (Tills et al. 2013). They showed the similarities in the timing and sequence of development by using a bio-imaging system to record simultaneously the development of hundreds of embryos. This reminds us of the first videos filmed in real time of the entire embryonic development of certain organism in the 1920s, in the vein of descriptive embryology (focusing on developmental stages).

Once the distinction established – between "developmental stages" and "developmental events" – it can be noted that within a "developmental sequence" (a sequence of developmental events in time) the timing of development might be affected differently.

1- Reiss identifies what he calls "transformation sequence": a "particular transformation sequence of states or events in the development of a specific feature (such as a morphological element or cell lineage)" (after Velhagen 1997, p. 204). He explains that, it may be a qualitative difference in the observed sequence of a mutant compared to the wild type (e.g., a sequence of events "A -> B -> C -> D " in a wild type may become, in a mutant, "A -> C -> F ": a change in the arrangement of the events). From this change, the timing from event A to event C may differ (see Fig. 15.7a).

2- Reiss observes that the "event sequences" themselves may describe "the relative sequence of events between distinct transformation sequences." More precisely, the "event sequences" describe "the order of developmental events in an individual or a collection of individuals (usually a species)" (e.g. in cranial ossification: 1.palatine ossifies -> 2.pterygoid ossifies -> 3.maxilla ossifies). In mutants, transformations sequences alter sequences relative to each other (e.g., the previous sequence 1->2->3 becomes 1-->2---->3 in a mutant, the number of '-' representing period between events) (Reiss 2003, p. 359–360) (see Fig. 15.7b).

However, the timing of development is not limited to the timing of "developmental sequences". There is another well-known way to characterize time in development, which is "developmental duration". While, "developmental sequence" subdivides development in events and compares the periods which separates them, "developmental duration" focuses on the general timing of development and compares the relative appearance of events within this general timing. Therefore, within "developmental duration", Reiss distinguishes two categories.

3-There is first what he calls: "the relative timing of the events of transformation sequences with respect to some intrinsic time scale of development." In this case, "even if sequence is conserved [e.g., the development of all neurons in drosophila lasts around 30 hours, from A to D], mutants may show events that occur earlier or later with respect to such a time scale" (e.g., expression of the molecule of Rhodopsin 5 and 6 in the eye of drosophila might shift from t = 15 h to t = 25 h, difference between A & B interval in Fig. 15.8c top and down). Such modification is usually also "a change in event sequence, because the duration of an interval in one transformation or event sequence [. . .] is used as an index of the intrinsic time scale with respect to which the relative timing of another has changed" (Reiss 2003, p. 360) (Fig. 15.8d). D-Second, Reiss identified "the overall rate of development with respect to absolute time." In this case, mutants "show the same sequence and relative timing events, but the absolute length of development is scaled uniformly longer or shorter" (Reiss 2003, p. 360) (Fig. 15.8d). In any of the previously identified situations, what matters about time scale analysis is not the change of specific morphological characters but it is the change of sequence of events, in time.

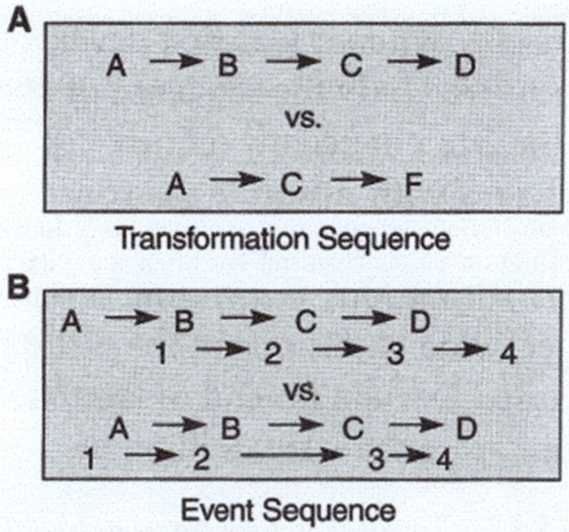

Fig. 15.7 Two aspects of development timing, time as sequence, from J. Reiss (2003). (**a**) Transformation sequence and (**b**) Event sequence

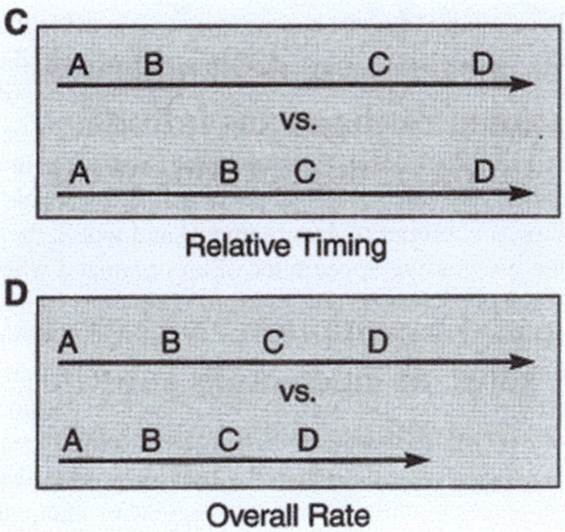

Fig. 15.8 Two aspects of development timing, time as duration from J. Reiss (2003). (**c**) Relative timing and (**d**) Overall rate

In the foreword of the book *Development Time and Timing*, Gottlieb summarizes how contemporary developmental biologists see time in development, as it follows:

> Undoubtedly, time and timing are inextricable aspects of developmental analysis, but, at least to the present writer, the temporal dimension operates more like a framework within which the "causes" of development operate, so that time and timing play a different role than a developing organism's interactions with physical and social objects, the flow of hormones, the expression of genes, the activity of neurons, and the like. (Turkewitz and Devenny (eds.) 1993, p. x)

Gottlieb made, then, a perfect diagnostic of the current problem about time for developmental biologists: time had become only "a framework within which the 'causes' of development operate," and the framework itself remained often neglected from the analysis. At least, in his opinion, its causal implications, and thus its explanatory power remained neglected. I would like to argue that – since it is possible not only to described time in development but also to use it as a way to "pattern" developmental process – the timing in development might explain development without necessarily causing it (opposing here causal explanations often understood as mechanistic explanation and "pattern explanations", similar to Graham Nerlich "geometrical explanations"). Indeed, even if in biology the focus on causation and mechanisms has often avoid the project of developing other models of explanation, it has also been argued, more generally, that "completely subsuming explanation into causation loses connections with some important issues" (Woodward 2014). In the following section I will show how "timing as a cause of development" and "timing as an explanation of development" may be decoupled.

15.3.2 Towards a New Approach of Time in Development

In a certain way, one can say that it is possible to think about time in development as one thinks about time in music. Stravinsky once said that "the phenomenon of music is given to us with the sole purpose of establishing an order in things, including, and particularly, the co-ordination between *man* and *time*."[18] Likewise, one can think about development as a process in which several aspects of time itself are co-ordinated and such a co-ordination would describe a precise structuration of time (i.e. proper to development) and would therefore be part of the explanation of the progressive appearance of an organized whole.

In music, there are different elements, which help to offer a precise characterization of general musical patterns: pitch (governs melody and harmony), rhythm (and its associated concepts tempo, meter and articulation), dynamics, and the sonic qualities of timbre and texture. Whereas, tempo is defined as the speed or the pace of a piece, rhythm is defined as the regular, repeated pattern of movement or sound. In other words, rhythm is the relationships between the parts, which make musical sentences equally spaced and regular or unequally spaced and syncopated. Therefore, the rhythm is the arrangement of sounds and silences over time. Similarly in developmental process, the developmental events are like the sounds in music, the sequence of developmental events like the rhythm, and the developmental duration like the tempo.

Recently, Janina Wellmann argued in a paper entitled "Folding into being: early embryology and the epistemology of rhythm" that from the nineteenth century, organic world was apprehended as being subject to a certain "rhythmic" structure. She explained that "unlike the continuous flux of time, [such a view implied] the restriction of the flux in favour of a rule" (Wellmann 2015, p. 19). More precisely, she showed that envisioning the living world in rhythmical terms informed music

[18] Stravinsky (1962, p. 54), quoted in DeLone and Wittlich. (Eds.) (1975, Chap. 3).

and aesthetic theories at the turn of the eighteenth to the nineteenth century. Consequently, music, in our case, may not be the analogy that underpins our conception of development but this could be the other way around: the observations of development that have influenced theorization in music.

Following this view, what is interesting here, in my opinion, is that in music underlying "timing mechanisms" have often been studied in order to give an account of "musical time," "time of music," or again, "music as a art of time" (Alperson 1980; Kramer 1988). In other words, "timing mechanisms" have been used to explain and theorize about time in music. Likewise, a return to the study and to the precise description of underlying "timing mechanisms" (and their articulation) in development could help to explain developmental process, as a process in time.

Assessing the connections between the different timing mechanisms within music has been a way to move beyond the sole descriptive stage, to understand time in music and even to explain music specificity. It could lead to similar results in development. About time in development, Reiss argued:

> Understanding the underlying timing mechanisms is essential if we are to move beyond the descriptive stage in our study of the evolution of developmental timing. Much criticism of studies of heterochrony has focused on their purely descriptive nature: how can heterochrony be a "mechanism" of evolution if it is only a result of underlying mechanistic changes? In particular, an understanding of [underlying timing] mechanism should allow us to understand the connection among the distinct aspects of the developmental time [. . .] (Reiss 2003, p. 363).

Reiss showed, for instance, that the concept of "positional information"[19] introduced by Lewis Wolpert in 1969 has been extremely useful in order to start thinking about these kind of underlying mechanisms. By describing the mechanisms cells might use in the spatial patterning of development, it became also possible to think about temporal information: an internal timekeeper would act as an intrinsic time scale used in the temporal patterning of development (Banerjee and Slack 2002). However, other models, likely to interact with one another, have been proposed: *hourglass*[20] (Tadros and Lipshitz 2009) and *clock*[21] models (Johnson and Day 2000; Pourquié 2003a; Tessmar-Raible et al. 2011; Hubaud and Pourquié 2014) (Fig. 15.9 Reiss p. 364 A et B). The studies on these temporal

[19]The concept refers to the fact that "cells acquire positional identities as in a coordinate system and then interpret this information according to their genetic constitution and developmental history" (Wolpert 1994) (e.g. the concept allows to understand cellular differentiation of cells genetically identical).

[20]The hourglass model refers to a model in which "a constant decay or buildup of products from an initial time point is used to control timing, with threshold levels of the product being used as cues to initiate developmental events" (Reiss 2003, p. 364).

[21]This notion of clock is different from Gould's "clock model". In the present case, the model is used to describe differentiation processes (e.g. somitogenesis in vertebrates: the process by which somites, blocks of mesoderm that give rise to a variety of connective tissues, are formed). The model describes processes of differentiation over time as the result of oscillating expression of particular genes.

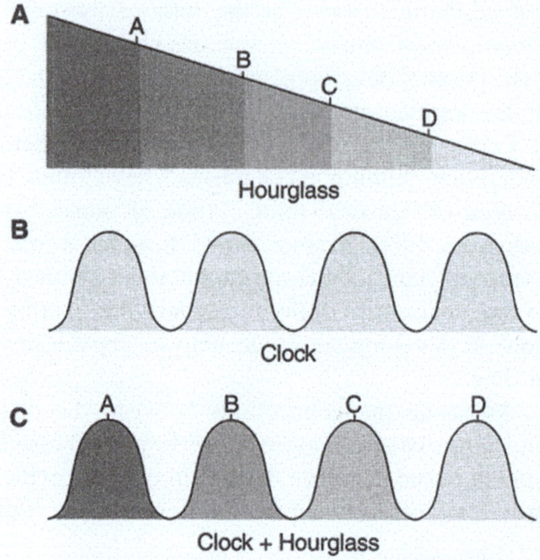

Fig. 15.9 Representation of three time development mechanisms: (**a**) hourglass, (**b**) Clock and (**c**) Combination of hourglass and Clock system (From Reiss 2003, Fig.3 p. 364)

aspects of development remain few and quite underdeveloped (see Pourquié 2003b).

Indeed, the fact that biologists often encountered difficulties when they try to account for time in a unified manner lead them to avoid addressing time-dimension in their explanation of developmental process. The reason is that time never really appears in a unique way, and for a single purpose, within developmental process. Consequently, time often appears to some biologists unmanageable and/or chaotic "and thus immune to analysis and understanding" (Gottlieb 1993).

In general, set of causes often replaces any attempt to give precise description of time in development. However, one could consider that the developmental process does not necessarily refer to a sole linear conception of time. Time in development – like time in music – should not necessarily be seen as a linear time (e.g. Bock von Wülfingen et al. 2015, p. 4, argued for instance that Stockhorst (2006) had shown how multiple concepts of time – linear, progress oriented, circular – had already overlapped during the eighteenth century).

Time could be characterized and analyzed from different perspectives without necessarily appearing chaotic. A precise characterization of the underlying timing mechanisms and of their connections could even help to "unfold time". However, it seems that the analysis and characterization of time in biology remains largely to be done, and that such characterization should not be limited to a linear view about time in development.

In the introduction of their book *Development Time and Timing*, Turkewitz and Devenny suggested that time should be seen as "formative for development" since it could be "a way of *relating* developmental events" (my emphasis). The timing "refers to the temporal relationships between the components of development that change during the course of the development of the organism." In this respect, they viewed timing "as formative in the sense that the effect of change on the organism in any component is determined by the status of the other components that comprise the organization of the individual and by the differential rates of change of the various components of development." Therefore, "the developmental time at which an event occurs will ultimately influence or shape both the trajectory of development and its outcome." Through this complex definition, it appears that the authors did not necessarily consider only a single timing – those of the developmental process itself – but also other characteristics of the time, which occur during development and could influence the whole timing of the process. Therefore, since they included the possible interaction of different aspects of time in development (including different dimensions of time), it appeared difficult for them to say if time in development, should, or should not, be seen as "truly causal" (Turkewitz and Devenny 1993, p. x, 10).

Such indeterminacy concerning the causal status of time in development relies, in my opinion, on a certain degree of confusion between two possible levels of analysis. The first level concerns the actual "causal status of time" in development whereas the second concerns the "explanatory status of time" in development. Turkewitz and Devenny claimed, that time was both "a way of relating developmental events" and something "formative for development". Whereas, the first affirmation has to do, in my opinion, with the causal status of time, the second affirmation has to do with the explanatory status of time. Another way to illustrate the distinction is to refer to Nerlich's proposal (1979) already mentioned above for which explaining phenomena requires appealing to "geometric properties of space itself, which requires an ontic commitment to space and does not reduce to a causal explanation in terms of material objects and relations among them" (for a more recent approach on topological explanations, see Huneman 2015). Similarly, considering the structure of time as an explanation of development would not be tantamount to considering causal mechanisms occurring over time as an explanation of development. Therefore, I argue it is possible to make a clear distinction between the two levels and to address separately the causal status of time in development (e.g. by looking at the effects of heterochrony on development[22]) and the explanatory status of time in development (e.g. by looking at the interaction between different aspects of time in development, such as the interaction between clock models and hourglass models).

Indeed, in biology an explanation is usually given by pinpointing the underlying mechanisms of a phenomenon. In development, the more often, biologists focus on spatial mechanisms (e.g., how a cell is spatially oriented relatively to another group of cell; or how a molecule/protein interact with another one, etc.) These spatial mechanisms necessitate referring to causal relationships because the interactions of

[22]Recently some studies have been developed which focuses on the timing of transcription and its importance for development, particularly concerning "transcriptional synchrony" (the fact that many genes are transcripted exactly at the same time) (e.g. Lagha et al. 2013). Few years ago, studies have been focusing on the isolation of "heterochronic genes" to assess developmental timing (e.g. Moss 2007, Poethig 2009, Frasch 2008).

the elements, within these mechanisms, are ongoing in a linear dimension. What is the difference with timing mechanisms?

As, I have tried to show, the way to apprehend time in development is not unique. There are different ways to explore it, but there are also different kinds of times within the developmental process that interact with each other (clock models and hourglasses models), leading to the developmental timing. The possible analogy between time in development and time in music highlights the fact that the precise characterization of time in development should be a first step for a precise understanding of its explanatory role.

For instance, in his book, *The Time of Music* (1988), Jonathan Kramer argued that in order to arrive at some current notions of musical time it is necessary to tackles, among other things, the dual aspect of time, that of being and becoming (i.e. the ongoing time goal-directed linearity to moment and the "vertical" time). He gathered these different aspects of time with respect to the more general terms "linear" and "nonlinear". As Ingrid Arauco explained "linear time in music is linked to process and the fulfillment of a listener's expectations in succession". Therefore, it obeys the rules of linear causality (sequential pattern, cause precedes effect and direct link between cause and effect). "Nonlinear time is by contrast non-processive" (Arauco 1990). Kramer explained that "nonlinear principles may be revealed gradually, but they do not develop from earlier events or tendencies" (Kramer 1988, p. 20). Textural consistencies and durational proportions are examples of nonlinear aspects of a musical piece. Therefore a nonlinear sense of time is not bound to succession (or causation) but to the meanings of gestures of tonal music which create that particular time in a piece, no matter where they occur. Kramer showed, in his study, that it is possible also to analyze the interaction between the processive and the gestural meanings of time, the linear and the nonlinear time.

Concerning development, a similar analysis would necessitate not to look at *what time an event occurs* (i.e. when in the succession of events over time an event occurs) but to look at *what kind of time an event A occurs compared to the kind of time another event, B occurs* (i.e. to look at the manifestations of time through its timings). It would then become possible to identify some common patterns of timing in development. Hourglass or clock mechanisms are examples of these common patterns, which have already been described. A further analysis would imply to draw the links between the linear dimension of time in development and its non-linear dimension (e.g. to watch when – over developmental process – hourglass mechanisms or clock mechanisms are expressed). From this point, it might be possible to identify some development timing patterns, which could be used as elements of comparison.

Indeed, the identification of such patterns could allow us to draw links between specific characteristics of time during development and developmental stages without focusing on characters (and without being confronted, for instance, to the problem of identification of a conserved stage for multiple species with the same body plan, when different organs may develop at different times in different species). The identification of a pattern of development timing would be more informative than any other stage based on character growth identification.

It seems that, through this analysis, time and timing of development, without necessarily being seen as a cause of development, could however appear as fully explanatory in the developmental process, since time would reveal some underlying mechanisms which could not be revealed by any spatial factors.

Finally, such an approach also shows that a precise analysis of developmental timing could lead to reassess our conception of developmental explanation by showing that such an explanation is not necessarily causal. Furthermore, developmental timing analysis shows that it is possible to provide an explanation of development that is not based on characters' analysis and thus neither on genes analysis alone. It is thus more than likely that such an approach might change the way one think about the relationship between ontogeny and phylogeny.

15.4 Conclusion: Explaining Evolution with the Timing of Development

In the introduction of this chapter, I suggested that evolutionary explanations and developmental explanations could be distinguished. At the present, it seems to me that not only such a distinction is needed (as Ernst Mayr already famously argued in 1961) in order to highlight specificities of both processes, but it also appears that within developmental explanations, causal and non-causal explanations should be distinguished. This will increase the possibilities of explanations for developmental process of course, but also for evolutionary process. Indeed, for many years a specific focus has been given to causal explanations leading to the idea that developmental and evolutionary explanations were apart.[23]

In this chapter, I have tried to give an explanation to this phenomenon based first on an historical analysis of time in development. I have shown that since basically the beginning of the twentieth century time has disappeared from the agenda of biologists (see also Bock von Wülfingen et al. 2015). The reasons of such oversight must be related to two conceptual shifts in the history of developmental biology: First, the stable epistemological shift introduced by de Beer from the notion of developmental stages to the notion of character and, second, the definitional shift introduced by de Beer, popularized by Gould and supported by Alberch and al, about the meaning of heterochrony, from the idea of a change in development timing to the idea of a change in characters' growth and size. This oversight of time as a decisive element in the analysis of development led to a specific focus on causal factors and to a mismeasure of the plurality of time dimensions in development. In other words, only the linear dimension of time has been taken into account

[23]See Huneman's Chap. 14 in this volume.

(a dimension in which causation could operate and make disappear time analysis in favor of space analysis).

Consequently, in order to offer a renewed view of time in development and another type of explanation for development, I have suggested one should look at the timings of development (i.e. the different shapes that time can take during the historical event of development, depending on whether it is considered as a sequence or as a duration, cf. Reiss 1989). I have proposed an analogy with the analysis of time in music including the difference between rhythm and tempo (a distinction which could have been itself influenced by the first observations of development in the eighteenth century (Wellmann 2015)). A few authors have started to describe some of these patterns of development timings in the literature (e.g. Wolpert 1969; Pourquié 2003a; Reiss 2003), however a great deal of work will remain to be done (Pourquié 2003b). I have argued that the characterization of those patterns would allow defining new developmental stages (i.e. redundant patterns occurring in different species during their developmental process), which would not be primarily based on characters' growth and size and defined along a causal linear dimension of time but would allow to explore different types of explanation (e.g. non-causal) for developmental process. This could help, for instance, to explain why developmental processes while they are similar in their succession of events may appear different/unique (as well as each piece of music has its typical gestures).

Finally, it appears that with such an approach, a decoupling of explanations – the developmental one and the evolutionary one – will not necessarily signify that one of the processes cannot help explaining the other one (i.e. that the two fields remain apart). In the present case, I have shown that the shapes time can take (oscillatory, rhythmical, sequential etc.) constitute important aspects of developmental process and that understanding the articulations of these different patterns during the historical situation of the developmental process might help to unfold development und thus explaining it. Furthermore, it has become now possible to follow the development of multiple organisms simultaneously (synchronicity) and for several generations (diachronicity) (Tills et al. 2013). These works could allow deciding on what patterns of development timings to focus on for evolutionary purpose, by targeting what are the heritable timing patterns during development. In other words, instead of looking at arbitrary stages, or at the growth of characters that might develop at different moments of development, one could look at how developmental timing patterns are transmitted over successive generations.

Ernst Mayr has introduced a distinction between the functional biologist and the evolutionary biologist that has remained important until today concerning explanatory frameworks[24] – even if it has started to be discussed for a couple of year (Ariew 2003; Rosenberg 2008; Laland et al. 2011). However, the problem of Mayr's distinction is not that he separates "developmental" approaches from "evolutionary" approaches but rather that within his explanatory framework Mayr

[24]See Ricqlès's Chap. 11 in this volume.

focuses mainly on causal explanations. Biologists too, both in developmental biology and evolutionary biology, have tended to focus on causal explanations through the identification and clarification of biological mechanisms (e.g. Craver and Darden 2013).

We may today offer another way to think about explanations in biology by introducing the multiple dimensions of time specific to the developmental process in biology (i.e. "the timing of development"). It seems to me that if developmental biology can actually help enlightening evolutionary biology, it is primarily by looking at non-causal explanations, including nonlinear time analysis, because if evolutionary processes may causally explain developmental processes, the reverse is simply not true.

Acknowledgments I would like to thank Christophe Bouton and Philippe Huneman for their fruitful comments on previous versions of this manuscript as well as Michel Vervoort for his precious advices and references. This work was supported financially by the "Who Am I?" Laboratory of Excellence (ANR-11-LABX-0071) funded by the French government through its "Investments for the Future" Program operated by the French National Research Agency (ANR) under grant no. ANR-11-IDEX-005-02, as well as by the ANR Program EXPLABIO (Explanations in Evolutionary Biology) under grant no ANR-13-BSH3-0007.

References

Abzhanov, A. 2013. Von Baer's law for the ages: Lost and found principles of developmental evolution. *Trends in Genetics* 29 (12): 712–722.

Alberch, P. 1985. Problems with the interpretation of developmental sequences. *Systematic Biology* 34 (1): 46–58.

Alberch, P., S.J. Gould, G.F. Oster, and D.B. Wake. 1979. Size and shape in ontogeny and phylogeny. *Paleobiology* 5 (3): 296–317.

Alperson, P. 1980. "Musical time" and music as an "art of time". *The Journal of Aesthetics and Art Criticism* 38 (4): 407.

Amundsen, R. 2001. Adaptation and development. In *Adaptationism and development*, ed. S.H. Orzack and E. Sober. Cambridge: Cambridge University Press.

Amundson, R. 2005. *The changing role of the embryo in evolutionary thought: Roots of evo-devo*. Cambridge: Cambridge University Press.

Arauco, I. 1990. Review the time of music by J Kramer. *College Music Symposium* 30 (1): 153–156.

Ariew, A. 2003. Ernst Mayr's 'ultimate/proximate' distinction reconsidered and reconstructed. *Biology and Philosophy* 18 (4): 553–565.

Banerjee, D., and F. Slack. 2002. Control of developmental timing by small temporal RNAs: A paradigm for RNA-mediated regulation of gene expression. *Bioessays* 24 (2): 119–129.

Bateson, P., P.P.G. Bateson, and P. Gluckman. 2011. *Plasticity, robustness, development and evolution*. Cambridge: Cambridge University Press.

Beatty, J. 1994. The proximate/ultimate distinction in the multiple careers of Ernst Mayr. *Biology and Philosophy* 9 (3): 333–356.

Beer, G.R. (de) 1930. *Embryology and evolution*. Gloucestershire: Clarendon Press.

Beer, G.R. (de) (2007[1940]). *Embryos and Ancestors*. Oxford: Caven Press.

Bock von Wülfingen, B., C. Brandt, S. Lettow, and F. Vienne. 2015. Temporalities of reproduction: Practices and concepts from the eighteenth to the early twenty-first century. *History and Philosophy of the Life Sciences* 37 (1): 1–16.

Calcott, B. 2006. *Major transitions in biological organisation*. At the Australian National University, Made Available in DSpace on 2011-01-04. https://digitalcollections.anu.edu.au/handle/1885/49391.

Craver, C.F., and L. Darden. 2013. *In search of mechanisms: Discoveries across the life sciences*. Chicago: University of Chicago Press.

Davidson, E.H. 1991. Spatial mechanisms of gene regulation in metazoan embryos. *Development* 113 (1): 1–26.

DeLone, R., and G.E. Wittlich, eds. 1975. *Aspects of twentieth-century music*. Englewood Cliffs: Prentice-Hall.

Dettlaff T.A., Igntatieva G.M. and Vassetzky S.G. (1987). The Problem of Time in Developmental Biology. Its Study by the Use of Relative Characteristics of Development Duration. *Sov. Sci. Rev. Ser. F. Physiol., Gen., Biol.* 1. pp. 1–88.

Fox Keller, E. 1996. *Refiguring life: Metaphors of twentieth-century biology*. New York: Columbia University Press.

Frasch, M. 2008. A matter of timing: MicroRNA-controlled temporal identities in worms and flies. *Genes & Development* 22 (12): 1572–1576.

Gluecksohn-Schoenheimer, S. 1938. The development of two tailless mutants in the house mouse. *Genetics* 23 (6): 573–584.

Gottlieb, G. 1993. Forword. In *Developmental time and timing*, ed. G. Turkewitz and D.A. Devenny, ix–x. Hillsdale: Lawrence Erlbaum Associates.

Gould, S.J. 1977. *Ontogeny and phylogeny*. Harvard: Harvard University Press.

Gould, S.J. 1988. The uses of heterochrony. In *Heterochrony in evolution*, ed. M. L. McKinney, 1–13. Springer US. Retrieved from http://link.springer.com/chapter/10.1007/978-1-4899-0795-0_1.

Haeckel, E.H. 1866. *Generelle Morphologie der Organismen allgemeine Grundzuge der organischen Formen-Wissenschaft, mechanisch begrundet durch die von Charles Darwin reformirte Descendenz-Theorie von Ernst Haeckel: Allgemeine Entwickelungsgeschichte der Organismen kritische Grundzuge der mechanischen Wissenschaft von den entstehenden Formen der Organismen, begrundet durch die Descendenz-Theorie (Vol. 2). Verlag von Georg Reimer*. Retrieved from http://books.google.com/books?hl=fr&lr=&id=RxmNSuKRp54C&oi=fnd&pg=PR26&dq=E+Haeckel+1866&ots=Lf2OIeXsKw&sig=nupCPpSKeNFVnjGUzf9CVL9HJdo.

Hall, B.K. 1999. *Evolutionary developmental biology*. New York: Springer.

Hall, B.K., and W.M. Olson. 2003. *Keywords and concepts in evolutionary developmental biology*. Cambridge, MA: Harvard University Press.

Hamburger, V., and H.L. Hamilton. 1951. A series of normal stages in the development of the chick embryo. *Journal of Morphology* 88 (1): 49–92.

Hanken, J. 2015. Is heterochrony still an effective paradigm for contemporary studies of evo-devo? In *Conceptual change in biology*, ed. A. Love, 97–110. New York: Springer. Retrieved from.

Hubaud, A., and O. Pourquié. 2014. Signalling dynamics in vertebrate segmentation. *Nature Reviews Molecular Cell Biology* 15 (11): 709–721.

Huneman, P. 2015. Diversifying the picture of explanations in biological sciences: Ways of combining topology with mechanisms. *Synthese*: 1–32.

Huxley, J. 1932. *Problems of relative growth*. New York: L. MacVeagh, The Dial Press. Retrieved from http://archive.org/details/problemsofrelati00huxl.

Johannsen, W. 1911. The genotype conception of heredity. *The American Naturalist* 45 (531): 129–159.

Johnson, M.H., and M.L. Day. 2000. Egg timers: How is developmental time measured in the early vertebrate embryo? *BioEssays* 22 (1): 57–63.

Keibel, F. 1910. *Normentafeln zur Entwicklungsgeschichte der Wirbelthiere*. Vol. Hft.11. Jena: Fischer. Retrieved from http://www.biodiversitylibrary.org/item/35263.

Kramer, J.D. 1988. *The time of music: New meanings, new temporalities, new listening strategies*. New York: Schirmer Books.

Lagha, M., J.P. Bothma, E. Esposito, S. Ng, L. Stefanik, C. Tsui, J. Johnston, et al. 2013. Paused Pol II coordinates tissue morphogenesis in the Drosophila embryo. *Cell* 153 (5): 976–987.

Laland, K.N., J. Odling-Smee, and S.F. Gilbert. 2008. EvoDevo and niche construction: Building bridges. *Journal of Experimental Zoology Part B: Molecular and Developmental Evolution* 310 (7): 549–566.

Laland, K.N., K. Sterelny, J. Odling-Smee, W. Hoppitt, and T. Uller. 2011. Cause and effect in biology revisited: Is Mayr's proximate-ultimate dichotomy still useful? *Science* 334 (6062): 1512–1516.

Mayr, E. 1961. Cause and effect in biology. *Science* 134 (3489): 1501–1506.

Moss, E.G. 2007. Heterochronic genes and the nature of developmental time. *Current Biology* 17 (11): R425–R434.

McKinney, M.L. 1999. Heterochrony: Beyond words. *Paleobiology* 25(2): 149–53. doi:10.1017/S0094837300026464.

Müller-Wille, S., and H.-J. Rheinberger. 2012. *A cultural history of heredity*. Chicago: University of Chicago Press.

Navis, A.R. 2007. Stages of chick development. *Embryo Project Encyclopedia*. Retrieved from http://embryo.asu.edu/pages/stages-chick-development#sthash.leTBR20X.dpuf.

Needham, J. 1959. *A history of embryology*. New York: Abelard-Schuman.

Nerlich, G. 1979. What can geometry explain? *British Journal for the Philosophy of Science* 30 (1): 69–83.

O'Connell, L. 2013. The Meckel-Serres conception of recapitulation. *Embryo Project Encyclopedia*. Retrieved from http://embryo.asu.edu/pages/meckel-serres-conception-recapitulation#sthash.NNU7SA1p.dpuf.

Poethig, R.S. 2009. Small RNAs and developmental timing in plants. *Current Opinion in Genetics & Development* 19 (4): 374–378.

Pourquié, O. 2003a. The segmentation clock: Converting embryonic time into spatial pattern. *Science* 301 (5631): 328–330.

Pourquié, O. 2003b. A biochemical oscillator linked to vertebrate segmentation. In *Origination of organismal form: Beyond the gene in developmental and evolutionary biology*, ed. G.B. Müller and S.A. Newman, 183–194. Cambridge: MIT Press.

Raff, R.A., and G.A. Wray. 1989. Heterochrony: Developmental mechanisms and evolutionary results. *Journal of Evolutionary Biology* 2 (6): 409–434.

Reiss, J.O. 1989. The meaning of developmental time: A metric for comparative embryology. *American Naturalist* 134: 170–189.

Reiss, J.O. 2003. Time. In *Keywords and concepts in evolutionary developmental biology*, ed. B.K. Hall and W.M. Olson, 358–367. Cambridge, MA: Harvard University Press.

Richardson, M.K., J. Hanken, M.L. Gooneratne, C. Pieau, A. Raynaud, L. Selwood, and G.M. Wright. 1997. There is no highly conserved embryonic stage in the vertebrates: Implications for current theories of evolution and development. *Anatomy and Embryology* 196 (2): 91–106.

Richardson, M.K. 1995. Heterochrony and the phylotypic period. *Developmental Biology* 172: 412–421.

Rosenberg, A. 2008. *Darwinian reductionism: Or, How to stop worrying and love molecular biology*. Chicago: University of Chicago Press.

Schmitt, S. 2004. *Histoire d'une question anatomique : La répétition des parties*. Paris: Muséum National d'Histoire Naturelle.

Smith, K.K. 2001. Heterochrony revisited: The evolution of developmental sequences. *Biological Journal of the Linnean Society* 73 (2): 169–186.

Stockhorst, S. 2006. Zeitkonzepte. Zur Pluralisierung des Zeitdiskurses im langen 18. Jahrhundert, Das achtzehnte Jahrhundert. *Zeitschrift der Deutschen Gesellschaft zur Erforschung des achtzehnten Jahrhunderts* 30: 157–164.

Stravinsky, I. 1962. *Autobiography*. New York: W.W. Norton & Co., Inc.

Tadros, W., and H.D. Lipshitz. 2009. The maternal-to-zygotic transition: A play in two acts. *Development* 136 (18): 3033–3042.

Tessmar-Raible, K., F. Raible, and E. Arboleda. 2011. Another place, another timer: Marine species and the rhythms of life. *Bioessays* 33 (3): 165–172.

Tills, O.S., D. Rundle, and J.I. Spicer. 2013. Parent–offspring similarity in the timing of developmental events: An origin of heterochrony? *Proceedings of the Royal Society of London B: Biological Sciences* 280 (1769): 20131479.

Turkewitz, G., and D.A. Devenny, eds. 1993. *Developmental time and timing*. Hillsdale: Lawrence Erlbaum Associates.

Velhagen, W.A. 1997. Analyzing developmental sequences using sequence units. *Systematic Biology* 46 (1): 204–210.

Virchow, R.L.K. 1860. *Cellular pathology as based upon physiological and pathological history: Twenty lectures delivered in the Pathological Institute of Berlin during the months of February, March, and April, 1858*. Birmingham: Classics of Medicine Library.

Von Baer, C.E. 1827. *De ovi mammalium et hominis genesi epistolam (etc.)*. Leop. Voss. Retrieved from http://books.google.com/books?hl=fr&lr=&id=Ns5YAAAAcAAJ&oi=fnd&pg=PT9&dq=von+Baer+(1827&ots=BqtyPqiOux&sig=3jIfun4lj-bNbqNQQmSZjUpeMUA.

Waddington, C.H. 1940. The genetic control of wing development in Drosophila. *Journal of Genetics* 41 (1): 75–113.

Wallace, B. 1986. Can embryologists contribute to an understanding of evolutionary mechanisms? In *Integrating scientific disciplines: Case studies from the life sciences*, ed. W. Bechtel, 149–163. Springer. Retrieved from http://link.springer.com/content/pdf/10.1007/978-94-010-9435-1_9.pdf.

Wellmann, J. 2015. Folding into being: Early embryology and the epistemology of rhythm. *History and Philosophy of the Life Sciences* 37 (1): 17–33.

West-Eberhard, M.J. 2003. *Developmental plasticity and evolution*. Oxford: Oxford University Press.

Wolff, C.F. 1774. *Theoria generationis*. Halae, ad Salam: Typis et sumtu Io.

Wolpert, L. 1969. Positional information and the spatial pattern of cellular differentiation. *Journal of Theoretical Biology* 25 (1): 1–47.

Wolpert, L. 1994. Positional information and pattern formation in development. *Developmental Genetics* 15 (6): 485–490.

Woodward, J. 2014. Scientific explanation. In *The Stanford encyclopedia of philosophy*, ed. E. N. Zalta. Retrieved from http://plato.stanford.edu/archives/win2014/entries/scientific-explanation/.

Author Index

C. Bouton, P. Huneman (eds.), *Time of Nature and the Nature of Time*, Boston Studies in the Philosophy and History of Science 326,
DOI 10.1007/978-3-319-53725-2

C

D

E

F

K

L

M

Subject Index

C. Bouton, P. Huneman (eds.), *Time of Nature and the Nature of Time*, Boston Studies in the Philosophy and History of Science 326,
DOI 10.1007/978-3-319-53725-2

The manufacturer's authorised representative in the EU is Springer
Nature Customer Service Centre GmbH, Europaplatz 3, 69115 Heidelberg,
Germany. If you have any concerns regarding our products, please
contact ProductSafety@springernature.com

Printed and bound by CPI Group (UK) Ltd, Croydon, CR0 4YY
05/10/2026
02228127-0011